Empiricism, Naturalism and the Metaphysics of Biology

Sandy C. Boucher

Empiricism, Naturalism and the Metaphysics of Biology

Taking a Stance

palgrave
macmillan

Sandy C. Boucher
School of Humanities, Arts and Social Sciences
University of New England
Armidale, NSW, Australia

ISBN 978-3-032-00554-0 ISBN 978-3-032-00555-7 (eBook)
https://doi.org/10.1007/978-3-032-00555-7

This Palgrave Macmillan imprint is published by the registered company Springer Nature Switzerland AG.
The registered company address is: Gewerbestrasse 11, 6330 Cham, Switzerland

To my parents, David and Diana Boucher

Preface

I formally began work on this book a year and a half ago, but in some ways it represents the culmination of work I have been engaged in for over two decades on metaphilosophy and the philosophy of science. In 2002, when I was a graduate student at the University of Connecticut, two books came out that may not have seemed at first glance to have much to do with one another, but which in different ways fired my intellectual imagination and influenced the course of much of my subsequent research: *The Empirical Stance* by Bas van Fraassen, and *The Structure of Evolutionary Theory* by Stephen Jay Gould. It struck me that ideas and themes from the two works could be combined to develop a general account of the large-scale perspectives, orientations and worldviews that shape and are shaped by evolutionary biology, something of which I have been trying, with varying degrees of success, to convince my colleagues ever since. This is the research project of which the current work is the result.

The more immediate impetus for this book came from the belated recognition on my part that the concept of a philosophical stance had in the last decade or so been developed in two quite different ways: in my work, in the idea of what I call a metaphysical stance; and in the work of several other theorists, in the idea of an epistemic stance. In my earlier work I had tried to shoehorn these into a single, unified conception. It was the realisation that this was the wrong approach, and that we needed to recognise two related, but fundamentally distinct, ways of elaborating the notion of a stance, each useful for different purposes, and each applicable to different topics in the metaphysics of biology, that suggested to me the idea of a book on the subject.

Armidale, NSW, Australia
October 2024

Sandy Boucher

Acknowledgements

For advice, encouragement and inspiration, thank you to Anne Hiskes, Patrick Fleming, Howard Sankey, Greg Restall, Daniel Stoljar, Curtis Forbes, Anjan Chakravartty, John Wilkins, Neil Thomasson, Marc Ereshefsky, Rose Novick, Terry Sullivan, Alan Scott, Erol Aslan, Caleb Hazelwood, Fiona Utley and David Coady. Thanks especially to Adrian Walsh for help and support and many stimulating conversations about metaphilosophy. Thanks to Arunaa Devi, Production Editor at Palgrave Macmillan, and to an anonymous reviewer for extremely helpful and generous feedback and suggestions. Above all a huge thank you to my family, Felicity, Leo and Claudia.

The material in this book has been presented at the Australasian Association of Philosophy Conference, the International Society for the History, Philosophy and Social Studies of Biology Conference, the Australian National University Philosophy of Biology Workshop, the University of Tasmania Philosophy Seminar, the University of New England Humanities Seminar Series, the Australasian Association for the History, Philosophy and Social Studies of Science Conference, the Australasian Philosophy of Biology Postgraduate Conference, a Workshop on Metaphysical Anti-Realism in Sydney, and the University of New England Philosophy and Friends Work in Progress Seminar. Thank you to those in attendance for very helpful feedback.

Thank you to several publishers for granting me permission to reuse previously published work. Chapter 3 incorporates material from 'An argument for global realism about the units of selection', *Biology and Philosophy* 38 (5): 1–22, 2023. Chapter 6 includes material from 'Biological teleology, Reductionism, and verbal disputes', *Foundations of Science* 26 (4): 859–880, 2021. Chapter 8 is based on 'Functionalism and structuralism as philosophical stances: van Fraassen meets the philosophy of biology', *Biology and Philosophy*, 30 (3): 383–403, 2015. Chapters 7, 9 and 10 incorporate material from 'An empiricist conception of the relation between metaphysics and science', *Philosophia*, 47 (5): 1355–1378, 2019. Chapters 9 and 10 include material from 'Stances and epistemology: values, pragmatics and rationality', *Metaphilosophy*, 49 (4): 521–547, 2018.

Competing Interests The author has no competing interests to declare that are relevant to the content of this manuscript.

Contents

About the Author

Sandy C. Boucher is Senior Lecturer in the Philosophy of Science at the University of New England, Armidale Australia.

Abbreviations

IBE	Inference to the best explanation
MN	Metaphysical naturalism
MoS stance	Metaphysics of science stance
NC	Naturalistic constraint on metaphysical inferences
PE	Punctuated equilibrium
SAI	Species as individuals thesis
SSR stance	Standard scientific realism stance

List of Figures

List of Tables

1

Introduction: Philosophical Stances and the Metaphysics of Biology

This book is an exploration of the metaphysics of biology from a particular metaphilosophical standpoint, that which employs the concept of a *philosophical stance*. The book thus exists at the intersection of metaphilosophy, philosophy of science, and the metaphysics of biology.

In recent years there have been exciting developments in the philosophy of science and philosophy of biology with respect to issues such as naturalised metaphysics (Slater and Yudell 2017), the metaphysics of biology (Dupré 2021), and meta-level questions about how we go about, and how we *should* go about, adopting and defending views on scientific ontology, especially in relation to scientific realism and empiricism, and the role played in this by philosophical stances (Chakravartty 2017). There have been several important, independent debates on these and related topics that have engaged leading researchers in the field in the last 10 years. In this book I propose to bring together some of these ideas and approaches to develop and defend a coherent meta-level picture of the metaphysics of biology that engages with elements of empiricism and naturalism. Such a project has the potential to be illuminating in two directions: in applying recent theories and insights in metaphilosophy to specific first-order issues in the metaphysics of biology we get a clearer

S. C. Boucher, *Empiricism, Naturalism and the Metaphysics of Biology*,
https://doi.org/10.1007/978-3-032-00555-7_1

picture of the value and fruitfulness of those theories and insights; and we thereby provide a new perspective on those specific cases that may be valuable to, and possibly help us make progress on, the first-order debates themselves.

The philosophical stance idea has been widely discussed in the philosophy of science in the last couple of decades, but the discussion has mostly been fairly abstract. In this book I intend to *apply* the idea to a number of existing topics, ideas and debates, all drawn from the metaphysics of biology, to explore whether it may provide some metaphilosophical insight and enlightenment in these areas. I will put the idea through its paces, test its usefulness, and consider whether and how it may help to guide us through and make better and more unified sense of a number of topics of contemporary interest.

1 Philosophical Stances

Van Fraassen's concept of a philosophical stance first emerged in the context of concerns about empiricism being self-undermining (1995, 2002). Suppose that empiricism is a factual belief, possibly concerning humans' sources of information about the world. The empiricist holds that all beliefs about factual matters are empirical and contingent. So the empiricist must hold that empiricism itself, if a factual belief, is empirical and contingent. It is hostage to future empirical evidence, which may show it to be false, perhaps by uncovering alternative sources of humans' information about reality. But there is at the very least something unsettling about this admission for the empiricist, since it was empiricist principles that led the empiricist to concede that empiricism could easily turn out to be false. For the empiricist, empiricism seemingly must be both unquestioned presupposition or 'dogma', *and* tentative, vulnerable empirical hypothesis. This is an untenable situation. As Mohler (2007, 210) nicely expresses it, van Fraassen is claiming that '[Empiricism] must not be subject to the challenge it itself grounds'.

This worry was a result of construing empiricism as a factual belief, van Fraassen suggests. The alternative is to construe empiricism is a 'stance':

> ...a philosophical position can consist in something other than a belief in what the world is like. The alternative is a stance (attitude, commitment, approach) ... What empiricists have shared over the centuries... has not most obviously been a set of beliefs... [empiricism is] an attitude, or rather a cluster of attitudes, a philosophical stance. (van Fraassen 1995, 86, 83)

Alspector-Kelly (2001, 421) notes that 'Empiricism is appropriately characterised by the attitudes of respect for science, suspicion of the comfort that explanation-by-postulate brings, and a call to experience as a safeguard against theoretical flights of fancy...' Notice that none of these things—respect, suspicion, a call—is a belief.

The key features of stances in van Fraassen's sense are:

- They are not reducible to beliefs; they are *adopted* rather than believed, like an approach or policy.
- They are *non-propositional*; so they are not to be thought of as true or false.
- They are largely *values-driven*. One adopts a stance which coheres well with one's values, both epistemic and non-epistemic.
- They are pragmatically justified in part in terms of their fruits; one adopts a stance at least partly on the basis of the *consequences* of doing so.

van Fraassen's paradigm examples of philosophical stances are empiricism, materialism, naturalism and secularism (2002). There has been much discussion in the literature about the nature and function of stances, (e.g. Boucher 2014, 2018a, 2018b, 2019; van Fraassen 1994, 1995, 1996, 2002, 2004a, 2004b, 2004c, 2007, 2011, and references therein), but for the moment we can make do with the above characterisation.

So characterized, the concept of a philosophical stance is quite flexible, and open-ended. And indeed the idea has, over the last decade or two, been developed in two quite different ways, something that has not generally been noticed: in the idea of an *epistemic stance*, and the idea of a *metaphysical stance*. We will see that these two elaborations of the nature and function of a stance, while they take us in quite different directions, can each be put to work in productive ways in the metaphysics of biology.

2 Epistemic Stances

Epistemic stances, according to Forbes, sit 'at a "meta-epistemological" level… above the level of epistemological theories and philosophies of science, which in turn sit above the level of more concrete, ground level facts (about our world's actual ontology and laws of nature, for example). Each level serves as a guide for developing beliefs and claims about things at the lower level, such that epistemic stances inform our development of theories and epistemologies of science, and these in turn inform our ontological beliefs on the basis of contemporary science' (Forbes 2017, 3329–3330). The realist stance is one such epistemic stance. The realist stance is opposed to the empiricist stance.[1] Traditional forms of scientific realism fit naturally within the realist stance, but the stance is broader than the typical beliefs with which we tend to associate scientific realism ('unobservable scientific entities exist', 'our best scientific theories are true or approximately true' etc.), involving non-belief elements such as values, attitudes and goals. The realist stance should be understood above all as involving or endorsing the demand for *explanation*, in particular explanation of the observable in terms of the unobservable (Chakravartty 2017). The empiricist rejects this demand.[2] As a number of theorists have pointed out, the empiricist and the realist endorse different *epistemic policies*: policies or strategies for forming factual beliefs (e.g. Teller 2004). This locates the disagreement at a different point from factual beliefs. It is not so much that the empiricist and the realist have different beliefs about the

[1] In the literature the realist stance is sometimes referred to as the 'metaphysics stance' or the 'metaphysical stance', but I will be reserving the term 'metaphysical stance' for something quite different, so to avoid confusion I will use 'realist stance', which is also not uncommon in the literature. Chakravartty (2024, 1311) claims this difference in usage is 'substantial', but I regard it as purely terminological.

Empiricists can of course be *common-sense* realists. The realism of the realist stance is the stronger form associated with scientific and metaphysical forms of realism.

[2] Aside from the empiricist and realist stances, Chakravartty (*ibid.*, 208–10) identifies a third epistemic stance, which he terms the 'deflationary' stance: the stance that recommends a non-factualist, pragmatist or quietist attitude (rather than one of realist commitment or empiricist suspension of judgement) towards metaphysical claims: one that does not, unlike the other stances, interpret ontological claims on face value. I shall set this stance aside for the purposes of my discussion, as Chakravartty does for the most part. See Henschen (2024) for a defence of a Kantian version of the deflationary stance in light of Chakravartty's arguments.

world or about science, but that they endorse different policies for generating factual beliefs. And these different policies reflect divergent goals, commitments, and values.[3]

The thought is that we can gain insight into what is fundamentally driving the debate around scientific realism by taking a step back from the characteristic beliefs and claims endorsed in the debate and considering, at a deeper level, the more basic value-laden *stances* that are underlying, generating and motivating those beliefs and claims. On this view the realist is a realist not principally because she thinks the evidence supports beliefs X, Y, and Z, but because she endorses a particular stance defined in terms of the value placed on explanation, or certain characteristic forms of explanation. The empiricist is an empiricist because she endorses a different stance defined in terms of other kinds of values.[4]

As Chakravartty (2017) notes—following Wylie (1986), Swoyer (2008) and others—the realism-empiricism dispute can be thought of in terms of the conflicting aims of securing explanatory power and minimizing epistemic risk. The realist thinks the benefits of explanatory power that come with committing to certain types of entity outweigh the epistemic risks associated with such commitment (given their 'distance' from the empirical evidence and their consequent lack of empirical vulnerability); the empiricist on the other hand thinks the epistemic risks of such commitment outweigh whatever explanatory power it may confer (Pils 2023).[5] The empiricist recommends venturing only with great caution

[3] The idea that realism is more a set of attitudes than a factual belief was articulated by Crispin Wright (1987). Realism, he suggested, involves a kind of compromise between the attitudes of modesty/deference (the world exists independently of us) and presumption/self-assurance (we can come to know this world).

[4] As Bryant notes (2021), we can easily identify what could be called 'epistemic stances' other than those concerning scientific ontology that are our focus here. Stances with regard to scientific ontology are just one kind of epistemic stance—one set of values and policies for belief and knowledge-generation—among others. Thus Pincock characterises epistemic stances quite broadly as 'collection[s] of policies that settle how evidence supports new knowledge claims' (Pincock 2024, 1). But I will stipulate that, for the purposes of this book, by 'epistemic stance' we shall mean 'stance with respect to scientific ontology'.

[5] Indeed the empiricist typically does not value explanatory power at all, thus it is perhaps misleading to say she believes it is *outweighed* by the need to minimise epistemic risk, as if the empiricist is *trading these off* against one another. The realist on the other hand does value both explanatory power and minimising epistemic risk, so can more accurately be described as trying to strike a balance between them, as trading one off against the another.

beyond the empirical evidence in matters of ontology, and regards the value the realist imputes to explanations in terms of the unobservable as illusory.

This is often connected with William James' point that in deciding what to believe we have two aims, to believe the truth and to avoid error (1956, 17–18). But these aims conflict: if we only cared about believing the truth we would be overly credulous, believing lots of falsehoods as well as truths; if we only cared about avoiding error we would be overly sceptical, believing very little, and thereby missing out on lots of available true beliefs. Typically in our epistemic lives we value both of these aims, and try to strike a balance between them, but the precise balance we strike will not be a matter of objective correctness or rational requirements, but a matter of values. The scientific realist and the anti-realist empiricist both value believing truth and avoiding error, but strike different balances: the former's willingness to commit to the existence of unobservable entities of science and the latter's unwillingness to do so is a reflection of the fact that in this context, the realist is willing to risk the possibility of error in order to potentially get the benefits of believing the truth, while the empiricist is not willing to risk the possibility of error even if that means risking missing out on available truth beliefs. Neither side is irrational; their dispute is values-based. This Jamesian view is a reflection of his *epistemological voluntarism* that both van Fraassen and Chakravartty endorse, which I shall discuss below.

Thus, one of the key advertised virtues of the epistemic stance approach is its ability to account for persistent rational disagreement in debates such as the scientific realism debate, by referring it to the existence of rival underlying, equally-rational epistemic stances, embodying conflicting non-belief elements: values, attitudes, goals, commitments, and the like. This leads in turn to the possibility of pluralism, relativism or quietism about the rival stances and the disputes they give rise to.

2.1 Naturalised Metaphysics, Epistemic Stances and Voluntarism

Chakravartty (2017) following van Fraassen, recommends a view that stresses stances and values. But he departs from van Fraassen in that he bases this on an analysis of different forms that *naturalised metaphysics* may take. According to the latter metaphysics ought to be sensitive to and grounded in our best empirical science (Ladyman and Ross 2007; Ladyman et al. 2013; Slater and Yudell 2017), and empirical inquiry more generally. As such it is contrasted with the a priori, intuitions-based, armchair metaphysics that dominates contemporary analytic metaphysics. What constraints this 'norm of naturalised metaphysics' imposes on metaphysical theorising is, Chakravartty suggests, not obvious. He argues that on the naturalist approach, 'metaphysical inferences' from empirical science and empirical inquiry to metaphysical conclusions are methodologically central (*ibid.*). But the metaphysical inferences one is willing to endorse are a function of one's epistemic stance. Those who adopt the empiricist stance endorse quite limited and cautious metaphysical inferences, that don't transcend the observable.[6] Those who adopt the realist stance are wont to endorse more adventurous and extravagant inferences that commit them to the existence of entities, properties and processes that go well beyond what we can observe. But these stances are, he suggests, largely a function of one's values and, in accordance with voluntarism, may be equally rationally permissible.[7] It is important to note that for Chakravartty, 'metaphysical inferences' are not straightforward empir-

[6] This suggestion, that even empiricists such as van Fraassen are engaged in a kind of naturalised metaphysics, in that, to the extent that they are not solipsist-phenomenalists, they are committed to the existence of some entities and properties that go beyond immediate sense-experience (2017, 53–54), is at odds with the self-conception of such empiricists, who believe themselves to be rejecting metaphysics entirely. Clearly Chakravartty is operating here with a very broad notion of metaphysics, according to which even my inference that the table in front of me exists counts as a metaphysical inference. Somewhat confusingly, one may endorse such metaphysical inferences without thereby adopting what he calls the metaphysics *stance*. See Psillos (2021) for a critique of Chakravartty's notion of metaphysical inference as too broad and liberal, implying that virtually every scientific or common-sense explanatory inference is a metaphysical inference. But for Chakravartty, that is precisely the point.

[7] Borge (2024) connects Chakravartty's view with the recent literature on the epistemology of peer disagreement.

ical inferences from observable evidence or data to metaphysical conclusions. Such reasoning 'has a significant a priori dimension', in that it typically involves considerations relevant to the attempt to *explain* the data, such as simplicity, coherence, fruitfulness etc. (Chakravartty 2017, 35).

Chakravartty's focus is scientific ontology, and in Part I this shall be our theme. 'Ontology' primarily concerns the question of what exists (Quine 1953). But inseparable from this is the question of what the things that exist are like (Humphreys 2013, 44–45). After all, committing to the existence of some entity is not of much use or interest unless we have some idea of what it is that we are committing to, i.e., of what its nature is, what properties it has, and how it is related to other entities. For example, in Chap. 4 we shall discuss realism and anti-realism about natural kinds, i.e. the question of whether our ontology should include natural kinds. But not surprisingly, this question has been bound up with the question of what natural kinds *are*, i.e., what they are like, what their nature is, and how they are related to other types of entity.

Chakravartty's discussion suggests a taxonomy of four broad views (i.e. stances) on how far we are permitted to extend belief with respect to ontology (in Chap. 2 we will have need to modify this taxonomy but it shall serve us adequately as a starting point for our discussion):[8]

1. To the observable only = Empiricism e.g. constructive empiricism.
2. To the explicit but unobservable entities[9] of science (electrons, protons, genes etc.). = Standard scientific realism without any 'metaphysics of science'.
3. To the implicit entities of science (laws, causation, natural kinds, properties, dispositions, modality etc.) = Metaphysics of science.[10]

---

[8] Chakravartty distinguishes between the explicit and implicit subject matter of science in his (2017). I shall discuss the distinction in more detail in Chap. 2. The distinction between the four stances is my own, and I am not suggesting Chakravartty would necessarily endorse it, though it seems to me to be implicit in his analysis.

[9] For convenience I shall use the term 'entity' throughout, but I shall be using it in a broad sense, where it may include forces, processes, types of causation, and so on. An entity in my sense need not be a 'thing' or 'object'. Chang (2022) adopts a similar usage.

[10] We shall see that there are stronger and weaker versions of the metaphysics of science stance, and that the stronger versions arguably shade into the armchair metaphysics stance; see Chap. 2, Sect. 2.1.

4. To non-scientific metaphysical entities (haecceities, quiddities, substances, Platonic forms, grounding relations, truthmakers, hylomorphic compounds, monads, necessary beings, the absolute, etc.) = Traditional armchair metaphysics.

There is, Chakrarvartty suggests, a *continuum*—though one we have out of convenience packaged into several discrete views—from metaphysical inferences very closely connected to the empirical evidence, through to metaphysics completely divorced from any grounding in empirical or scientific evidence.[11] Many want to get off the 'slippery slope' at some point. Naturalised metaphysics is indeed defined as the view that we must get off before reaching 4. But where? Where we get off depends, he suggests, on judgments of epistemic risk—how much of this we are willing to tolerate to get the benefits of explanatory power, for instance—which in turn is a function of value-laden epistemic stances, as we outlined above.[12]

A few points of clarification.

Firstly, only the first three views above count as potentially naturalised metaphysics, for Chakravartty. Once we adopt 4, we are positing metaphysical entities with no apparent connection at all with the sciences; the norm of naturalized metaphysics is violated.

Secondly, all of these views except empiricism are examples of the realist stance.

Thirdly, each stance will typically include the permissions of the stances weaker than itself: the standard scientific realism (SSR) stance will license belief in observables, the metaphysics of science (MoS) stance will license belief in observables and the explicit entities of science, and so on.

Fourthly, I am defining the SSR stance as a stance that rejects the MoS stance, i.e. one that does not license beliefs in the implicit subject matter of science. If someone accepts both the explicit and implicit entities of

[11] See Robus (2015) for a critique of this picture that focuses on the theory-ladenness of experience.

[12] Magnus (2018) notes a similarity between Chakravartty's notion of epistemic risk and the concept of inductive risk at work in discussions on science and values, where it is non-epistemic as well as epistemic values that contribute to the rationality of belief or suspension of judgement; see Douglas (2000).

science I will interpret them as endorsing the MoS stance and rejecting the SSR stance. The stances exclude one another: one cannot be part of the other. Of course many theorists who accept the MoS stance would think of themselves as *also* accepting something like the SSR stance, indeed would think of the MoS stance as providing the necessary metaphysical underpinnings of the SSR stance. But we can define a stance that licenses belief only in the explicit, not the implicit, entities of science: I am stipulating that this is what we shall mean by the SSR stance. *This* stance is rejected by those who embrace the MoS stance. Understanding the latter in this way will make our discussions neater and clearer: for instance, we don't think of the SSR stance as including the empiricist stance, despite the fact that the former usually licenses belief in observable entities. The stances exclude one another. While this may be partly a verbal question, if the relation between the MoS stance and the SSR stance can be made to mirror the relation between the latter and the empiricist stance, it will help to keep things clear.

Fifthly, 'standard scientific realism' is often used in the literature to refer to scientific realism in its traditional form, the view that our best scientific theories are true or approximately true, and unobservable, theoretical entities exist mind-independently. More recent versions of scientific realism such as 'entity realism', which is committed to the existence of entities but not the truth of theories (Cartwright 1983, Hacking 1983), and 'structural realism', which is committed to the existence of structure but not entities (things or objects)[13] (Ladyman 2014), would not count as standard scientific realism in this sense. I am using 'standard scientific realism' to mean the stance according to which beliefs in the existence of the explicit but unobservable entities of science—but not beliefs in implicit entities of science—are permissible.[14] For me both standard scientific realists in the traditional sense—those that are committed to the truth of theories and the existence of entities—and entity realists, would count as adherents of either the SSR stance, if they reject inferences to the implicit entities of science, or the MoS stance, if they

[13] 'Ontic' structural realists think the entities don't exist, only structure, while 'epistemic' structural realists think we only have *knowledge* of structure; they are agnostic about the existence of the entities (Ladyman 2014).

[14] In Chap. 2 I shall say more about what the idea of 'permission' here involves.

accept them. Structural realism requires either the SSR stance, or the MoS stance, depending on whether the 'structure' that is the focus of their realism is considered part of the explicit or implicit subject-matter of science, something on which I don't have firm views.[15] The fact that structural realists reject (or are agnostic about) scientific 'entities' in the sense of objects or things does not put them outside of these stances since, firstly, I am using 'entity' broadly, so that it includes 'structure', and secondly, the stances *permit*, but do not *compel*, belief in entities in the narrower sense.

Sixthly, I am using 'metaphysics of science' to designate a stance with respect to scientific ontology. It is more commonly used to refer to a philosophical field. And one could, presumably, contribute to this field, and thus be 'doing' metaphysics of science without adopting what I am calling the metaphysics of science stance. One could, that is, engage in debates about the existence and nature of the implicit entities of science from the standpoint of a different stance, say, the empiricist stance. I will suggest in Chap. 4, for instance, that the epistemic account of natural kinds is (or at least can be interpreted as) an expression of the empiricist stance, yet it is still clearly a view within the *field* of the metaphysics of science. This distinction between metaphysics of science as stance and as field of inquiry will be important to keep in mind in what follows.

Seventhly, naturalised metaphysics, for Chakravartty, takes its cue not necessarily from science *simpliciter*, but from *empirical inquiry*, which is neither necessary nor sufficient for scientific inquiry: not necessary, since the sciences incorporate a priori and highly speculative components (Chakravartty 2017, 81; Robus 2015, 848), and not sufficient, since empirical inquiry is not limited to scientific inquiry.[16] Of course much of science counts as empirical inquiry—the most reliable form we possess—and much empirical inquiry is scientific, such that we will not go far wrong if we think of naturalised metaphysics as indexed to science, and I

[15] Whether or not this structure is part of the implicit subject matter of science, some prominent structural realists, such as Ladyman and Ross (2007), clearly embrace the MoS stance in their commitment to the reality of things like objective modality, as I discuss further in Chap. 2.

[16] Slater (2017, 74) similarly suggests that the naturalist ought to insist that metaphysical claims not conflict with well-established facts from any source; while this will often be science, it need not always be.

will often speak that way in this book. But this does not always hold, hence Chakravartty's suggestion that the inference from my seeming to see a tree in front of me to the conclusion that there is indeed a tree in front of me counts as a rudimentary metaphysical inference in accordance with naturalised metaphysics, even though there is nothing 'scientific' involved. That is what motivates the continuum above from empiricism through to various forms of scientific realism: as we move along it we encounter increasingly adventurous and risky forms of inference from—i.e. epistemic 'distance' from—observation and empirical evidence, not necessarily from science. The thought is that the metaphysical inference to the tree in front of me is slightly risky (given the possibility of scepticism about the external world); the inference to the existence of electrons is more risky; and the inference to the existence of Platonic forms is more risky still. For this to be coherent as a single spectrum of epistemic risk—as I believe it is—the quantity that is increasing must be distance from empirical evidence, not science *per se*. This is not uncontroversial; some friends of naturalised metaphysics want to index it to science only (e.g. Bryant 2021, Arenhart and Arroyo 2021). The primary reason for preferring Chakravartty's conception is, in my view, the coherence, explanatory value and fruitfulness of the continuum, when applied to different issues; the chapters in Part I will help make the case for this claim.

Eighthly, Chakravartty's continuum represents a way of thinking about naturalised metaphysics in which posits 'closer' to science and empirical evidence are, other things being equal, more naturalistically respectable than those further down the 'slippery slope': believing in Platonic forms is, on this picture, more dubious from a naturalistic perspective than believing in bosons. While this is the sense of naturalism we shall primarily be working with, we shall see there is another, non-equivalent form that features in some of the debates I shall be discussing (and which will become important in Chap. 4)—what I will call 'conceptual naturalism', to contrast it with the more ontology-oriented form Chakravartty is concerned with.[17] According to conceptual naturalism, when one is offering

[17] I refrain from calling this latter form 'ontological naturalism', to avoid confusing it with a quite different doctrine that shall be discussed in Chap. 6.

a philosophical account or definition of some scientific or scientifically-relevant concept—'reduction' or 'natural kind' or 'species' or whatever—one should be guided by, and not depart too radically from, the way scientists in the relevant field think about and deploy the concept, or related concepts, and surrounding theories. One may violate conceptual naturalism in this sense by offering an account of some concept which departs too drastically from scientific theory and practice—the classic accusation of philosophy dictating to science—even if one is not thereby *positing entities* with questionable naturalistic credentials, so not violating naturalism in Chakravartty's ontology-oriented sense. Conceptual naturalism interacts in complex ways with epistemic stances. We shall see, for instance, in Chap. 4 that the desire to honour conceptual naturalism has led some theorists of species and natural kinds away from the realist stance, or, more precisely, has led them to embrace positions that do not *require* the realist stance.

Finally, a word about epistemic risk. Chakravartty suggests that beliefs that are less empirically vulnerable are, *ceteris paribus*, more epistemically risky. This might seem to get things backwards; does not greater empirical vulnerability entail *greater* epistemic risk? This is the sense of 'risk' we find in Popper, when he claims that scientific theories are more risky to the extent that they are more falsifiable: here risk is understood in terms of the likelihood of being empirically disproven.[18] Unfalsifiable claims count as 'safe' and non-risky on this view. Chakravartty on the other hand, as I interpret him, is understanding epistemic risk not in terms of the likelihood of getting things wrong where this is shown by the evidence, but in terms of the possibility of *getting things wrong but not knowing it*. A metaphysical claim that is not falsifiable—say, 'God exists'—is risky in this sense because if it was wrong there would be no way of telling—at least, no empirical way. What we risk is not having to reject our beliefs because we recognize they have been empirically falsified, but continuing to hold false beliefs because there is no way their falsity can be recognized. And conversely a highly empirically vulnerable claim is less risky in this sense

[18] In fact there is confusion about this in the literature: Arenhart and Arroyo (2021, 2), and Robus (2015, 850), mistakenly construe Chakravartty's notion of epistemic risk as Popper's opposite notion.

because if false, its falsity is likely to be brought to light. After all, persisting in error of which one is ignorant is presumably less desirable than having one's errors corrected by the evidence.[19] Clearly Popper's sense of risk and Chakravartty's are each legitimate; we need not choose between them. But it is Chakravartty sense that we shall be working with here.[20] It is in fact the one that has always been operative in the debate over scientific realism, in which the realist's beliefs, transcending the observable, have been understood to be more epistemically risky—more a matter of 'sticking one's neck out' epistemically—than the more cautious commitments of the empiricist.

I do depart somewhat from Chakravartty's construal of epistemic risk however. He suggests that just as empirical vulnerability reduces epistemic risk, so too does explanatory power: a hypothesis that is risky due to its distance from the empirical evidence may have its risk reduced by its explanatory power (*ibid.*, 87). The way I understand epistemic risk however, it has just one dimension, not two: it increases with distance from the evidence, but *it itself* does not decrease with an increase in explanatory power: the hypothesis in question retains the same degree of risk, it is just that this risk can be *offset*, or *outweighed*, by its explanatory power. A claim about invisible, intangible fairies that grant wishes, and a claim about non-actual possible worlds are, let us say, equally distant from the empirical evidence, score equally on empirical vulnerability (very low), and are thus, in my view, equally epistemically risky; but in the latter case, the explanatory power of the entities may *outweigh* (rather than *reduce*) this risk (which is not the case for the fairy hypothesis), such that the hypothesis about possible worlds may, depending on one's stance and epistemic commitments, be, all things considered, worthy of belief and a potential item of knowledge. Whether or not this is just a

[19] Popper, of course, agrees with this. The riskiness he associates with empirical falsifiability is he thinks desirable: the riskier a theory is, i.e. the more falsifiable it is, the better.

[20] Humphreys (2013, 71–2) suggests that beliefs based on ordinary perception, as well as speculative metaphysical claims based on intuition, are in our age *more* epistemically risky than the deliverances of science, which would imply that the appropriate stance today for those who wish to minimise epistemic risk is actually the SSR (or possibly MoS) stance, not the empiricist stance. I am dubious however that the belief in the existence of scientific entities like bosons or species is less risky than common sense beliefs based on sense-experience, like 'there is a table in front of me' (think of the Moorean facts).

terminological difference between Chakravartty and myself I'm not certain, but it seems to me that my construal is the only way to make sense of the thought (endorsed, apparently, by Chakravartty) that the realism-empiricism debate is at bottom a question about *balancing*, or *trading off*, the values of explanatory power and the minimization of epistemic risk. Only if epistemic risk is independent of explanatory power, and unaffected by it, can we make sense of the idea that the two values compete, and one may be traded off against the other, or prioritized over it. In running into the burning building I am risking my life, but I may judge that the potential benefit—saving the life of the baby trapped inside—outweighs the risk, makes the risk worth taking, such that all things considered, I ought to try to save the baby. But this benefit does not *reduce* the risk I run. The risk is independent of and unaffected by the potential benefits.

Stance Voluntarism

As I noted above Chakravartty follows van Fraassen in endorsing *voluntarism* about stances.[21] Voluntarism in epistemology generally is the view that the constraints on rationality are minimal and relatively easy to satisfy, essentially that one is being rational so long as one is not being irrational, where to be irrational is to violate logical or probabilistic rules of consistency and coherence, or, a more pragmatic constraint, to sabotage or undermine oneself by one's own lights or standards.[22] Rationality is

---

[21] I discuss voluntarism here in relation to epistemic stances, but it applies equally to metaphysical stances.

[22] There are questions, broached in the recent literature, about whether these minimal constraints on rationality, when applied to stances, are too permissive (van der Merwe 2024), or not permissive enough (Bryant 2021); and whether there might be other, stance-neutral criteria for assessing an epistemic stance: for instance, whether a stance that recommends a policy that gives rise to preponderance of false beliefs and invalid inferences may be regarded as *ipso facto* a defective epistemic stance, given that the point of an *epistemic* stance after all is to produce knowledge and justified belief (Chakravartty 2017); and whether this criterion can be assimilated to the minimal voluntarist conditions for rationality (as Chakravartty thinks, *ibid.*) or is independent of them. While recognising the importance of these issues, fully engaging and resolving them is beyond the scope of this work. I cannot hope to provide a complete defence of the picture I am presenting in Part I of naturalised metaphysics, epistemic stances, and voluntarism (though I attempt to defend aspects of it in Chap. 2); my aim is rather primarily to outline and then *apply* this framework, to see what

thus merely 'bridled irrationality'. As van Fraassen puts it, on the voluntarist view, when 'evidence and reason reach their limit', we may just 'make up our minds' (2002, 91–2). On this view rationality is about permission not obligation. Rational beliefs or stances are permitted, not compelled, beliefs or stances (Steup 2011, 23). So we have quite a low bar on a stance counting as rational. If rationally binding rules of inference do not determine which stance to adopt, then one may choose between equally-rational options on pragmatic or values-based grounds.[23]

The role of values in belief or stance choice is typically thought to be central to the voluntarist conception, as is the idea of a certain degree of freedom to choose among rational options (we shall see in Chap. 2 however that whether the notion of *free autonomous choice* is really essential to voluntarism may be questionable). But voluntarism in epistemology is traditionally associated with the stronger claim that many or all of our beliefs are under the direct control of the will, something that seems implausible at least with respect to a great many of our beliefs. The form of voluntarism we are considering need not be committed to such a claim. As van Fraassen notes, 'voluntarism in epistemology does not have to involve the idea that we can believe at will. That making up one's mind involves some choice along the way, or that there is an inescapable role played by value judgements in managing our opinion, does not imply that it is simply a matter of free choice made 'at will'…' (2007, 351).

Moreover, as Chakravartty notes, we can construe stance choice in voluntarist terms without having to go full-voluntarist will respect to the beliefs generated on the basis of or against the background of a stance (2017, 219–220; 2024, 1311). One's voluntaristic choice of a stance might influence the types of beliefs that are options for one, but the beliefs one subsequently adopts can and should be adopted on standard evidential grounds. This kind of contrast is similar to one Foley draws

---

work it might do in helping us to understand and analyse various issues and debates in the metaphysics of biology. That can help to prove its value and justify its adoption. See Elder (2019) for a useful recent defence of, and application of, the epistemic stance (and voluntarist) approach.

[23] I discuss the basis of stance choice, and the question of whether and how a stance may be rationally evaluated, in Boucher (2018a). In Boucher and Forbes (2024), we argue that epistemic stances such as realism and empiricism are or should be adopted on the basis of both one's values, *and* the pragmatic and epistemic benefits they give rise to, or are believed to give rise to.

with respect to the effect of pragmatic or ethical factors on belief-justification (1991, 1994, 2000[24]). In picking wild mushrooms for dinner, the evidential standard that my belief that a mushroom is poisonous has to meet to count as epistemically justified is pretty low, since it doesn't matter much if I am wrong about this. But the standard of evidence that my belief that a mushroom is non-poisonous must meet to be justified is much higher, since being wrong in this belief may have life-threatening consequences. Foley also makes the point that as finite human beings who must live our lives, and have many ends other than our epistemic ones, there are enormous practical constraints on the amount of time we can spend testing and confirming our beliefs. Only a small minority of the beliefs we hold will be deemed important or interesting enough to spend any real amount of time and resources testing; for those beliefs, the standards of epistemic justification will be much higher than for our other, less important beliefs. Pragmatic considerations help us to decide which beliefs, or potential beliefs, to test and try to confirm or disconfirm, and also how much time and energy to invest in doing so. Foley argues however that these non-epistemic factors contribute *indirectly* to the justification of beliefs. They constrain the process of belief formation, but the beliefs that are formed by the processes they so constrain are still held primarily for truth-linked, that is, epistemic, not pragmatic reasons. They specify the stringency of our standards of evidence; they do not replace evidence as the ground of belief. 'They … impose constraints on inquiry, but subject to those constraints one's aim will be to determine which beliefs are true, not which beliefs are useful.' (2000, 186) The same can be said about stances and beliefs concerning scientific ontology. Value-laden stances indirectly constrain and influence the process of belief formation and justification, but within those constraints, it is the evidence, not the stance, that ought to direct one what to believe. See Chakravartty (2017, 219–220). I will discuss stance voluntarism in more detail in Chap. 2.

This concludes our initial discussion of epistemic stances. The majority of recent work on philosophical stances has been concerned with

[24] See also the extensive recent literature on 'pragmatic and moral encroachment' (Kim and McGrath 2019).

epistemic stances, in the context of the realism-empiricism dispute in the philosophy of science. But in this book we shall be also exploring philosophical stances of a quite different kind: what I will call 'metaphysical stances'.

3 Metaphysical Stances

Van Fraassen's two paradigm examples of philosophical stances are empiricism and materialism. As we have seen, discussion of the empiricist stance, and its rival the realist stance, has led to the more general formulation of the notion of an epistemic stance. The case of materialism however points us in a quite different—and, I will, argue, neglected—direction.

Van Fraassen's argument for construing materialism as a stance draws on what has come to be called 'Hempel's Dilemma', having been first raised by Carl Hempel (1980). Materialism (or 'physicalism'[25]) is the view that, as it is often put today, the physical facts fix all the facts. But what is a physical fact? What is a physical object or property? What is it to be physical? The naturalist, reasonably, advises us to defer to our most fundamental, and successful, science of the physical, modern physics, to answer this question. But we then encounter a dilemma. If we look to present-day physics to tell us what it is to be physical, we face the problem that we have every reason, given the history of science, to expect that many or even most of the entities, forces and processes referred to in current physics will in the future be discarded, as theory develops and paradigms are overturned. In that case a physicalism tied to the ontology of current physics would turn out to be false: it would, implausibly,[26] die along with the entities of current physics to which it was yoked. Perhaps

[25] Sometimes 'physicalism' is construed as a more recent and permissive replacement for the more metaphysically restrictive 'materialism' (Wilson 2006; Brown and Ladyman 2009), but other times they are treated as synonymous, which I shall do here.

[26] As I have argued elsewhere (Boucher 2020) the problem with the current-physics horn is not just that in that case materialism would turn out to be false, but that it is very implausible that if the ontology of current physics was overturned, *this would make* materialism false, especially since the latter has apparently survived successive revolutions in physical theory in the past.

we should instead link physicalism to the ontology of some future, ideal or completed physics. The problem with this proposal is that it doesn't seem that there is anything very meaningful we can say about the content of this future, completed physics other than that it the science of the future that will provide us with the compete and correct ontology of the physical, give us the complete and correct theory of the physical, etc. It is easy enough to see that a physicalism tied to such a notion would be vacuous. As Hellman notes, '...we lack any general criterion of 'physical object, property or law' framed independently of existing physical theory' (quoted in Melnyk 1997, 623).

Van Fraassen accepts that Hempel's dilemma is fatal for the standard understanding of materialism as a factual belief or truth-apt proposition—something along the lines of 'the physical facts fix all the facts' or 'everything that exists supervenes on or reduces to the physical'. But he suggests that most discussions of the dilemma present a false dichotomy: either the dilemma is accepted as unanswerable and materialism is consequently dismissed (as meaningless or clearly false); or an attempt is made to respond to the dilemma by embracing one of its horns, and trying to take the sting out of it, to mix metaphors.[27] Van Fraassen claims however that we can accept that the dilemma shows that the prospects of materialism construed as a factual belief being non-trivially true are likely hopeless, without thereby dismissing or undervaluing the role and importance of materialism, *if we construe the latter not as a factual belief but as a stance in his sense*, that is, a cluster of values, attitudes, goals, commitments and policies, characterised in particular by a kind of ontological deference to physics as constituted at the time. So construed, materialism may play an important role in science and philosophy (*ibid.*, 60. See also van Fraassen 1996; Ney 2006; Dupré 1993, 91):

> I propose the following diagnosis of materialism: it is not identifiable with a theory about what there is but only with an attitude or cluster of attitudes. These attitudes include strong deference to the current content of science in matters of opinion about what there is. They include also an

[27] For example Melnyk (1997) and Vicente (2011) embrace the 'current physics' horn of the dilemma while Dowell (2006) embraces the 'future physics' horn.

> inclination (and perhaps a commitment, at least an intention) to accept (approximative) completeness claims for science as actually constituted at any given time. (2002, 60. See also his 1996)[28]

I am not concerned at this point with whether van Fraassen is correct in claiming, against a number of contemporary philosophers, that Hempel's dilemma establishes that materialism is unsalvageable as a factual belief.[29] I present his argument merely in order to exhibit *his* grounds for thinking of materialism as a philosophical stance. The key point for us is that if materialism is a stance it is seemingly not an *epistemic* stance, like the empiricist or realist stances. It is not an epistemic policy or strategy for generating knowledge in the same way. It does not in any very obvious way 'inform our development of theories and epistemologies of science'. It is not a standpoint from which the legitimacy of metaphysical inferences, or certain ontological beliefs, can be legislated, or at least, not in the way an epistemic stance is. It does not belong on Chakravartty's continuum. It is not characterised by a particular solution to the problem of how to balance the competing values of explanatory power and minimising epistemic risk. It is rather a *metaphysical position reinterpreted as a stance rather than a factual belief*.[30] It is what I shall call a 'metaphysical stance'.[31]

3.1 Empiricism and the Role of Metaphysical Stances in Science

While one does not have to be an empiricist to accept arguments such as van Fraassen's argument for construing materialism as a stance, the metaphysical stance approach—the interpretation of certain metaphysical

[28] As van Fraassen sometimes puts it, while the empiricist stance involves deference to or respect for the *methods* of science, the materialist stance involves deference to or respect for the *content* of science (van Fraassen 2002).

[29] I do think he is correct in this claim; see my (2020); but am not defending that here.

[30] Van Fraassen tends to see the empiricist stance and materialist stance as rivals. If my analysis is correct, they are not rivals in the way that the realist stance and empiricist stance are rivals. And indeed some, such as Ladyman (2011) accept both.

[31] Again, not to be confused with 'metaphysical/metaphysics stance' as it is used by writers like Chakravartty to refer to what I am calling the 'realist stance'.

views and theories as stances—is particularly useful for the empiricist, in particular when it comes to theorising the relation between metaphysics and science.

It is widely accepted that science and scientific theories are influenced by metaphysical assumptions, commitments and presuppositions. The old logical positivist view that there is a strict line separating (meaningful) science from (meaningless) metaphysics is no longer tenable. Yet empiricists have retained their anti-metaphysical attitudes, and still want to defend a view on science on which science is to a very considerable extent innocent of metaphysics, at least as the latter is traditionally understood (Chakravartty 2010). The metaphysical stance approach offers the empiricist a means of reconciling these seemingly contradictory claims.

Van Fraassen is well-known for his withering critique of contemporary analytic metaphysics (2002) as 'coherent nonsense', a critique that is largely shared, as we shall see, by naturalists such as Ladyman and Ross (2007), though the latter seek to preserve and reform, rather than entirely discard, metaphysics. But there is in van Fraassen's work on stances a more positive empiricist program with respect to the relation between metaphysics and science: 'I … lack sympathy for metaphysics, though not in general: only for pre-Kantian metaphysics—and then only if practised after Kant' (van Fraassen 1989, viii).

We can, I suggest, find resources in van Fraassen's work to defend the claim that the way to reconcile the undeniable role of metaphysics in science with the empiricist's hostility to metaphysics in general and its contamination of science in particular, is to see science and scientists as influenced by metaphysical positions and commitments reconceptualised as metaphysical stances. For instance, in his (1996), van Fraassen argues that metaphysical positions, such as materialism and naturalism, understood as stances shape the practice of science largely through their inspirational, motivational effects, rather than through their content as factual theses about the world. 'There is… a way in which materialism can play a role in shaping science… not through the content of its theses, entering as assumptions, but through attitudes for which these theses are mere codes' (1996, 151). He writes that if metaphysical speculation 'has great heuristic and inspirational value, let's encourage scientists to so far forget themselves as to constantly engage in it' (2004b, 181).

This is what I am calling a metaphysical stance. It is a working assumption of this book that both the epistemic stance approach and the metaphysical stance approach are potentially useful and instructive, when applied to different topics in the metaphysics of biology. I will return, at the end of the book, to the question of how the two paradigms may be related, after having, in the main part of the book, applied each of them to several issues in the metaphysics of biology.

3.2 A Generalised Explanation for Philosophical Impasse

There is a further reason for discussing metaphysical stances as well as epistemic stances. A central motivation and justification for the epistemic stance analysis is the 'impasse' or 'stalemate' character of the realism-empiricism debate in the philosophy of science. Forbes writes, '[a]fter 35 years of debate between scientific realists and constructive empiricists, philosophers of science seem to have reached an impasse… it would seem that the most we can say is that there are versions of scientific realism that are coherent, plausible, and reasonable philosophies of science; but at the same time, it would seem we can say the same for constructive empiricism…' (Forbes *ibid.*, 3328). Realists have been unable to persuade empiricists to abandon their views and vice versa; no convergence or consensus is in sight. The epistemic stance interpretation of the debate is intended to be an explanation for this situation. Thus in the literature van Fraassian stances are discussed almost exclusively in the context of this particular debate in the philosophy of science.

But there is an obvious objection to this. Surely this is the situation throughout philosophy! Philosophy in general is characterised by a lack of consensus, by theorists finding that they are unable, for the most part, to rationally persuade their opponents to change their minds, and by the elaboration of internally consistent, coherent, empirically adequate, opposing theories or views that persist over time.

There are three possible responses to this objection. First, one may try to claim that there *is* something special about the realism-empiricism dispute, that the stalemate or impasse character of it is more pronounced or problematic, that opposing views are more entrenched, argument less effective, than is the case elsewhere in philosophy. Thus Slater writes, '[t]he intractability of the scientific realism debate is legend. It is also rather remarkable, considering the impressive set of resources available for its resolution. How is it that we have not yet reached a firm consensus on this question? Doesn't this failure cry out for explanation?' (Slater 2021, 33). This is implausible. When we look at other areas of philosophy we find, it seems to me, precisely the same situation as the realism-empiricism dispute: rival detailed, coherent, carefully argued positions with respect to some topic are developed and defended, become entrenched, and can't be dislodged by argument. The debates do not move towards resolution. Think of physicalism vs. dualism in the philosophy of mind, internalism vs. externalism in epistemology, deontology vs. consequentialism in ethics, and many others.[32] The second response concedes this point (that there is nothing special in this respect about the realism-empiricism dispute), but still claims that the epistemic stance interpretation is a good explanation for impasse in only the realism-empiricist dispute. If E is a good explanation for some phenomena P having property F, the observation that other phenomena Q+ also have property F, but that E is not obviously an explanation for Q+ having F, does not necessarily mean that E is not a good explanation for P having F. Perhaps there is some other explanation for why Q+ have F. While this is a coherent view, it is, I'd suggest, unsatisfying. Unless we are furnished with some reason why the realism-empiricism dispute and the other disputes in philosophy are disunified in this way—why the explanation for impasse in the former case does not carry over to the latter—we are entitled to be sceptical of exceptionalism with respect to the realism-empiricism dispute and the philosophical stances that underlie it.

The third response—the one I favour—rejects exceptionalism. The realism-empiricism dispute in the philosophy of science is not the only dispute in which an impasse-like situation may be explained by linking views to underlying stances. Confining discussion of stances to this

[32] If it is claimed that in these disputes, argument and debate continues, whereas in the realism-empiricism dispute, argument has dried up as all sides recognise the hopeless impasse-situation, we need only point to the large number of books and articles that continue to appear defending realism or empiricism with what are at least thought to be rational and persuasive arguments.

particular dispute is a mistake.[33] We need to explore the role of stances elsewhere in science and philosophy. That is what I do in Part II. As well as epistemic stances—various forms of realism and empiricism with respect to scientific (or non-scientific) ontology—there are also metaphysical stances.[34] I am certainly not defending the strong claim that the lack of progress towards resolution in *every* dispute in philosophy—even every dispute between the sort of carefully worked out, coherent and entrenched positions I have mentioned—is a result of the positions being expressions of underlying stances. That is a bold claim that might be correct, but I remain agnostic on it here.[35] I will argue for the more limited claim that stances play an important role in science and philosophy beyond the realism-empiricism dispute in the philosophy of science, and I look at three disputes in Part II where this seems to be the case.[36] How much further this may be generalised is a question for future work.

4 The Metaphysics of Biology

In this book I will be interpreting a number of issues in the metaphysics of biology in terms of the stance framework(s) outlined above. The metaphysics of biology deals with questions such as: Are species real? Are species individuals or natural kinds? Do species or other biological kinds have essences, and what are these essences? Is there design and teleology in the natural world? Is there a fact of the matter about the level at which natural selection acts? Is natural selection a force? Is reductionism in biology correct, and if so, what form? Are there emergent properties in biology? What is biological individuality and what entities possess it? Is

[33] I agree that what I am calling *epistemic* stances do relate only to the realism-empiricism dispute with respect to scientific ontology (though I will suggest they apply more widely than the traditional scientific realism debate). But epistemic stances are not (*contra* much of the recent literature) the only kind of stances: there are also metaphysical stances.

[34] It is clear that van Fraassen always intended the stance idea to apply quite widely (hence his suggestion that materialism, naturalism, and secularism are stances, not just realism and empiricism); the restriction of it to the realism-empiricism dispute is a more recent feature of the literature.

[35] In my (2018a) I raise the possibility that the stance idea could, among other things, help account for the lack of progress in philosophy generally, but do not defend the claim in any detail.

[36] Having said that, the arguments I offer in Part II for treating the positions in question as stances are not primarily related to the impasse-character of the disputes. While that is in the background, the specific arguments I offer will be somewhat different.

function prior to structure or structure prior to function in the natural world? I shall be dealing with some (though not all) of these topics. I shall suggest that debates about some of them naturally fit the epistemic stance framework, and others naturally fit the metaphysical stance framework.

Why am I focusing on the metaphysics of biology? My work in this area over a number of years has convinced me that the stance approach—both the epistemic and the metaphysical stance approach—provides illumination with respect to a number of topics within it. But I do not claim that there is anything special about biology or the philosophy of biology that makes the stance approach *particularly* appropriate or fruitful with respect to it, and it seems doubtful that this would be the case. Still, there is a good reason for my exclusive focus on the metaphysics of biology. The majority of examples and applications in the literature on stances, and the relation between science and metaphysics, still come from the hard sciences,[37] so in focusing rather on biological examples and debates, I am offering a different and novel perspective, and enriching the discussion by bringing the metaphilosophical and metaphysics of science debates into contact with contemporary issues in the philosophy of biology.[38] For example, as I note in Chap. 2, there has been surprisingly little discussion of naturalised metaphysics in the context of the metaphysics of biology. Many advocating a naturalised metaphysics have a heavy physics-focus. So discussing naturalised metaphysics and epistemic stances in the context of biology provides something of a corrective or counterweight to the dominant physics orientation in the literature. My book can thus be seen as a contribution to a broader trend within philosophy of biology that seeks to correct and counter-balance the narrow and limiting focus on the hard sciences that has traditionally predominated in the philosophy of science and the scientific realism debate. The situation is improving, to be sure, but we still have a way to go before this physics-centrism is dislodged for good and all.

---

[37] For example, Elder (2019) applies the stance voluntarism approach to the debate about the existence of the multiverse in physics.

[38] As Triviño (2023, 51) also does: 'The debate in Metaphysics of Science concerning the interaction between metaphysics and science has been mainly approached from the perspective of the scientific discipline of physics. In this paper, I address this debate from a different framework, namely that of biology.'

5 Plan of the Book

The plan of the book is as follows. In Part I I will first explain, and offer a detailed defence of, the general picture of naturalised metaphysics and epistemic stances I briefly sketched above (Chap. 2). I shall then consider four issues in the metaphysics of biology (Chaps. 3, 4, 5 and 6), and argue that construing them in terms of the epistemic stance framework is fruitful in understanding the debates and the commitments of the positions therein. These are primarily matters of ontology, and of the metaphysical inferences that philosophers draw from science—so long, that is, as the stances in play are naturalistic.

As I noted, one of the advantages of the epistemic stance approach is its ability to explain persistent rational disagreement in matters of scientific ontology. These benefits can be seen when we apply the account to the metaphysics of biology. One key theme in Part I is that, while certain views on scientific ontology can and should be adopted on the basis of evidence and rational argument, this always takes place against the background of stance choice, either explicitly or implicitly. Only if certain stances have been adopted, will certain forms of evidence or patterns of argument be admissible. Arguments and evidence relating to the existence of certain entities will only be relevant if one has adopted a stance on which ontologically committing to entities of that kind is permissible, in a sense that I will explain in Chap. 2.

In Chap. 3 I shall look at the question of realism vs pluralist conventionalism about the units of selection, a central issue in the metaphysics of biology. It lends itself well, I will argue, to the epistemic stance interpretation. Realism about the units of selection as explicit entities of biology requires at least the SSR stance, while conventionalist pluralism about the units of selection at least sometimes is motivated by the values and attitudes characteristic of the empiricist stance. The debate between various forms of realism and pluralist conventionalism in this area is familiar, but has not been interpreted in terms of the epistemic stance, scientific ontology, and naturalised metaphysics framework; that is something I aim to do in this chapter. I shall present a detailed argument for realism about the units of selection, consider the ways in which it presupposes the realist stance, and explore ways the empiricist anti-realist might respond. The upshot will be the articulation of a kind of pluralism at a higher level: pluralism between the two equally rational epistemic stances, the realist and the empiricist stances, each yielding characteristic positions on the reality of the units of selection.

Chapters 4 and 5 address the metaphysics of biological kinds. Interpreting these issues in terms of epistemic stances is fruitful firstly in clarifying the nature of the disputes in meta-level terms, offering a fresh approach to an old problem, and secondly in opening up the space for a pluralism with respect to the value-laden epistemic stances that yield different first-order ontological views in these domains. I suggest in Chap. 4 that we can discern a spectrum of views from anti-realism with regard to species and kinds that typically takes the form of pluralism or nominalism, and is sometimes motivated by the empiricist stance, through to species or higher-taxa realism—requiring at least the SSR stance, the epistemic stance that licenses ontological commitment to the explicit subject matter of science—through to views that require the MoS stance, the stance that licenses belief in the existence of the implicit subject matter of biology, such as natural kinds, essences and dispositions—the position characteristic of the realist program associated with philosophers such as Alexander Bird (2007) and Brian Ellis (2001). The epistemic stance framing helps to uncover the metaphysical and epistemological commitments of the various positions taken in the recent debates concerning natural kinds. The epistemic account of natural kinds defended by Magnus (2012), Dupré (2002) and Ereshefsky and Reydon (2015) is, I argue, often motivated by the kind of rejection of metaphysical explanation characteristic of the empiricist stance, while the views associated with thinkers such as Boyd (1999) and Franklin-Hall (2015)—sometimes called weak realism, sometimes called anti-realism about natural kinds—involves, I argue, a somewhat unstable combination of the empiricist and realist stances. The stronger 'metaphysical' account of kinds defended by thinkers such as Bird (2018), purports to be an expression of the MoS stance, although it is one which, in its a priori speculations about the metaphysical status of kinds—whether they are universals, or are sui generis, etc.—arguably risks veering off the cliff of naturalised metaphysics into armchair metaphysics territory. This at-times conceptually confusing literature becomes clearer once we interpret it in terms of the approach that I recommend.

In Chap. 5 I look in detail at essences and essentialism in biology, with a focus on species essences. I outline and defend a particular account of what essences are, and use it to classify views on the metaphysics of species as essentialist or non-essentialist, in a way that runs counter to the way some of these views tend to be construed. This will, in turn, enable us to identify the underlying epistemic stances that are required for different views on the metaphysics of species.

Chapter 6 focuses on functions and teleology in biology. Views on the metaphysics of function, teleology and design in the natural world include strongly realist views, weaker realist views (naturalistic reductionism) and anti-realist (e.g. eliminativist) views. These are views in scientific ontology: they concern the metaphysical inferences we should draw from evolutionary science. Does the science license the inference that teleology and design exist in the natural world in a real, literal and irreducible sense, as emergentists such as Cameron (2004) claim? Or does it license rather the inference that teleology and design exist in the natural world in a real, literal but reducible sense (as Dennett (2017) argues)? Or should we rather infer from modern evolutionary theory that teleology and design do not form part of the natural world (as Ghiselin (1994) argues)? Again, we find that views on scientific ontology are largely a function, or outgrowth, of epistemic stances. Realism about teleology and functions as implicit entities of science requires at least the MoS stance. Naturalistic reductionism about functions requires only the weak form of this stance, while strong realism requires at least the strong form of the stance, and possibly the armchair metaphysics stance, depending on one's perspective, indeed, of one's stance. Full-blooded anti-realism about functions is sometimes motivated by the empiricist stance.

The upshot of Part I is that the disputes and positions we have looked at have more to do with values and attitudes, and less to do with evidence and argument, than is commonly supposed.

In Part II of the book I will consider three further issues in the metaphysics of biology (Chaps. 8, 9 and 10), and argue that construing them in terms of the metaphysical stance framework is fruitful, in both understanding, and making progress on, the debates. These are *not* usefully understood as matters of ontology. They are different types of metaphysical questions, having to do with the most general features of the natural world and the relations between them, but not, in any obvious way, with questions of existence. For the most part, my discussion in Part II concerns the role that metaphysical assumptions play in scientific practice. I will present two arguments for construing the positions in question as metaphysical stances, a historically-oriented one in Chap. 8, and an epistemically-oriented one in Chaps. 9 and 10.

Chapter 8 explores functionalism and structuralism in biology. A longstanding metaphysical dispute about the natural world is the 'function-structure' dispute: it concerns whether in the biological world function is prior to structure, or structure prior to function. This dispute can be interpreted as a dispute between rival metaphysical stances. These stances approach the natural world in a certain way, from a certain perspective, with a tendency to focus on, emphasise, and find meaning and significance in certain features, and de-emphasise others. This interpretation can explain the continuity of this metaphysical dispute across radical shifts in scientific theory: the dispute predates the Darwinian revolution, for instance. This is an example of the way in which the construal of metaphysical positions as stances allows the empiricist in particular to accept that metaphysical assumptions and presuppositions have an important epistemic function in scientific practice, while honouring empiricist scruples about traditional metaphysics. But one need not be an empiricist to accept the arguments I present.

A second dispute in the metaphysics of biology that is fruitfully interpreted as a clash of rival metaphysical stances is that between punctuationism—the view that change in the natural world typically exemplifies a punctuational pattern, i.e. long-term stability occasionally punctuated by rapid, large-scale change—and gradualism—the view that change in the natural world is typically gradual, uniform, and incremental. This is the topic of Chap. 9. These general views find expression in different scientific theories, for instance punctuated equilibrium and Darwinian gradualism as theories of tempo and mode in evolution. But they are more general and abstract than those more specific, empirical theories. I will argue that if construed as factual claims, punctuationism and reductionism are demonstrably false. But if understood as metaphysical stances, their manifest value and importance in science can be vindicated and explained.

In Chap. 10, the final set of metaphysical issues that I will argue can be fruitfully understood in terms of the metaphysical stance framework are those relating to holism and reductionism. My argument for construing holism and reductionism as metaphysical stances proceeds from the observation that *pluralism* is the appropriate position to take with respect to reductionist and holist explanations: reductionism is true of some systems, while holism is true of others. This undermines both holism and reductionism understood as general factual claims about nature. Thus, if there *are* interesting and important general reductionist and holist positions, which I believe there are, these cannot be broad factual claims or hypotheses about nature, but must rather be metaphysical stances. In this Chapter I also discuss a particular form of holism, *dialectics*. This is the Hegelian-Marxian doctrine, roughly, that reality is a differentiated totality in which the totality is in some sense primary (the whole is prior to the parts), that change is a constant and pervasive feature of reality, and that change takes place as the result of conflict and contradiction. It can, I argue, be interpreted as a metaphysical stance—a way of seeing the world or style of thinking—that may influence the practice of science in certain characteristic ways, and may thereby be pragmatically justified.

In the Conclusion I will consider the relations between epistemic stances, as discussed in Part I, and metaphysical stances, as discussed in Part II, their similarities and differences. And I shall outline what position the empiricist and the realist ought to take with respect to both the meta-level questions—should the relevant positions be construed as stances?—and the lower-level questions—what stance should we adopt?—for both epistemic and metaphysical stances.

References

Alspector-Kelly, M. 2001. Should the Empiricist be a Constructive Empiricist? *Philosophy of Science*, 68. 413–431.

Arenhart, J. R. B., & Arroyo, R. W. 2021. The spectrum of metametaphysics: Mapping the state of art in scientific metaphysics. *Veritas* (Porto Alegre), 66(1): 1–21.

Bird, A. 2007. *Nature's Metaphysics: Laws and Properties*. Oxford University Press.
———. 2018. The Metaphysics of Natural Kinds. *Synthese* 195: 1397–1426.
Borge, B. 2024. Disagreement about Scientific Ontology. *Journal for General Philosophy of Science* 55 (2): 229–245.
Boucher, S.C. 2014. What is a philosophical stance? Paradigms, policies and perspectives. *Synthese*, 191(10): 2315–2332.
———. 2018a. Stances and epistemology: values, pragmatics and rationality. *Metaphilosophy* 49 (4): 521–547.
———. 2018b. What is the relation between a philosophical stance and its associated beliefs? *dialectica* 72 (4): 509–524.
———. 2019. An Empiricist Conception of the Relation between Metaphysics and Science. *Philosophia*, 47 (5): 1355–1378.
———. 2020. Methodological naturalism in the sciences. *International Journal for Philosophy of Religion*, 88 (1): 57–80.
Boucher, S.C. and Forbes, C. 2024. The Pragmatic Turn in the Scientific Realism Debate. *Synthese*, 203 (4). 1–23.
Boyd, R. 1999. Homeostasis, Species, and Higher Taxa. R. A. Wilson (ed.), *Species: New Interdisciplinary Essays*. Cambridge, MA: MIT Press. 141–185.
Brown, R. and Ladyman, J. 2009. Physicalism, Supervenience and the Fundamental Level. *Philosophical Quarterly*, 59: 20–38.
Bryant, A. 2021. A Thousand Flowers on the Road to Epistemic Anarchy: Comments on Chakravartty's *Scientific Ontology. Dialogue* 60 (1): 1–13.
Cameron, R. 2004. How to be a Realist about Sui Generis Teleology yet feel at home in the 21st Century. *The Monist*, 87(1), 72–95.
Cartwright, N. 1983. *How the laws of physics lie*. Oxford: Clarendon Press.
Chakravartty, A. 2010. Metaphysics Between the Sciences and Philosophies of Science. In P.D. Magnus and J. Busch (eds.). *New Waves in Philosophy of Science*. Palgrave Macmillan. PDF: http://www.nd.edu/~achakra1/research.html, 1–17.
———. 2017. *Scientific Ontology: Integrating Naturalised Metaphysics and Voluntarist Epistemology*. Oxford University Press.
———. 2024. Resolving Debates about Scientific Realism: The Challenge from Stances. *Philosophy of Science* 91, 1306–1315.
Chang, H. 2022. *Realism for Realistic People: A New Pragmatist Philosophy of Science*. Cambridge University Press.
Dennett, D. C. 2017. *From bacteria to Bach and back: The Evolution of minds*. New York: W. W. Norton & Company.
Douglas, H. 2000. Inductive Risk and Values in Science. *Philosophy of Science* 67 (4): 559–579.

Dowell, J. 2006. The Physical: Empirical, not Metaphysical. *Philosophical Studies* 131: 25–60.

Dupré, J. 1993. *The Disorder of Things: Metaphysical Foundations for the Disunity of Science*. Cambridge MA: Harvard University Press.

———. 2002. Is 'Natural Kind' a Natural Kind Term? *The Monist*, vol. 85, No. 1: 29–49.

———. 2021. *The Metaphysics of Biology*. Cambridge University Press.

Elder, J. 2019. Defending stance voluntarism. *Philosophical Studies*, 176(11), 3019–3039.

Ellis, B. 2001. *Scientific Essentialism*. Cambridge University Press.

Ereshefsky, M. and T.A.C. Reydon. 2015. Scientific Kinds. *Philosophical Studies*, 172: 969–986.

Foley, R. 1991. Rationality, Belief and Commitment. *Synthese*, 89. 65–92.

———. 1994. Pragmatic Reasons for Belief. J. Jordan (ed.), *Gambling on God: Essays on Pascal's Wager*. Lanham, Md.: Rowman & Littlefield. 31–46.

———. 2000. Epistemically Rational Belief and Responsible Belief. R. Cobb-Stevens (ed.). *Proceedings of the Twentieth World Congress of Philosophy*. Bowling Green, OH: Philosophy Documentation Centre, Bowling Green State University. 181–188.

Forbes, C. 2017. A pragmatic, existentialist approach to the scientific realism debate. *Synthese* 194: 3327–3346.

Franklin-Hall, L. 2015. Natural Kinds as Categorical Bottlenecks. *Philosophical Studies*, 172 (4): 925–948.

Ghiselin, M. 1994. Darwin's language may seem teleological, but his thinking is another matter. *Biology and Philosophy* 9(4): 489–492.

Hacking, I. 1983. *Representing and Intervening: Introductory Topics in the Philosophy of Natural Science*. Cambridge University Press.

Hempel, C. 1980. Comments on Goodman's *Ways of Worldmaking*. *Synthese*, 45 (2). 193–199.

Henschen, T. 2024. Subatomic Particles, Epistemic Stances, and Kantian Antinomies. *Journal for General Philosophy of Science*, 55 (2): 247–268.

Humphreys, P. 2013. Scientific Ontology and Speculative Ontology. Don Ross, James Ladyman & Harold Kincaid (eds.), *Scientific Metaphysics*. Oxford University Press. 51–78.

James, W. 1956. The Will to Believe. In James, W. *The Will to Believe and other essays in popular philosophy*. New York: Dover. 1–31.

Kim, B. and McGrath, M. (eds.). 2019. *Pragmatic Encroachment in Epistemology*. Routledge.

Ladyman, J. 2011. The Scientistic Stance: The Empirical and Materialist Stances Reconciled. *Synthese*, 178 (1). 87–98.

———. 2014. Structural Realism. *The Stanford Encyclopedia of Philosophy* (Summer 2023 Edition), Edward N. Zalta & Uri Nodelman (eds.). URL = https://plato.stanford.edu/archives/sum2023/entries/structural-realism/

Ladyman, J. and D. Ross, with D. Spurrett and J. Collier. 2007. *Every Thing Must Go: Metaphysics Naturalized*. Oxford: Oxford University Press.

Ladyman, J., D. Ross and H. Kincaid (eds.), 2013. *Scientific Metaphysics*. Oxford: Oxford University Press.

Magnus, P.D. 2012. *Scientific Enquiry and Natural Kinds: From Planets to Mallards*. Palgrave Macmillan.

———. 2018. Cautious realism and middle range ontology. *Metascience* 27: 365–70.

Melnyk, A. 1997. How to keep the 'Physical' in Physicalism. *J Philos* 94(12): 622–637.

Mohler, C. 2007. The Dilemma of Empiricist Belief. B. Monton (ed.). *Images of Empiricism: Essays on Science and Stances, with a Reply from Bas C. van Fraassen*. Oxford: Oxford University Press. 209–228.

Ney, A. 2006. Physicalism as an Attitude. *Philosophical Studies*, 138: 1–15.

Pils, R. 2023. Scientific Realism and Blocking Strategies. *International Studies in the Philosophy of Science* 36(1): 1–17.

Pincock, C. 2024. Defending a Realist Stance. *International Studies in the Philosophy of Science*, 37 (1–2), 1–15.

Psillos, S. 2021. Scientific Ontology: Fact or Stance? *Dialogue* 60 (1): 15–31.

Quine, W.V.O. 1953. On What There Is. W.V.O. Quine. *From a Logical Point of View*. Cambridge, MA: Harvard University Press. 1–19.

Robus, O.M. 2015. Does Science License Metaphysics? *Philosophy of Science* 82: 845–55.

Steup, M. 2011. Empiricism, metaphysics, and voluntarism. *Synthese* 178: 19–26.

Slater, M. 2017. Naturalized Metaphysics and the Contention over the Ontological Status of Species. M.H. Slater and Z. Yudell (eds.). *Metaphysics and the Philosophy of Science: New Essays*. Oxford University Press. 55–80.

Slater, M.H. 2021. Extending the Ladder of Stances: Comments on Chakravartty's *Scientific Ontology*. *Dialogue* 60: 33–42.

Slater, M.H. and Z. Yudell (eds.) 2017. *Metaphysics and the Philosophy of Science: New Essays*, Oxford University Press.

Swoyer, C. 2008. Abstract Entities. T. Sider, J. Hawthorne, D. Zimmerman (eds.), *Contemporary Debates in Metaphysics.* Oxford/Malden/Carlton: Blackwell, 11–31.

Teller, P. 2004. What is a Stance? *Philosophical Studies*, 121 (2).159–170.

Triviño, V. 2023. Exploring the interactions between metaphysics and science: lessons from the metaphysics of biology. *Philosophica*, 31(1): 51–69.

van der Merwe, R. 2024. Stance Pluralism, Scientology and the Problem of Relativism. *Foundations of Science* 29 (3): 625–644.

van Fraassen, B.C. 1989. *Laws and Symmetry*. Oxford: Clarendon Press.

———. 1994. Against Transcendental Empiricism. T. Stapleton (ed.). *The Question of Hermeneutics.* Dordrecht: Kluwer. 309–335.

———. 1995. Against Naturalised Epistemology. P. Leonardi and M. Santambrogio (eds.). *On Quine.* Cambridge: Cambridge University Press. 68–88.

———. 1996. Science, Materialism and False Consciousness. J.L. Kvanvig (ed.). *Warrant in Contemporary Epistemology: Essays in Honor of Plantinga's Theory of Knowledge.* Lanham, MD.: Rowman and Littlefield. 149–181.

———. 2002. *The Empirical Stance.* New Haven, CT: Yale University Press.

———. 2004a. Précis of The Empirical Stance. *Philosophical Studies*, 121 (2). 127–132.

———. 2004b. Replies to Discussion on The Empirical Stance. *Philosophical Studies*, 121 (2). 171–192.

———. 2004c. Reply to Chakravartty, Jauernig, and McMullin, unpublished typescript of a talk delivered at a symposium on *The Empirical Stance* at the Pacific APA in Pasadena, CA.

———. 2007. From a View of Science to a New Empiricism. B. Monton (ed.). *Images of Empiricism: Essays on Science and Stances, with a Reply from Bas C. van Fraassen.* Oxford: Oxford University Press. 337–383.

———. 2011. On Stance and Rationality. *Synthese*, 178 (1). 155–169.

Vicente, A. 2011. Current Physics and 'The Physical'. *Br J Philos Sci* 62: 393–416.

Wilson, J. 2006. On Characterizing the Physical. *Philosophical Studies*, 131: 61–99.

Wylie, A. 1986. Arguments for Scientific Realism: The Ascending Spiral. *American Philosophical Quarterly* 23(3), 287–297.

Wright, C. 1987. *Realism, Meaning and Truth.* Blackwell.

Part I

Epistemic Stances

2

Naturalised Metaphysics, Scientific Ontology, and Epistemic Stances

In the Introduction I briefly sketched a view about naturalised metaphysics, scientific ontology and epistemic stances, deriving from some ideas in van Fraassen and Chakravartty in particular. In this chapter I shall develop and defend this picture in more detail, before applying it, in the subsequent four chapters, to some topics in the metaphysics of biology.

1 Implicit and Explicit Entities of Science, and Naturalised Metaphysics

Recall in the Introduction I distinguished between two versions of the realist stance: the standard scientific realism (SSR) stance, which licenses belief in, i.e. ontological commitment to, the explicit but unobservable subject matter of science but demurs when it comes to the entities that are part of the implicit subject matter of science—e.g. essences, causal powers, dispositions etc.—and the metaphysics of science (MoS) stance,

S. C. Boucher, *Empiricism, Naturalism and the Metaphysics of Biology*,
https://doi.org/10.1007/978-3-032-00555-7_2

which is willing to embrace both the explicit and the implicit entities of science, but rejects 'purely' metaphysical entities unknown to science.[1]

We need a clearer idea of what is meant by 'explicit' and 'implicit' entities of science. Chakravartty (2007, 68–9) distinguishes them in this way. There are entities the nature of which science sees it as its business to investigate—electrons, protons, genes, species, spacetime, etc.[2]—and entities that the sciences mention 'in passing' but whose nature science is not in the business of investigating—e.g. laws, causation, properties, modality. While the explicit subject matter of science clearly gives us candidates for scientific ontology, the entities mentioned just in passing also arguably give us candidates for scientific ontology, though this may be more controversial.

'Mentioned in passing' is not an altogether happy way of designating the latter class of entities however, since presumably for something to belong to this category it is not necessary that science refer to them *under that name*. For example, insofar as scientists refer to instances of natural kinds, essences, causal powers, or dispositional properties, it is rarely if ever under those names. Insofar as it is part of the business of science to investigate particular instances of these things—particular natural kinds, particular causal powers etc.—yet not part of its business to investigate the general nature of these categories—natural kinds in general, causal powers in general etc.—and the relations among them, they fall into the category of the 'implicit' entities of science.

1.1 Implicit Entities of Science and Non-naturalistic Metaphysical Entities

I take it the distinction between explicit and implicit entities of science, and between these and the entities of armchair metaphysics, is in a general sense relatively clear and uncontroversial. Arguably the entire

[1] In this book I shall be using 'ontological commitment to Xs' and 'realism about Xs' more or less interchangeably, that is, I shall assume that S is ontologically committed to X just in case S is a realist about X. And I shall understand realism primarily in terms of mind-independence.

[2] Although arguably questions of realism or anti-realism with respect to these entities are questions of ontology, and are thus philosophical questions.

metaphysics of science program (in the sense of a program not a stance) presupposes something like this latter distinction, inasmuch as it takes as its subject matter metaphysical questions that arise out of the sciences, not out of the philosophical armchair.

Yet when we consider the category of implicit scientific entities, a problem arises. Recall our taxonomy of four stances with their characteristic entities. There are seemingly paradigm cases for the characteristic entities of three of the stances: the empiricist stance (tables and trees), the SSR stance (genes and protons), and the armchair metaphysics stance (God, haecceities). There are likely to be borderline cases for these entities: the distinction between observable and unobservable entities is notoriously vague, for instance. But there are seemingly clear examples of entities that fall into each of these categories. When we consider the characteristic entities of the MoS stance, however—causation, laws, natural kinds, properties, essences, modality etc.—things are not so straightforward. The problem is that these entities can seemingly be postulated, and investigated, from either a naturalistic (i.e. MoS) or non-naturalistic (armchair) standpoint. Self-styled metaphysicians of science explore the nature of entities such as these; but so do those who pursue metaphysics in the traditional, a priori manner. Knowing merely that a philosophy article addresses the nature of causation, or properties, or modality, is insufficient for one to determine whether the article is a species of naturalistic metaphysics of science, or non-naturalistic armchair metaphysics.[3] Modality, and the related concept of essence, is a particularly clear example. Even the most naturalistic of naturalised metaphysicians, Ladyman and Ross (2007) argue that the sciences tell us about the modal structure of the world,[4] such that the nature of modality falls legitimately within the province of naturalistic metaphysics and scientific ontology; yet speculation and theorising about modality, in the form, for instance, of debates about possible worlds, has standardly been firmly associated with old-style metaphysics

[3] For instance, Dupré and Meincke (2020, Introduction) note that Mumford and Tugby (2013, Introduction) claim to be doing metaphysics of science, discussing questions about laws, causation etc., but that they appear to be pursuing traditional armchair metaphysical projects, rather than genuinely science-based metaphysics.

[4] 'The world has an objective modal structure that is represented by causal claims and laws in the special sciences' (Ladyman 2017, 154).

done from the armchair, with Lewis' modal realism being one of most celebrated, or notorious (depending on one's point of view) examples. Another example is 'process metaphysics'. Forms of process metaphysics—roughly the view that the basic, fundamental constituents of the world are processes, not enduring 'things'—have historically been defended in a fairly traditional armchair way, by Whitehead among others (Browning 1965)—indeed the view goes back to Heraclitus. Yet in recent years a notable form of process ontology, at least with respect to the natural world, has been defended in an unimpeachably naturalistic way. Dupré (2021), for instance, has argued that close attention to what our best contemporary biological science is telling us forces on us the view that the natural world is fundamentally composed of processes, not entities or substances. For instance, organisms, on this view, should be thought of as processes not things or substances (see Dupré and Nicholson 2018).[5]

Pressure can be put on the distinction between naturalistically respectable implicit scientific entities and non-naturalistic metaphysical entities in a somewhat different way. Recall the point made above that the sciences make reference to particular instances of natural kinds, essences, causal powers, and dispositional properties, even if they don't use those terms, and thus that these entities count as implicit entities of science. Traditional metaphysicians will also claim that the sciences make reference to particular instances (or instantiations) of at least some of the metaphysical categories they are interested in: universals, tropes, substances, etc., though of course not under those names. Thus if theorising about things such as these is to be dismissed as insufficiently naturalistic, it seemingly can't be by way of a simple criterion relating to whether instances of the category take their place as part of the subject-matter of the sciences.

To see this, we can compare the account of metaphysics in Paul's influential (2012). Metaphysics, she argues, 'describes the metaphysically prior categories and constituents of the physically fundamental entities, and in this sense describes features of the world that are more fundamental that those of the natural sciences' (Paul 2012, 5). That is, science inquires into the nature of the physically fundamental entities; but these

[5] Compare Armstrong's (1997) factualist view that the fundamental constituents of the world are states of affairs. This is defended mostly from the armchair. See below.

entities belong to general categories—'object', 'property', 'process', 'event', 'law', 'cause', 'modality' etc.—on the nature of (and relations between) which science is silent. It is the job of metaphysics to supply this deficiency. This is on the face of it quite similar to Chakravartty's account: both highlight the role of metaphysics as giving systematic accounts of the kinds and categories instances of which science makes reference to, but the nature of which it does not pause over. Paul would also no doubt be happy to accept Chakravartty's suggestion (2007) that such theorising concerns the 'ontological underpinnings' of contemporary science.

1.2 The Methodological Criterion

If we are to label the kind of metaphysics Chakrvavartty recommends as 'naturalistic' and the more traditional approach recommended by Paul as 'non-naturalistic'—the former an expression of the MoS stance, the latter of the armchair metaphysics stance—it seemingly has to be via considerations of *methodology*, not solely subject-matter. Naturalistic metaphysics takes science as its cue, its starting point, looking to what science seems to be telling us, even if just implicitly, about the nature of and relations between the categories and kinds in question (whereas for non-naturalistic metaphysics, theorising largely takes place in the armchair). If there is a topic on which science seems in principle unable to tell us anything whatsoever—e.g. Paul's example of bundle theory vs substance-attribute theory—the naturalised metaphysician takes no interest in it.

Naturalistic metaphysics thus emphatically rejects the claim, endorsed by Paul, that metaphysics is 'prior' to science. Of course, *ex hypothesi*, science does not tell us in detail what natural kinds, essences, causation etc. are. That is not its job. Philosophy has a major role to play in such inquiries. Nonetheless there is, according to the naturalised metaphysician, much that science is able to tell us implicitly about these things, if we know how to listen. To put it another way, our best current scientific theories may not tell us what exactly natural kinds, essences and the like are; but they constrain philosophical theorising about such things to a considerable degree. These kinds of influences and constraints imposed on metaphysical theorising by the content of science are not recognised (at least, not sufficiently) by traditional intuitions-based armchair metaphysics, at least according to the avowedly naturalised metaphysician. The topic of Chap. 4, natural kinds, is a case in point. One will not find many papers in science journals on the subject of the metaphysics of natural kinds and essences, unless a philosopher happens to get a philosophy paper published in one; yet many theorists suggest that science has much to tell us, implicitly, about the nature of natural kinds and essences, and that indeed one of the failings of recent, influential philosophical discussions of natural kinds that have had a big impact outside of the philosophy of science, such as Kripke and Putnam's work on natural kinds and natural kind terms, was precisely that it wasn't sensitive to developments in the sciences, especially biology, that have implications for what natural kinds and essences are like, and what natural kinds and essences there are.

So, given that often entities don't wear their naturalistic credentials on their sleeves, it may frequently be this admittedly imprecise methodological criterion, relating to whether one's metaphysical theorising is substantially influenced and constrained by, grounded in and informed by, our best current science, rather than a criterion based solely in the character or status of the postulated entities themselves (e.g. universals and tropes as inherently 'purely metaphysical' thus naturalistically disreputable entities) that allows us to distinguish between naturalistic and non-naturalistic metaphysics, between the MoS stance and the armchair metaphysics stance, between scientific and non-scientific ontology (Humphreys 2013; Chakravartty 2013, 32).

Another reason we need to appeal to methodology is that there are arguably cases of traditional, armchair metaphysics where no recondite, purely 'metaphysical' entities need be posited at all. For instance, in the classic debate about material constitution—concerning whether, for example, the statue and the clay it's made of constitute two objects and if so how is it that they seem to be colocated—seemingly the only entities that need to be involved are the familiar ones of the statue and the lump of clay themselves. Of course in solving the puzzle philosophers *may* posit 'metaphysical' entities such as hylomorphic compounds and the like, but this seemingly isn't necessary. One could simply argue, 'the statue and the lump of clay have different properties—for instance, the lump existed before the statue did and could continue to exist after the statue is destroyed—so, given Leibniz's law, they are distinct entities, and thus distinct entities can spatiotemporally coincide, despite what common sense tells us'. No special metaphysical entities are being invoked in this line of reasoning, assuming that 'property' is understood in a non-technical sense.[6] Thus, if this debate and similar ones are to be classified as pieces of armchair metaphysics, the grounds would have to be that since there is no possibility of scientific evidence being in any way relevant to solving the problems, even indirectly, the debates are and must be conducted at a level that is entirely a priori, from the armchair. No position in the debates enjoys any degree of empirical vulnerability.

The way Chakravartty expresses the relevant methodological criterion is by way of the concept of *metaphysical inferences* from science. Metaphysics done in a naturalistic spirit involves inferring—where this is understood in a quite wide sense, typically involving explanatory reasoning—one's metaphysics from our best current science; non-naturalistic metaphysics does not, or at least, not primarily. Naturalistic and non-naturalistic metaphysicians may both study modality, but a naturalistic approach to modality would involve (for the most part) drawing inferences from well-confirmed science regarding the existence and nature of various grades of necessity and possibility; a non-naturalistic approach would theorise and speculate (for the most part) from the armchair. The same goes for theorising about properties, causation, laws, natural kinds and so forth. While these things—or for that matter atoms and protons!—can theoretically be investigated a priori, with no (or minimal) engagement with the relevant science, they are, so the naturalist claims, best postulated, and theorised about, on the basis of metaphysical inferences from science, our most reliable source of objective knowledge about the world.

---

[6] Another example may be the problem of persistence over time. In the ship of Theseus puzzle, seemingly the only entities involved are the ships themselves and their component parts, and maybe certain relations such as the relation of identity. One could claim, perhaps, that in such examples what is at issue is really the nature of *objects, properties, relations* and the like *in general*, rather than ships, statues and lumps of clay. But at least on the face of it, problems such as these do not involve any special metaphysical items or require technical metaphysical terminology in order to be formulated (though solving them may be another matter). That's why we are easily able to get our first-year philosophy students to understand them using the terms and concepts of ordinary language.

Even uncontroversial instances of traditional armchair metaphysics at times involve the attempt to support certain views by citing scientific theories and evidence. For example, Schaffer's metaphysical system, combining priority monism with Aristotelian relations of grounding and dependence (2009, 2010), counts as traditional non-naturalistic metaphysics if anything does. The thesis that the whole is prior to the parts, for instance, has been criticised by Melnyk (2013, 80) on the reasonable grounds that that there is no scientific reason to suppose there is such a thing as 'priority' in this sense. Yet even Schaffer argues that priority monism is supported in part by quantum mechanics. (Russell defended pluralism by appealing partly to science (1956, 338–39) and partly to common-sense; see Della Rocca 2020, 284). When assessing the naturalistic credentials of a piece of metaphysics, the question is not whether there is *some* influence from science on the metaphysics, i.e. *some* metaphysical inferences from science are drawn, but *how much.*

There is also the related fact that, as Chakravartty points out, metaphysical inferences can be more or less *direct*: an inference to the existence of the table in front of me is more direct and immediate than an inference to the existence of electrons, which is more direct than an inference to the existence of natural kinds, which is more direct than an inference to the existence of universals (arguably). As I shall discuss in more detail in Chap. 4, claims about the existence and nature of natural kinds have, seemingly, been more responsive to science than claims about the existence and nature of universals, and thus have enjoyed a degree of empirical vulnerability that claims about universals typically lack. Arguably traditional essentialism about natural kinds has been refuted by the evidence concerning the role of natural kinds in science; whereas the realism vs. nominalism dispute with respect to universals is often given as a paradigm example of a metaphysical dispute where the rival views are strongly empirically equivalent (Ney 2012).[7]

---

[7] Of course, on a Quinean confirmation-holist picture, all beliefs are potentially revisable in light of experience, and the empirical evidence. But even Quine accepts that some beliefs are closer to the core, and further from the periphery, of our web of belief, than others, making it less likely they will be revised in response to empirical data that conflicts with the web overall (Chakravartty would say they are less empirically vulnerable). Put in Quinean terms, beliefs about universals are closer to the core of our web of belief than beliefs about natural kinds. See Sect. 1.3 below.

This is Chakravartty's continuum of deepening metaphysical commitment, increasing distance from science and empirical evidence, decreasing empirical vulnerability, increasing epistemic risk, and increasingly more tenuous naturalistic respectability. Even if one *aims* to theorise about, say, universals in a naturalistic spirit, drawing inferences from and offering explanatory hypotheses that are sensitive to, the relevant parts of science, it may be that claims about universals lack sufficient empirical vulnerability, and thus represent too great an epistemic risk, for the naturalistic metaphysician to be willing to commit to them. This theme shall be explored in Part I.

The way this is often expressed is that the weaker the connection between the science and the ontological beliefs, the more the latter are *underdetermined* by the former (Henschen 2021, Humphreys 2013). It may be that *all* ontological claims are underdetermined by science, but the *degree* of underdetermination increases the 'further' we get from the science: claims about natural kinds are more underdetermined by the scientific evidence than claims about electrons, and claims about universals are more underdetermined still (arguably). Bryant writes:

> [N]aturalized metaphysics carries lower epistemic risk precisely because it is less underdetermined than traditional metaphysics. Naturalized metaphysics is less underdetermined because it is constrained by science, and the scientific constraint narrows the range of theoretical options we may consider live. So, comparative underdetermination can be used to explain why naturalized metaphysics is epistemically preferable to non-naturalized metaphysics. (Bryant 2021, 7)

While I think this is broadly correct, it is overly simplistic in presenting a binary opposition: naturalised vs. non-naturalised metaphysics. We should, instead, recognise a spectrum, even within naturalised metaphysics, between greater and lesser degrees of underdetermination.

With this in mind, returning to the argument above for parity between natural kinds, laws etc. on the one hand, and universals, tropes etc. on the other, on the grounds that science is concerned with instances of both types of category (though not necessarily under those descriptions), we can say that while this might be true, the claim that, say, species S or chemical element E, is a natural kind, and the claim that species S or chemical element E is a universal, are not on a par epistemically, since they (arguably) represent metaphysical inferences of differing strength, speculativeness and hazard. The extent of epistemic hazard one is willing to countenance to get the putative benefits of deeper ontological commitment, is a function of one's epistemic stance.

It is one of the strengths of Chakravartty's account that it recognises that the extent to which metaphysical theorising is influenced by and grounded in science is often a matter of degree. Metaphysical inferences can be immediate or more indirect. And the question of how direct this influence needs to be for one's metaphysical theorising to count as sufficiently naturalistic is frequently contested, as we shall see in subsequent chapters. Chakravartty, as we've seen, makes the further point that such disputes are not ultimately factual, but reflect different epistemic stances, that in turn are the function of values. Thus stances don't just disagree about how far to extend our ontology; they disagree about whether certain extensions are consistent with naturalism.

1.3 Quine on Science and Metaphysics

The view that there is no clear, sharp, unambiguous boundary separating naturalistic from non-naturalistic metaphysics, that such differences are often a matter of degree, will no doubt recall to mind Quine's celebrated thesis that there is no sharp boundary separating science and metaphysics (1953b), that the two are continuous, such that the differences between them are a matter of degree. Metaphysical claims, for Quine, simply occupy a position closer to the core of our web of belief than scientific claims, being more conceptual and speculative, and thus are generally less likely to be revised in the light of experience, though such revision is not ruled out. As far as I am aware, Chakravartty does not attempt to justify his view in Quinean terms, and neither shall I. While there are (perhaps superficial) similarities between the claims, there are important differences. Chakravartty's claim is not really that *there are no* qualitative differences in kind between science and all of metaphysics, or between different forms of metaphysics, but that *whether there are* such differences, and precisely where the relevant boundaries are located, is typically a stance-relative value judgment with respect to which rational disagreement is possible (and often actual). Quine, in our terms, is *taking a stance* on this matter: the epistemic stance according to which even apparently quite recondite metaphysical claims and speculation are continuous with science, that they indirectly confront empirical evidence, holistically, along with the rest of our web of belief, so enjoy a degree of indirect empirical vulnerability, and are thus, in this respect at least, epistemically legitimate (and Quine's views have often been appealed to by those who would seek to defend the legitimacy of metaphysics in the face of sceptical and naturalist doubts[8]). Chakravartty, on the other hand, is merely claiming that this *is* a stance, rather than a rationally compelled belief; and that there are rival, equally-rational stances that deliver quite different verdicts concerning the legitimacy of certain forms of metaphysics.

2 The Four Stances Again

In the Introduction I gave a taxonomy of four stances defined in terms of their attitude to the question of how far to extend our ontology. The above discussion indicates that we need to modify this taxonomy somewhat to reflect methodological considerations.

We begin with our categories of entities:

1. Observables
2. The explicit but unobservable entities of science (electrons, protons, species etc.)
3. The implicit entities of science (laws, causation, natural kinds, properties, dispositions, modality etc.)
4. Non-scientific metaphysical entities (haecceities, quiddities, substances, grounding relations, hylomorphic compounds, monads, potentialities, necessary beings, etc.)

[8] Quine's 'On What There Is' (1953a) is usually credited as the seminal document that opened the door to the analytic metaphysics that would flourish in the post-positivist period. Much of this counts as armchair metaphysics in our sense, but armchair metaphysics that often seeks to justify itself in Quinean and quasi-naturalist terms, that is, in terms of the supposed similarities it shares with speculative science (its use of inference to the best explanation, for example). For discussion see the papers collected in Chalmers et al. (2009).

We can now define our four stances again:[9]

Empiricist stance

Permits inferences only to observables. (Minimal metaphysical inferences)

Realist stances

Naturalist realist stances[10]

SSR stance

Permits metaphysical inferences to the explicit but unobservable entities of science, but not to implicit entities of science.

MoS stance

Permits metaphysical inferences to the implicit (and explicit) entities of science.

Non-naturalist realist stance

Permits inferences to or belief in non-scientific metaphysical entities.

To the extent it theorises about implicit entities of science, it is not primarily by way of metaphysical inferences from science and empirical evidence, but *a priori.*

---

[9] See Arenhart and Arroyo (2021) for a similar taxonomy of views on the relation between science and metaphysics. Their categories: empiricism; full-blooded naturalism; full-blooded metaphysics; and various moderate-naturalist views that seek to integrate science and metaphysics, map reasonably closely onto my categories of the empiricist stance, the SSR stance, the armchair metaphysics stance, and varieties of the MoS stance, respectively. There are some differences between their mapping of the terrain and mine. Theirs is less concerned with ontological commitment to certain types of *entities*. Moreover, although they refer to their taxonomy as a 'spectrum', they tend to present the different approaches as fairly sharply distinguished, whereas I allow more for borderline cases and differences of degree. The biggest difference however is that despite being influenced by Chakravartty, they do not adopt the epistemic stance approach, and this vitiates the explanatory power of their (otherwise insightful) analysis. So the distinction I draw between the empiricist stance on the one hand, and all forms of the realist stance on the other, based on values such as explanatory power, does not feature in their taxonomy. They also do not employ the voluntarist framework, so the key notion of *permissibility* with respect to scientific ontology, which I outline in more detail later in this chapter, is missing from their account.

[10] More precisely, these stances are *intended* to be naturalist stances. As we have seen, how far along Chakravartty's continuum one may travel before one's claim to be pursuing naturalised metaphysics loses credibility is a stance-relative judgement. From the standpoint of the SSR stance, the MoS stance may fail to be sufficiently naturalistic, thus can't be called a 'naturalist' stance in a non-tendentious way. If I refer to the MoS stance as a naturalist stance from now on this should be understood to mean 'intended to be' a naturalist stance.

2.1 Stronger and Weaker Forms of the Metaphysics of Science Stance

I shall talk at times of stronger and weaker versions of the MoS stance. These cannot be as sharply distinguished as, say, the MoS and the SSR stance, or as clearly defined; the difference is a matter of degree. While both endorse inferences to implicit entities of science, and both aspire to be naturalistic, proceeding largely on the basis of metaphysical inferences from science, the weaker version is more cautious, less willing to venture very far beyond the science and posit inflationary metaphysical items, than the strong version. We will see indeed that from the standpoint of the weak MoS stance, the strong version risks severing the connection between metaphysics and science, and thus violating the norm of naturalised metaphysics. It risks, that is, turning into the armchair metaphysics stance. But this judgement, again, is stance-relative and open to dispute.

2.2 The Standard Scientific Realism Stance

In the Introduction I said that I am defining the SSR stance as a stance that rejects the MoS stance, i.e. one that does not license beliefs in the implicit subject matter of science. For instance, one who adopts the SSR stance may, if the arguments in favour of the view(s) are judged to be cogent, accept that electrons, protons and quarks and the like exist, but demur when it comes to countenancing and speculating about entities at a 'deeper' level: laws, properties, dispositions, structure, modality, and the like. This lack of commitment to entities at a more fundamental metaphysical level could take the form either of an agnosticism regarding such entities, or an outright denial of their existence, mirroring the two empiricist options regarding the explicit entities of science—agnosticism (as in constructive empiricism) and instrumentalist eliminativism.[11]

---

[11] That is, those who accept the SSR stance will adopt a characteristically empiricist attitude towards the deeper metaphysical level and the entities that reside there.

Perhaps the view in the recent scientific realism debate that comes closest to this position is the 'localist' version of realism, according to which the local first-order scientific evidence for the existence of certain explicit entities of science—the evidence adduced by scientists themselves, and forming the basis of *their* ontological commitment—is sufficient evidence for the realist philosopher to commit to the entities (Achinstein 2002, 2010; Maddy 2007). Distinctively philosophical arguments for realism, such as the No Miracles Argument, are unnecessary. For instance, to the extent that Jean Perrin's celebrated experiments in the early twentieth century conclusively established, to the satisfaction of the scientific community, the reality of atoms and the truth of the atomic hypothesis, philosophers should believe that atoms exist and atomic theory is true.[12] Philosophy, with its philosophical reasoning, has little to add to this. On this view, '(t)here simply is no room for philosophical debate outside of science to settle questions about whether a particular theory is true, whether certain entities exist, or what we have good reason to believe' (Wolff 2019, 486). This points us to the fact that there may be disagreement within a stance concerning the directness or immediacy of the metaphysical inferences that take us from the scientific evidence to ontological commitment: the localist holds that in a sense the required inferences have already been drawn *within* science, and there are no further inferences or philosophical arguments required, it just remains for the realist philosopher to endorse what the scientists already believe; while the more traditional proponent of the SSR stance thinks that in addition to the scientific evidence we need philosophical arguments, such as the No Miracles Argument, to conclusively establish the realist view.

It would be a mistake however to suppose that the localist view amounts to nothing more than, as it were, repeating what scientists already believe and assert, such that no philosophical position or epistemic stance is involved. Arenhart and Arroyo (2021) make this mistake in suggesting that the localist view—what they call 'full-blooded naturalism' (*ibid.*, 4)—eschews metaphysics entirely. To the extent

[12] See van Fraassen (2009) for an alternative, empiricist interpretation of this episode that is frequently invoked by realists as decisive evidence for their view.

that the view is a version of scientific realism, recommending that we commit to the reality of explicit but unobservable scientific entities, it is an instance of the realist stance, and must therefore be implicitly underlain by the values characteristic of that stance. As van Fraassen taught us, given the availability and coherence of the empiricist alternative, any view that involves realist commitment is thereby *metaphysically* committed, whether its proponents like it or not. After all, the only grounds for believing in the existence of many of the unobservable entities embraced by the localist is their ability to explain the relevant observable phenomena, so it is hard to see how such a realist view could fail to be underwritten by a stance among whose central features is that it accords a high value to explanatory power in matters of ontology.[13]

While I am not aware of the localist position being explicitly linked to a rejection of belief in the implicit entities of science, such a rejection would seem to go hand-in-hand with the localist's claim that only the first-order scientific evidence is relevant to questions of scientific realism and ontological commitment, since we do not tend to find scientists conducting experiments with the aim of establishing the existence of implicit entities such as causal powers, natural kinds, essences, objective modality and so on. If they did, the entities would be among the entities the nature and behaviour of which science sees it as its business to investigate, and so they would, by definition, be among the explicit, not implicit, entities of science. There are no equivalents of Perrin's experiments for entities such as causal powers and natural kinds *as such*. Certainly defenders of the MoS stance will endeavour to draw metaphysical inferences from science to the existence of such entities, and may even claim that such inferences are relatively direct and unproblematic. Some philosophers of biology argue that it clearly follows from well-established parts of evolutionary

[13] Moreover, FitzPatrick (2013) argues convincingly that in making the case for localism, Achinstein (2002, 2010) in fact helps himself to certain philosophical assumptions and arguments that ought to be illicit on the localist view. Simply 'reciting first-order evidence', it turns out, is *not* sufficient for SSR. Philosophy is always involved. And, we might add on behalf of the metaphysicians of science, once philosophical assumptions and reasoning have been allowed in, why may we not appeal to them in drawing stronger metaphysical inferences to the existence of implicit entities?

biology, for instance, that species have historical essences, and thus that essences (or at least species-essences) exist (see Chap. 5). But it is typically philosophers, not scientists *qua* scientists, who draw such inferences, and philosophers *qua* philosophers who take on the ontological commitments. Compare Platonism about numbers: Platonists argue it follows immediately from uncontroversial mathematical statements such as 'there is a prime number greater than 5' that prime numbers exist, and hence that numbers exist. But it is the philosopher of mathematics, not the mathematician *qua* mathematician, who typically draws such inferences, and who takes on the ontological commitments.[14] Mathematicians *qua* mathematicians can be neutral on all questions of ontology. Thus I suggest we should interpret the localist view as a version of the SSR stance, recommending we commit to those explicit entities of science the existence of which has, it is reasonable to believe, been established by empirical scientific means, but stop there. In keeping with this interpretation, the localist view in some forms is linked with strong forms of naturalism, see for instance Maddy (2007). The thought is that the naturalist should commit only to the entities the reality of which has been established by science itself, not to entities belief in which requires distinctively philosophical forms of reasoning.

Whether such a stopping point is philosophically defensible is of course open to question. One key line of argument pursued in recent years by those would recommend the MoS stance is that such a stopping point is unmotivated and the resulting form of realism impoverished, that if scientific realism is to be fully articulated and made plausible, and defended successfully against its rivals, it needs to be grounded in firm metaphysical foundations; we need, as the title of Chakravartty's (2007) book put it, a 'metaphysics for scientific realism'.[15] '[T]he internal coherence of realism as a philosophical position depends on having not only a plausible account of the unobserv-

[14] Granted, numbers are among the explicit, not the implicit entities of mathematics. But the philosophical character of Platonism regarding numbers makes this example analogous to philosophical inferences to the existence of implicit scientific entities.

[15] See also Ellis (2001) and Bird (2007).

ables described by scientific theories, but also some understanding of the most central metaphysical items one invokes in giving such an account' (Chakravartty *ibid.*, loc. 2547). This includes properties, dispositions, natural kinds and the like. Ladyman (2018) argues similarly that a successful defence of scientific realism requires inflationary ontology, in particular an ontology of objective modality (natural necessity).[16]

One likely motivation—though it may fall short of being an actual argument—for the stopping point would be the naturalist's desire to get off the 'slippery slope' in good time before one falls into the metaphysical abyss. After all, the argument that stopping before metaphysics get more heavy-duty is unmotivated and leaves us with an impoverished ontology can be multiply iterated; it was the reason given by scientific realists for going beyond the observable to commit to the unobservable but explicit entities of science; it is offered by advocates of the MoS stance to justify going beyond these explicit entities and embracing implicit ones; it is given by those who defend still stronger forms of realism to support going beyond implicit entities of science to embrace at least some traditional metaphysical entities unknown to and seemingly remote from science such as universals, quiddities, and possible worlds. If naturalist or empiricist scruples constrain one to try to avoid pursuing this reasoning into such murky metaphysical depths, the argument that 'if you've come this far, there are no grounds for not going further' needs to be resisted at some point. Why not at the point where implicit entities make their appearance?

Of course, the empiricist will interject, why even go this far? Van Fraassen has long argued, as it were from the opposite direction, that the idea that one can accept explicit but unobservable entities without plunging into the metaphysical depths is a vain hope. Scientific realism, in all its forms, is fundamentally an expression of the metaphysical impulse, the impulse to posit unobservable entities, properties and processes lying behind and explaining the appearances. Scientific realism without metaphysics is like Hamlet without the Prince, such that if one dislikes

[16] See also Berenstain and Ladyman (2012).

metaphysics, one would do well to reject realism entirely and embrace empiricist anti-realism (see also Jaksland 2022).

Chakravartty considers the worry that once we've countenanced implicit entities, we're on the slippery slope towards the metaphysical abyss: 'Equipped with an understanding of properties, structures, and particulars, is the realist now required to give an account of the metaphysical basis of this understanding, and so on and so forth, ad infinitum?' (*ibid.*, locs. 2554–2555). He claims this is not a big worry, as the realist can simply stop giving explanations at this point. As we all know, explanations must come to an end somewhere. A possible reply to this is that it overlooks the fact that the reasons he and others give for embracing implicit scientific entities—the need to give a deeper and more satisfactory metaphysical explanation of science and produce a more coherent and defensible overall realist position—are precisely the reasons that can be given for going still further, and embracing traditional metaphysical items—universals, quiddities, possible worlds and the rest. Once the argument procedure of positing metaphysical-explanatory items to account for and ground the items already in our ontology is permitted, the metaphysical genie is arguably out of the bottle.

We need not adjudicate this dispute here. It is a key part of my (and Chakravartty's) argument that such disputes are in any case ultimately stance disputes, and not likely to be susceptible of objective rational resolution. The point I wish to make here is simply that even if one does think the SSR-without-MoS resting point is unstable and unmotivated, we can still define the stance that recommends we commit only to the explicit entities of science (for which we have good evidence), resisting the urging to go further in the direction of inflationary metaphysics.

3 Epistemic Stances and Ontological Commitment

I have said that different epistemic stances can be defined or characterised in terms of the attitudes they engender towards certain forms of inference and types of belief concerning scientific ontology, or ontology more generally. This needs to be clarified in several ways.

3.1 Permissions and Prohibitions

Stances *permit* one to commit to entities of a certain type, in a sense to be clarified below, but they do not *compel* one to believe in all entities of that type, or any particular entities of that type. The empiricist is permitted to believe in everyday, observable entities, but she is not thereby compelled to believe in all everyday, observable entities, or any particular everyday, observable entities. Even more clearly, the SSR stance permits one to believe in the explicit but unobservable entities of science, but one is not thereby compelled to believe in all explicit but unobservable entities of science, or in any particular explicit but unobservable entities. And the same goes for the MoS stance and the implicit entities of science. *Which* entities from the class of entities to which one's stance permits one to commit one actually does commit to will depend on judgments concerning the strength of evidence and argument (scientific or philosophical) for the existence of certain entities. If one adopts the SSR stance, one may think the evidence supports belief in electrons but not dark matter, for instance. This is a familiar point about the commitments of scientific realism. The realist need not, and should not, be so excessively credulous or worshipful of science that she uncritically embraces each and every newly proposed unobservable scientific entity, including highly speculative or heterodox ones. She can be more judicious with her ontology, assessing the first-order evidence for the reality of certain postulated entities just as the scientist does, as well as any relevant local philosophical arguments. And in the case of the MoS stance, the realist may of course think the arguments and evidence support the existence of laws but not causation; of natural kinds but not essences; and so on.

I am here agreeing with those who recommend that the realist abjure global philosophical arguments for realism about unobservable entities in general, such as the No Miracles Argument (e.g. Fitzpatrick 2013). Realism about entities needs to be more local, and assessed on a case-by-case basis (Bird 2022; Magnus 2012). But that need not mean the realist will only assess the first-order scientific evidence in a naïve and philosophically uninformed way (*contra* naturalistic localism, discussed above). In between such first-order evidence and global arguments such as the No Miracles Argument lie specific criteria that realist philosophers have offered for determining when realist commitment is and is not appropriate. Chakravartty's semi-realism (2007), for instance, recommends realists commit only to the concrete relations of 'detection properties' (properties 'connected via causal processes to our instruments and other means of detection' (*ibid.*, loc. 740)), while Eronen (2019) suggests realism is justified with respect to entities that are robust in the sense that their existence is supported by multiple lines of independent evidence,[17] and Strevens (2018) argues that realist commitment is more secure when the entities in question are able to explain a range of different phenomena. These criteria are, however, internal to the realist stance, a point I shall return to in Chap. 4.

Whether one who has adopted a stance has good reason to believe in some putative 'permitted' entity, then, depends on whether she judges that there are good arguments and evidence for the existence of said entity, and this isn't settled by the prior adoption of the stance. It is true that whether certain *forms of argument*, as well as certain types of ontological belief, are legitimate may also, at least sometimes, be a stance-dependent question. But even if it is, there is still a 'gap' between the adoption of a stance that licenses certain types of belief and forms of argument, and the actual acceptance of certain beliefs on the basis of certain arguments. If a stance S licenses argument form A and beliefs of type B, it doesn't of course follow that there must *be* an argument, or a convincing argument, of form A that supports a particular belief of type

[17] This is, he says, a sufficient but not a necessary condition for realist commitment; we may be justified in believing in entities, such as experiential qualities, that are not robust in this sense (*ibid.*, 2345).

B, or that one who has adopted S must endorse such an argument. For example, the realist stance licences the argument form 'inference to the best explanation (IBE)' and beliefs in unobservable entities; it doesn't follow that the realist must think that there is a good IBE in favour of believing in unobservables in general, or in particular cases (though of course most realists do in fact think that).

Of course, it would be an odd sort of realist who decided that in *no* case did the evidence support belief in the explicit entities of science (for SSR) or the implicit entities of science (for MoS). In practice if one adopts an epistemic stance one is likely to believe in the existence of a fairly broad range of entities commitment to which is characteristically permitted by that stance.[18] If one claimed to be standard scientific realist but didn't believe in any, or believed in hardly any, of the explicit entities of science it would be reasonable to question in what sense one really was a realist. But so long as the stances only involve *permission* to commit, such a position is coherent. The point here is just that value-laden stances carve out space for certain types of belief, for certain ontological commitments. But how one who accepts a stance ultimately populates that space with particular beliefs and commitments will be a matter of evidence, argument, and epistemic considerations generally.

What this means—and this will be important in subsequent chapters—is that if someone is, say, a realist about some implicit entity of science, this may be good evidence that she has adopted the MoS stance. But not conclusive evidence, since as we've seen one may believe in the existence of implicit entities of science while adopting the armchair metaphysics stance. Similarly if one believes in some explicit but unobservable entity of science one may have adopted either the SSR stance or the MoS stance. But in both cases we can at least say this person has rejected the weaker stances that prohibit such belief. If someone is an anti-realist about some implicit entity of science however, this is *not* good evidence—or is at best quite weak evidence—that they have *not* adopted the MoS

[18] Devitt suggests that the standard scientific realist holds that tokens of *most* scientific types exist mind-independently (2011). This may be too strong however. If someone held that tokens of all the types referred to in fundamental physics existed, but tokens of no types in any other science—chemistry, biology, the social sciences—existed, they would intuitively still be a standard scientific realist, even though they did not hold that tokens of most scientific types exist.

stance. To know whether they reject this stance we would need to know whether their anti-realism about the relevant entities derives in part from their rejection of characteristic values and attitudes associated with that stance, and their acceptance of values and attitudes associated with weaker stances.

As I noted in the Introduction, typically the stronger stances will include the permissions of the weaker stances as a subset of their permissions: the SSR stance will permit commitment to observable entities, and the MoS stance will permit commitment to both observable entities and unobservable explicit entities of science (see Table 2.1). This does not mean that one who accepts the SSR stance also accepts the empiricist stance, or that one who accepts the MoS stance also accepts the SSR stance (and the empiricist stance), as noted in the Introduction. As I am defining them, the stances do not nest in this way, they exclude one another: one who accepts the SSR stance thereby rejects empiricism, as that is the stance that does *not* permit belief in unobservable explicit entities of science, and one who accepts the MoS stance thereby rejects the SSR and empiricist stances, as they are the stances that do *not* permit belief in implicit entities of science.

Table 2.1 The four categories of entities, and the epistemic stance(s) on which inferences to the existence of the entities is permissible

Category of entity	Stance(s) on which inferences to the existence of the entities is permissible
Observables	Empiricist stance SSR stance MoS stance Non-naturalist realist (armchair metaphysics) stance
Explicit, unobservable scientific entities	SSR stance MoS stance Non-naturalist realist (armchair metaphysics) stance
Implicit scientific entities	MoS stance Non-naturalist realist (armchair metaphysics) stance
Non-scientific metaphysical entities	Non-naturalist realist (armchair metaphysics) stance

Of course, one could, and some do, embrace a position according to which *only* the explicit theoretical entities of science exist (or some of them at least), while commonsense observable entities do not. There are bosons and fermions but not tables and chairs, for example. We can still view this as an expression of the SSR stance as I have construed it: on that stance one is *permitted* to believe in ordinary, everyday objects, but need not. Similarly one could hold that both the explicit and implicit entities of science exist, but commonsense observable entities do not, e.g. there are bosons and fermions, and natural kinds, laws and causal powers, but not tables and chairs. This could fit within the MoS stance, which, like the SSR stance, permits but does not compel belief in everyday commonsense objects.[19] One *could* define a stance on which one is only permitted to believe in explicit theoretical entities of science, or the explicit and implicit entities, and is forbidden to believe in commonsense objects. But I am not aware of anyone arguing for a view that would be profitably interpreted in terms of such a stance, so a principle of Occam's razor for stances—don't multiply stances without necessity—would suggest that we construe the 'scientific entities exist but everyday objects do not' position in terms of the stances we already have on the table—SSR and MoS.

If one adopts stance S, in what sense is one prohibited from committing to the characteristic entities of the stances stronger than S? When we say the empiricist stance rejects unobservable entities, or the SSR stance rejects implicit entities,[20] in what does this rejection consist? One possibility is that such belief is deemed *rationally impermissible* by the stance. While this is a possible view, it is not van Fraassen's or Chakravartty's (or my) view with respect to stances. As we've seen, they, and I, defend voluntarism about stances, the view that a stance is rational so long as it is not irrational, where to be irrational is to violate logical or probabilistic rules of consistency and coherence, or to sabotage oneself: self-defeat by one's own lights. On this view, it is very likely that the stances we are discussing here are all rational, as are the ontological commitments they characteristically license. Thus the empiricist should not regard the SSR

[19] It would be very odd however for one who accepted the MoS stance to reject the explicit entities of science.

[20] This rejection can take the form of agnosticism—suspending judgement—rather than outright denial.

stance as irrational, nor any particular belief in the existence of explicit but unobservable entities of science. (At least, the latter are not irrational *in virtue of* being beliefs in the existence of explicit but unobservable entities of science. They may, according to the voluntarist, be irrational for other reasons, for instance if they are, perhaps when conjoined with some other beliefs, logically or probabilistically inconsistent or incoherent.) The voluntarist-empiricist rejects beliefs about unobservables not because they are irrational (*qua* beliefs about unobservables), but because they, and arguments for them, conflict with the values, attitudes and epistemic policies associated with the empiricist stance, and express rival values, attitudes and epistemic policies. Van Fraassen's classic arguments for empiricism, and against realism (1980), did not aim to show that the latter was rationally impermissible, but simply that it was not rationally obligatory: it is optional; as is empiricism (Lipton 2004, 150). Both views are rational; and where rational meets rational, values and attitudes decide.

For instance, ontologically committing to the unobservable explicit entities of science typically is a consequence of valuing the explanatory power that such commitments afford, even if this involves considerable epistemic risk. It involves or presupposes commitment to inference to the best explanation (IBE) as an obligatory rule of inference. These are values and commitments that the empiricist *qua* empiricist rejects. Thus, the empiricist rejects beliefs in the unobservable explicit entities of science not because the evidence doesn't support such beliefs (making them irrational or epistemically unjustified) *per se*, but because the evidence supports such beliefs, and they count as epistemically justified, only in a framework and against a background that includes IBE as a required rule of inference, explanatory power as a value that trumps other values, and so on, in other words, assuming the realist framework or stance (Psillos 2011). Likewise, the advocate of the SSR stance, while she may accept IBE and value explanatory power in moderation, regards belief in the existence of the implicit entities of science as a bridge too far, in that it expresses a commitment to these things that is intemperate, and a disregard for epistemic risk that is cavalier. She values explanatory power more than does the empiricist, but values minimising epistemic risk more than does the advocate of the MoS stance. She is troubled by the distance from empirical evidence of beliefs in implicit entities.

Let us then distinguish two forms of permissibility and two forms of prohibition:

A belief B is *R-permissible* for S iff B (and the means by which it is justified) is consistent with the rules of rationality.

A belief B is *R-prohibited* for S iff B (and the means by which it is justified) violates the rules of rationality.

A belief B is *V-permissible* for S iff B (and the means by which it is justified) is consistent with the values and attitudes characteristic of the stance S accepts.

A belief B is *V-prohibited* for S iff B (and the means by which it is justified) violates the values and attitudes characteristic of the stance S accepts.

For the empiricist, beliefs in explicit but unobservable entities of science are R-permissible (not R-prohibited), but are V-prohibited (not V-permissible). For the standard scientific realist beliefs in the implicit entities of science are R-permissible (not R-prohibited), but are V-prohibited (not V-permissible).[21] Such, at least, is the voluntarist view that I am assuming here. I shall defend it against objections below.

This way of understanding stance prohibitions and permissions gives us a means of responding to a tempting critique of the stance approach to scientific ontology. According to this objection, it is only rational to believe that entities exist or fail to exist, or to remain agnostic, on the basis of the evidence. Thus if we decline to believe in a type of entity because such a belief is V-prohibited by our stance, and the latter is accepted for non-evidential reasons, we are being irrational: our beliefs are not based on the evidence.[22] This objection overlooks the fact that, as I noted above, the very idea of believing in *certain types* of entity on the basis of *certain types* of evidence (or forms of argument) only makes sense

[21] Van Fraassen (2004a, 173, 176) also suggests there can be a kind of 'pragmatic incoherence' (as opposed to logical incoherence) involved in accepting a stance whilst holding beliefs that go against the spirit of the stance, or whose rejection is naturally or characteristically associated with one's stance.

[22] See Pincock (2024) for a version of this objection. I address Pincock's critique of voluntarism below.

against the background of the adoption of a stance or framework (Psillos *ibid.*). The question whether the evidence supports belief in, say, electrons, only arises once we have already adopted a stance according to which it is epistemically legitimate to infer from certain sorts of scientific evidence to belief in explicit but unobservable scientific entities. So it is misleading to present the grounds of scientific ontology as being a matter or evidence *or* stance choice. Rather, questions of evidence only come into play given the prior adoption of a particular stance. Evidential considerations are downstream from stance choice. The fact that this is not always made explicit, that theorists may *suppose* they are basing their ontological beliefs on a direct comparison of those beliefs with reality, independent of epistemic stances, only shows that what van Fraassen calls 'false consciousness' with respect to stances and beliefs, is widespread. If a theorist thinks the evidence supports belief in the existence of electrons, they have implicitly adopted the realist stance, according to which such evidence and such inferences are legitimate, whether they are aware of this or not. On this view, the position of one who declines to believe in electrons on the basis of the V-prohibitions of the empiricist stance is not, as the above objection suggests: 'I reject belief in electrons since such a belief is prohibited by my stance, irrespective of the evidence'. It is, rather: 'I do not endorse the stance according to which the question of whether the evidence supports belief in electrons is a legitimate question. Thus the question whether the evidence supports beliefs in electrons just doesn't arise for me.'

3.2 Stances and Views

I have been talking about *people*, the stances they adopt and the beliefs they hold. In subsequent chapters we shall be encountering *views*: e.g. the epistemic interpretation of natural kinds, or the etiological account of biological functions, that have implications for scientific ontology. If a view implies that some explicit but unobservable entity of science exists, I shall say that, other things being equal, it is an expression or manifestation of the realist stance, and that it requires at least the SSR stance. If a view has the implication that some explicit but unobservable entity of science *does not* exist, I shall say only that it is *not, in itself*, an expression or manifestation of the realist stance, *not* that it is (necessarily) an expression or manifestation of the empiricist stance, since, as we've seen, more is required for this than just anti-realism. Likewise, if a view implies that some implicit entity of science exists, I shall say that, other things being equal, it requires (at least) the MoS stance, but if a view has the implication that some implicit entity of science does not exist, I shall say only that it is not, in itself, an expression or manifestation of the MoS stance, not that it is (necessarily) an expression or manifestation of a weaker stance. Someone who accepts a particular stance in general can still endorse a view that is not *in itself* an expression or manifestation of that stance. One may accept the MoS stance, for instance, yet endorse a view about natural kinds that implies that natural kinds do not exist, such that that view is not in itself an expression or manifestation of the MoS stance. As we've seen, to accept a stance one need not be a realist about *all* the entities the stance characteristically permits one to believe in. The anti-realist views become associated with the weaker stances only when the grounds for proposing or endorsing them are bound up with the values, attitudes and commitments characteristic of the weaker stances: if, for instance, an anti-realist view about natural kinds is accepted in part because natural kinds are implicit entities of science, and belief in such entities is V-prohibited by one's stance.

I am claiming that certain views or beliefs *require* certain stances: realism about explicit but unobservable entities of science requires at least the SSR stance, for example. Is this in conflict with the widely-held view that holding certain beliefs is *insufficient* for adopting particular stances, since for this one must also accept the characteristic values and attitudes of that stance, and one may hold the belief without endorsing these?[23] There is no conflict here, since when I claim that certain views or beliefs require certain stances, the 'require' here has to do with V-permissions, not R-permissions. We have seen that according to the voluntarist view I am endorsing, for the empiricist, belief in unobservable entities is R-permitted. Thus, as far as rationality is concerned, the empiricist is permitted to hold such beliefs, and therefore, in this sense, holding the beliefs is not

[23] I endorse this view in my (2018b).

sufficient for endorsing the realist stance, as the orthodox view contends. Yet such beliefs are V-prohibited for the empiricist, and so in that sense, the beliefs do require a stance on which they are V-permitted, such as the SSR stance; we may say they V-require, but do not R-require, at least the SSR stance. All references to beliefs or views or accounts 'requiring' certain stances should henceforth be understood in terms of V-permissions and V-prohibitions.

4 Stance Voluntarism

4.1 Pincock on the Obligatoriness of the Realist Stance

In the previous section I spelled out the sense of permission and prohibition that is involved in the voluntarist conception of stances. But should we accept stance voluntarism? Pincock (2024) argues we should not; more precisely, that the realist should not accept voluntarism about her realist stance. If the realist holds that, in accordance with voluntarism, there are not theoretical grounds, but only pragmatic grounds, for accepting the realist stance, she has to accept that her realist commitments are not rationally compelling but rather entirely optional, and ultimately arbitrary. Her belief in, say, unobservable entities like electrons, is voluntarily chosen and thus doesn't amount to knowledge.

We have seen in the previous section one way the voluntarist can respond to this kind of argument. It makes the mistake of counterposing questions of evidence to questions of values as if these are *competing* grounds for belief. It overlooks the fact that questions concerning what beliefs are supported by what evidence only arise once a certain stance has been adopted. Thus Pincock imagines the realist, and one who rejects the realist stance, considering the same evidence for the existence of unobservables, with the realist accepting that the only reason that she, but not her opponent, thinks the evidence supports their existence is that she has freely adopted the realist stance. But the realist and one who rejects the realist stance do not consider the same evidence: as I noted above, the question of whether the evidence supports the existence of unobservables just doesn't arise for the latter.

But it is worth examining Pincock's arguments a little further. Pincock suggests that if a belief (in, say, unobservables) is only acceptable given one has adopted the realist stance, and this adoption is—and is *recognised* as being—optional and unforced, this means the belief itself must be (and must be recognise as being) optional and fall short of knowledge. But this doesn't follow. It is perfectly consistent, the voluntarist claims, for S to freely adopt a stance based on S's values, but then, once the standards of evidence and forms of argumentation licensed by that stance are in play for S, to regard certain beliefs that are the output of these as rationally compelling for S. Part of what one commits to when one accepts a stance is precisely to regard beliefs that are generated on the basis of the forms of argument and standards of evidence permitted or encouraged by that stance as potentially epistemically-rationally justified, and as potential items of knowledge. *Epistemic* stances are above all policies for knowledge-generation. Reflecting that one's choice of stance was optional need not, it would seem, undermine one's confidence in beliefs formed on the basis of the commitments of the stance, or lead to the kind of pragmatic incoherence Pincock fears. As Baumann nicely expresses this view:[24]

> Stances are … second-order practical attitudes concerning (the acquisition of) first-order beliefs about the world. Even if first-order belief-voluntarism is problematic, it is not entailed by second-order attitude-voluntarism. The stance constitutes a framework within which beliefs can be assessed as correct, true or rational. The arbitrariness of a stance or framework is not necessarily inherited by the beliefs resulting from inquiry within that stance or framework. (2011, 33)[25]

Pincock's alternative for the realist is to reject stance voluntarism: the realist ought to regard her stance as rationally obligatory. The way to do this, Pincock suggests, is to advance an argument for the epistemic

[24] Baumann in fact goes on to criticise this view (*ibid.*).

[25] While this is a generally accurate statement of the view in question, I would object to the suggestion that stance choice is 'arbitrary', as should be clear from the remainder of my discussion.

reliability of IBE (commitment to IBE is the defining feature of the realist stance) that shows it to be rationally obligatory. The problem with this is that, as Pincock admits, such an argument is not stance-independent, but rather presupposes the realist stance: it cannot hope to persuade one who rejects the realist stance, but is rather a way for someone who is already a realist to rationally convince themselves of the superiority of their realist stance. But this seems to just concede the point to the voluntarist, who is happy to accept that once the realist has adopted the realist stance, she can provide arguments internal to the stance that are rationally convincing (to the realist), for being a realist, or a particular kind of realist (Chakravartty 2024, 1312). As Psillos argues in his (2011) paper,[26] if we accept the realist framework, we can construct an argument that presupposes that framework for believing in unobservables: such a belief is indispensable for achieving the realist's goals. But such an argument will be compelling only for those who already accept the goals and values of the realist framework. The latter, he suggests, is freely adopted, as the voluntarist claims.[27]

Pincock faces a dilemma, then. If the arguments he offers for the reliability of IBE, and thus the rational obligatoriness of realism, are meant to be stance-independent, then we have a traditional argument for

[26] In other respects Psillos is not a proponent of the sort of voluntarism Chakravartty defends; see his (2021).

[27] This notion of a stance-internal justification plays in important role in Elder's defence of stance voluntarism (2019) against an argument of Baumann's (2011), that is not dissimilar to Pincock's. Baumann claims that it is irrational to adopt an epistemic stance while acknowledging that one does not do so for epistemic reasons. What he means by 'epistemic reasons' is a claim to the effect that the stance is truth-conducive; it 'is the right one for the purposes of inquiry' (*ibid.*, 29). Elder convincingly responds to this (and similar) objections by distinguishing stance-internal from stance-external contexts. She notes that the stance voluntarist *can* accept that an epistemic stance needs to be justified on the basis of epistemic reasons, but these will be *internal* to and *presuppose* the stance. The only external epistemic reasons relevant to stance choice are those pertaining to the rationality-constraints on a stance, which on the voluntarist conception are minimal and underdetermine stance choice. Thus in the external context stance choice is governed predominantly by one's values, not epistemic reasons; but to convict the voluntarist of irrationality on this score is to beg the question, as the voluntarist position is precisely that there is nothing irrational about adopting on the basis of one's values an optional stance that satisfies the minimal constraints on rationality. Hence when Baumann and others claim that on the voluntarist view, stances are not adopted on the basis of epistemic reasons, they are either referring to the internal context, in which case what they allege is false, or the external context, in which case it is correct but irrelevant. See also Surovell (2019) and Teller (2011).

realism as a thesis, and it's unclear why we need to talk about *stances* at all. The realist stance without voluntarism just collapses into traditional thesis-realism. But the whole motivation for the 'stance-turn' in this area, a turn that Pincock seemingly wants to endorse, is the recognition that, after several decades of debate, we are at an impasse—the realist cannot convince the empiricist to abandon her position and vice versa. The stance idea in its voluntarist form is meant to be a partial explanation of, and partial solution to, this predicament. As Chakravartty says, the persistent rational disagreement we find with respect to realism and scientific ontology is plausibly understood as deriving from disagreement at the deeper level of rival epistemic stances, that embody conflicting values, goals, commitments etc., but are likely to be equally rational.

If, on the other hand, the arguments for the reliability of IBE, and thus the rational obligatoriness of realism, are internal to the realist stance, then Pincock is not saying anything the voluntarist disagrees with. This is in fact the horn of the dilemma Pincock favours. His view, then, seems to boil down to the claim that the realist's stance is mandatory for the realist. But that has an air of triviality. Has anyone ever questioned that the realist ought to be a realist?[28]

What the above suggests is that being a non-voluntarist about stances is not an option. The stance view and the voluntarist conception of stance choice are inseparable. At the risk of using non-voluntarist-sounding language, once we adopt the stance view, voluntarism is mandatory, not merely permissible. As Lipton (2004) and others have made clear, the shift van Fraassen recommends from thesis-empiricism to stance-empiricism is precisely the shift from thinking of empiricism as rationally mandated (by facts about, for instance, our sources of information about the world) to thinking of it as non-obligatory but permissible, as one rational stance among others that those with empiricist values and attitudes ought to adopt. Without voluntarism, the stance idea loses its point and purpose.

[28] Chakravartty (2024, 1312) argues that Pincock illicitly conflates 'choosing a rational stance' with 'choosing a stance that is rationally obligatory', and in supposing that the former must involve the latter, he thereby begs the question against the voluntarist.

4.2 Rief and Scott on Historicising Stance Choice

A very different critique of stance voluntarism is suggested by the sociologists Rief and Scott (2025).[29] They make use of the notion of a stance, in particular the opposition between the empiricist and 'metaphysical' (i.e. realist) stances, in characterising the commitments of Germanic social theorists and sociologists of knowledge in the interwar period. In particular, they discuss a version of the empirical/empiricist stance as it features in the work of the Marxist philosopher, social theorist and historian of science (and associate of the Vienna Circle) Edgar Zilsel. Zilsel offered a Marxist analysis of the development of science in the early capitalist period in terms of the interactions between social groups that had previously been relatively isolated from one another, such as craftsmen and engineers, scholars and scientific theorists. Capitalist economic conditions and social relations forged the links between social groups and social forces (bringing together, for example, those with theoretical knowledge and those with practical skills), that were a precondition of the advancement of science. These developments were associated with the historic triumph of the empirical worldview, the 'scientific spirit'—something Rief and Scott identify with the empirical stance. Yet this account seems to suggest a quite different basis for the adoption of a stance than we get on the voluntarist view.

> The contemporary notion of stance, as developed by van Fraassen and like-minded philosophers, is framed in individualist-liberal terms. It is a personal *commitment*; we *adopt* a stance according to our inclinations and preference, and/or our distaste for the alternative. … [I]n Zilsel, the empirical stance, or what he called the 'scientific spirit', is historicized. It is not a *choice* but a social transformation, or, expressed a little differently, it is a choice that can only be made under specific economic, historical, and technological conditions. In other words, Zilsel has a *sociological* understanding of what has come to be called a stance. (*ibid.*)

[29] Thanks to Alan Scott for helpful discussions on this topic.

Set aside the fact that what Rief and Scott are calling the empirical stance (the 'scientific spirit') is rather weaker and more amorphous than what we have been calling the empiricist stance, with its V-prohibitions and V-permissions with respect to ontology. (Arguably this 'scientific spirit' is neutral as between empiricism and naturalistic scientific realism: in Rief and Scott's discussion, the sociological opponents of what they are calling the empirical stance are not scientific realists in our sense, but the conservative, idealist and organicist metaphysical thinkers of the time such as Otto Spann, who in our terms can be thought of as partisans of the armchair metaphysics stance.) They are nonetheless making an important point about stance choice. One need not, and perhaps typically does not, simply adopt a stance as a free and isolated individual; rather stance choice takes place in the context of social, cultural and ideological forces and influences.[30] We may be as unable to freely 'choose' to adopt a stance as we are unable to adopt all our beliefs from scratch: many of our beliefs are inherited from our culture or family or society, and this is likely to be the case for our stances too. Those who were responsible for entrenching scientific culture and method in the early-modern period seemingly did not adopt the empirical stance as a free, conscious decision, rather such a stance was the almost-inevitable ideological or intellectual counterpart of the 'specific economic, historical, and technological conditions' prevailing in that context.

Yet none of this, it seems to me, need undermine the basic stance voluntarist view.[31] Despite the connotations of the name 'voluntarism', the voluntarist view (*contra* Bryant 2021) is consistent with stance choice often, or typically, being constrained or deeply influenced by social or cultural factors. The voluntarist need not be a liberal-individualist. The two core claims of voluntarism are 1. That stance choice is a function of

[30] This will especially be the case for those sympathetic to Marxism, like Zilsel and myself.

[31] Though it may be inconsistent with Chakravartty's verson of the view; see his (2017, 221), where he emphasises the free choice of a stance over the 'passive, unconscious absorption of ambient values', and suggests that the 'sociological' account of stance adoption is 'unpromising', given that stance choice seems to be underdetermined by such background influences. See Bryant (2021) for objections to Chakravartty's version of voluntarism; I claim these objections need not apply to all forms of voluntarism, if we relax the 'free autonomous choice' aspect of it.

values (understood broadly) rather than evidence, and 2. That the conditions stances need to satisfy in order to count as rational are minimal: any stance that does not violate logical or probabilistic rules of consistency and coherence qualifies as rational. No stance will be rationally obligatory and many will be rationally permitted. The two claims are related in that 1 makes sense in the light of 2: if different stances are equally rationally permissible, one will choose between them on the basis of one's values.

With respect to 1, a stance is adopted, explicitly or implicitly, on the basis of one's values, but many of our values are likely to be inherited or unconsciously imbibed from our society or culture rather than freely chosen. This is perhaps particularly true for non-epistemic values. In Part II we shall consider Darwin's commitment to the 'gradualist stance', a metaphysical stance related to a set of social, political and ethical values extolling the virtues and desirability of gradual change, progress and evolution in human societies and institutions, values that Darwin certainly did not freely choose in splendid isolation, but rather imbibed from his Victorian society. It may even be true that this stance could only become an option under the 'specific economic, historical, and technological conditions' prevailing at that place and time (or similar places and times). For now I remain agnostic about whether this is the case, but if it was, this would not conflict with voluntarism. Even epistemic values may often be a matter of collective or community commitment or tradition rather than individual free choice. Recognising these facts should make stance choice, even on the voluntarist conception, appear less individualistic, and more 'historicised'.

With respect to 2, it is important to remember that this concerns *rational* permissibility and obligatoriness. For those who adopted the empirical stance in the period Zilsel is talking about, rival stances, to the extent that they counted as rational, were, according to voluntarism, options for them in the sense of being rationally permitted (the empirical stance was not rationally obligatory), but the rivals may not have been serious options for them in practice given their values: they were R-permissible but V-prohibited. Similarly, for Darwin the rival stance to the gradualist stance, the punctuationist stance, may have been R-permissible but V-prohibited. Thus the voluntarist can, I claim, accept the historicization of stance choice recommended by Rief and Scott.[32]

4.3 Voluntarism, Empiricism and Anti-realism

In the recent scientific realism debate, the form that empiricist anti-realism has taken has generally been the agnostic, or 'suspend judgement' version, as in van Fraassen's constructive empiricism, which recommends, if one shares the empiricist outlook, suspending judgement about the existence of the unobservable entities of science, rather than outright denial of their existence. This form of anti-realism is particularly congenial to voluntarism, indeed it may be thought it is the only form consistent with it (Chakravartty seems to think so, 2017). The empiricist agnostic does not take *any* stance on the existence of the relevant entities, which fits with the voluntarist idea I have been rehearsing that considering arguments and evidence for the existence or non-existence of the relevant entities is just not something the empiricist need be moved to do, given her stance and values. Positively asserting the entities *don't* exist, on this line of thought, involves as much metaphysical commitment, the same degree of 'sticking one's neck out' epistemically, as asserting that they do, and presumably could only result from the kind of attention to the relevant arguments and evidence that I have suggested the empiricist abjures. Suspending judgment is a weaker, less committal stance than either of these, as per the empiricist's epistemic caution and wariness of risk.

Yet in Part I I am going to be considering 'atheist', not just agnostic versions of anti-realism, and suggesting that some of them can be expressions of the empiricist stance.[33] I think this is justified. As a purely histori-

[32] As an anonymous reviewer noted, if stances reside at the meta-level, the questions raised in this section about the extent to which our values are freely chosen reside at the meta-meta-level—the level above stances. Arguably discussion of epistemic stances needs to take this level into account to a greater degree than hitherto.

[33] Bryant (2021) also wonders why Chakravartty only considers agnostic, and neglects atheist, anti-realist stances, though she agrees with Chakravartty that allowing the latter would give rise to worries about incoherence for the voluntarist; I am trying to assuage those worries.

cal point, while constructive empiricism is considered the most promising version of empiricism on the market today, during other historical periods 'atheistic' forms such as instrumentalism and conventionalism have been popular, and have been the form that empiricism has tended to take. In other words, the empiricist stance is a broad church; one expression of it is constructive empiricism; but it can be, and has been, expressed in instrumentalist and conventionalist (i.e. atheist) terms as well. Denying that the latter are expressions of the empiricist stance would seem unmotivated. I shall argue in subsequent chapters that to the extent that the atheistic versions of anti-realism I shall be considering involve characteristic empiricist values and attitudes, they should be seen as expressions of the empiricist stance, no less than agnostic views.[34] One legitimate way of limiting one's epistemic commitment to the observable and avoiding explanatorily-justified inflationary metaphysics is to deny that unobservable entities exist. The empiricist rally-cry of 'down with all explanatory-metaphysical posits!' is possibly better satisfied by atheism than agnosticism.

The atheist view can be justified in terms of Occam's razor. Arguably when it comes to ontology, the default position should be atheism: we should assume that entities don't exist in the absence of positive evidence that they do. And arguably this applies even when we are not considering, and are not *willing* to consider, the evidence for or against their existence. If that is correct the empiricist who does not consider the evidence or arguments for or against the relevant unobservable entities should hold that the entities do not exist, rather than suspending judgment. But I don't need to take a stand on whether atheism with respect to unobservable entities is in all cases the appropriate empiricist attitude; all I have hoped to show is that we should not insist that empiricism must in all cases take the form of agnosticism.[35]

Furthermore, the empiricist-atheist can still be a voluntarist. She can deny the reality of the entities while accepting that the realist stance is

---

[34] One can be an atheist anti-realist without accepting the empiricist stance if one's anti-realism is not motivated by empiricist values and attitudes.

[35] The distinction between atheism and agnosticism in this context has always been a bit slippery anyway; even van Fraassen at times expresses his empiricism in terms of the latter rather than the former: 'theoretical entities as *fictions* facilitating systematic account' (1980, 204; emphasis added).

rational. She does not consider the arguments or evidence for the existence of the entities because she does not value explanatory power, and she denies their existence, perhaps due to Occam's razor as above, but (if she is a voluntarist) she accepts that one who does value explanatory power, and in considering the arguments and evidence comes to the conclusion that the entities exist, is not being irrational, so long as she satisfies the minimal voluntarist conditions for rationality. Clearly one can hold not-P rather than suspending judgement, while denying that those who hold P are being irrational. This is not the same as holding *both* P and not-P, which of course would be irrational.

This concludes our discussion of stance voluntarism.[36] I shall turn now to look more closely at naturalised metaphysics. The concept of a naturalistic approach to metaphysics is a central notion in Part I of this book. There has been much debate in recent years concerning what it could mean for metaphysics to be naturalised, and whether and how such an approach may be justified. Thus, it will be well to have a brief look at some of this discussion.

5 Recent Discussions of Naturalised Metaphysics

I have followed Chakravartty in interpreting naturalised metaphysics in terms of metaphysical (typically explanatory) inferences from theories, concepts and evidence in science to ontology broadly understood, shaped by epistemic stances. It may be worth considering some other conceptions of naturalistic metaphysics and metaphysics of science in the literature, and seeing how they relate to the one I have presented.

[36] For discussions of stances and voluntarism see also Teller (2011), Chakravartty (2011), Steup (2011), Silva (2021), and Bhakthavatsalam (2023). Kusch (2020) discusses the relation between stances, voluntarism and relativism.

5.1 Science-first vs Metaphysics-first Approaches

Some distinguish between two forms of scientific metaphysics (Guay and Pradeu 2020).[37] The first form looks to science to help us answer traditional metaphysical questions. This approach is continuous with conventional armchair metaphysical projects. It is *scientific* metaphysics to the extent that it is claimed we may be assisted in resolving these traditional problems by attending to developments in the sciences; but it is emphatically a 'metaphysics-first' approach (Guay and Pradeu *ibid.*); the science is there as a resource to be drawn on by metaphysicians to aid them in their antecedently-existing and antecedently-worthwhile metaphysical inquiries. The second form takes a more determinedly 'science-first' approach and is characteristically pursued by philosophers of science rather than metaphysicians: its starting point is the theories, concepts and categories of science itself, and it seeks to articulate a picture of the metaphysical structure of the world (or parts of it) that current science seems to be giving us, with very little deference to any traditional metaphysical projects or problems. One prominent version of this is the 'radically naturalistic metaphysics' defended by Ladyman and Ross (2007); this is 'a metaphysics that is motivated exclusively by attempts to unify hypotheses and theories that are taken seriously in contemporary science' (*ibid.*, 1).

There is clearly something to this distinction. For instance, Dupré and Meincken (2020, Introduction) note that we find the two approaches in recent work on biological individuality. Traditional metaphysicians, especially those advocating neo-Aristotelianism, interested in the nature of identity have in recent years drawn on work in the biological sciences to try to resolve or at least make progress on general and antecedently-existing metaphysical problems to do with identity and persistence over time, while philosophers of biology have tended to consider the issue of biological identity or individuality on its own terms, as a specific problem that arises *within* biology and is pressing partly for scientific reasons (for instance Clarke 2016, 894) notes that without a theory of biological individuality biologists have no systematic and consistent way of counting

---

[37] See Triviño (2023) for a similar distinction; see Stanford (2017) for a somewhat different taxonomy.

offspring, and thus no way of measuring fitness). The example of process metaphysics, mentioned above, also illustrates this distinction. The claim that 'biological science supports the process metaphysics view' can be made on the basis of a metaphysics-first or science-first approach.

Yet the way the distinction is sometimes presented is confusing. For instance, Guay and Pradeu (2020) suggest that the metaphysics-first approach is characterised by the fact that it appeals to science to help address questions to do with things like causation, laws, and individuals. Yet this is also something that the science-first approach does: Ladyman and Ross draw conclusions from physics concerning the existence or non-existence of individuals, for instance, as well as other 'traditional' metaphysical entities or phenomena such as modality. Thus Stanford (2017), *contra* Guay and Pradeu, includes Ladyman and Ross among those who think science can answer traditional metaphysical questions. As we've seen, theorists who clearly count as pursuing the science-first approach may often commit to, and theorise about the nature of, entities that are also part of the subject matter of traditional armchair metaphysics: properties, laws, individuals, causation, modality, processes etc. As we've seen, it is partly the methodology used, not only the entities themselves, that determine whether a piece of metaphysics is naturalistic (science-first) or non-naturalistic. Guay and Pradeu (*ibid.*) also classify Chakravartty among the metaphysics-first crowd (as does Triviño 2023), which I think is disputable, as he is much more a philosopher of science than a classic metaphysician, and, as we've seen, although he argues that science needs a (partly a priori) metaphysical underpinning that it is the job of philosophers to provide, he at least *aims* to do this metaphysics in a broadly naturalistic, science-first manner, and is sceptical of traditional a priori and armchair metaphysical projects that are too 'distant' from scientific-empirical evidence.

I'd suggest that rather than necessarily seeing two distinct projects here (science-first and metaphysics-first), we should recognise a continuum of more or less scientifically-informed metaphysics, and categories different approaches in terms of our taxonomy of stances. The science-first approach is characterised by the prominence of metaphysical inferences from science, where these may or may not be inferences to the existence and nature of 'traditional' metaphysical entities. Yet, as we've seen, such inferences can be more or less direct, and be accompanied by a greater or lesser degree of a priori or non-empirical forms of reasoning. If the inferences become too indirect, the underdetermination too great, and the a priori reasoning comes to predominate, the connection to the science may be judged to have snapped, such that the claim that the piece of metaphysics in question is an expression of the naturalistic 'science-first' project may be doubtful—it may be indistinguishable from non-naturalistic 'metaphysics-first' approaches. Yet how indirect metaphysical inferences need to be, and how prominent a priori reasoning has to be, for us to deem that the tie to the science has been severed, is subject to rational disagreement and will depend on one's stance.

'Metaphysics-first' approaches will often be expressions of the armchair metaphysics stance, not necessarily because they deal in traditional non-scientific metaphysical entities, though of course they might, but because to the extent that they are informed by science this is limited or superficial, or the science appealed to is outdated[38] or not well-understood,[39] and the metaphysical inferences from science are secondary and subordinate to purely a priori or non-empirical reasoning.

Consider the work of the Australian metaphysician David Armstrong. Armstrong had a great interest in and respect for science, saw his philosophy as 'this-worldly' and scientifically-oriented, and thought that we must look to natural science to tell us, for example, which universals exist (1997, 25). Yet his metaphysical system was developed, as he admits, for the most part a priori, with limited input from science. Indeed he admits

---

[38] Ladyman and Ross (2007) note metaphysicians' tendency to assume classical physics when speculating about the fundamental nature of reality, and ignore the fact that modern quantum mechanics and relativity theory have rendered this classical picture obsolete (*ibid.*, 17–27). The same point is made by Wilson (2008), who complains about metaphysicians who ignore the content of the sciences that are most relevant to answering the questions they're interested in, and instead proceed by the a priori construction of self-consistent classical possible worlds. They assume classical accounts of matter when what they're interested in is really quantum phenomena. But, he claims, even if we grant them the aim of constructing merely possible classical worlds, they still tend to ignore the areas of contemporary science that are needed to achieve *this* aim. See also Humphreys (2013).

[39] Gesturing towards science without sufficient engagement with or grasp of the relevant theories or concepts is of course something philosophers may often be guilty of; see Kitcher (2023) for a critique of the superficial understanding of evolutionary theory involved in much recent philosophical work on 'evolutionary debunking arguments'.

that the factualist hypothesis he defends in his (1997), according to which states of affairs are ontologically fundamental, being a purely philosophical view defended with philosophical arguments, enjoys a relatively low 'epistemic credit rating' as compared with 'science-based' theses such as physicalism (*ibid.*, 9–10). Put in our terms, his ontological views and commitments may be, as he himself is aware, too distant from science, and too epistemically risky, to be serious candidates for objective knowledge from a naturalist point of view.

If however, the input from science to one's metaphysics becomes prominent enough, and is the product of serious and sustained engagement with and a sophisticated grasp of, parts of contemporary science, the metaphysics may have a legitimate claim to be regarded as naturalistic, and an expression of the MoS stance.

No doubt there are borderline cases, and, again, how much and what kind of input from science is needed for a piece of metaphysics to count as naturalistic will be subject to legitimate, stance-based disagreement, but there do seem to be fairly clear and uncontroversial cases on either side. For instance, few, I take it, would suppose that the fact that Schaffer appeals to facts about quantum entanglement in his defence of his priority-monist metaphysics is sufficient to qualify this as a piece of naturalistic, scientific metaphysics, given the predominance of traditional a priori reasoning in his arguments for the view (2009, 2010), while few, I take it, would deny naturalistic status to the claim that the indistinguishability of particles in quantum mechanics implies either that quantum particles are not individuals, or that our standard conception of individuality needs to be revised (Guay and Pradeu, *ibid.*).

Thus I think our stance-based taxonomy can subsume and clarify the distinction between the science-first and metaphysics-first approaches.

5.2 Metaphysics in Science Vs. Metaphysics for Science

Triviño (2022, 2023) distinguishes two forms of metaphysics of science: metaphysics *in* science and metaphysics *for* science, with a particular focus on metaphysics in and for biology. Metaphysics *in* science concerns the 'metaphysical commitments and implications' of scientific theories and practices (2023, 57) and thus is similar to the sort of scientific metaphysics I have been discussing. Metaphysics *for* science, on the other hand, is more about the *application* of metaphysical concepts and distinctions, often developed in a priori philosophical contexts, to make sense of the metaphysics of particular sciences.[40] For instance, Epstein (2015) suggests[41] that philosophers of social science can and should make use of concepts, tools and distinctions current in cutting-edge metaphysics (e.g. the notions of grounding and anchoring) in order to better understand the metaphysics of the social world. Similarly, philosophers of biology have long helped themselves to concepts such as supereminence, causation, dispositions, emergence, and multiple-realis ability, to understand the metaphysics of biology. These notions are, as Triviño says (following French and McKenzie 2012, 2015), part of the already-existing 'metaphysical toolkit' metaphysicians of science may draw upon to make progress on understanding the metaphysics of particular sciences (Triviño 2023, 60; see also Arenhart and Arroyo (2021, 14–15) for discussion). This is an important form of scientific metaphysics, but not one that I will be spending a lot of time on in Part I, partly because it doesn't, for the most part, concern ontology, at least not directly.[42] Whether, say, fitness is supervening on the physical, or functions are multiply-realizable by different structures, or social groups are 'grounded' in facts about individuals, are somewhat different questions from the sort of ontological questions with which I am mainly concerned in Part I, questions concerning which entities science gives us reason to believe *exist*, and the sorts of epistemic stances required for different degrees of ontological commitment. That is, the former are not *realism* questions in quite the sense that is my focus in Part I.

---

[40] Triviño says this is about the clarification of scientific concepts (2023), but I wouldn't put it quite that way; it is, I would suggest, more about understanding the metaphysical facts. Also, the name 'metaphysics *for* science' is a little misleading: it's not clear the *science itself* needs this kind of metaphysical inquiry in order to perform its tasks adequately. It is primarily philosophers of science that this kind of work is 'for'; but let's not quibble about names.

[41] This is my example, not Triviño's.

[42] I shall have more to say about projects such as these in Part II.

What is more directly relevant to our discussion in Part I however, is that Triviño argues that since with respect to both metaphysics in biology and metaphysics for biology, we find a mix of a priori and empirical elements—armchair theorising along with close contact with science—the categories of naturalistic and a priori (armchair) metaphysics (or, as she puts it, the metaphysics-first and the science-first approaches) do not obviously fit, and so she proposes a different category, 'applied' metaphysics, for both (2023). I think this is premature. As I explained above, the stance approach recognises a spectrum, from clearly naturalistic, through to clearly non-naturalistic (entirely a priori) metaphysics, with some fuzziness in between. The label 'naturalistic' is appropriate to the degree that metaphysical inferences from, or otherwise close contact with, science, rather than purely a priori reasoning, are prominent. In examples like the ones Triviño gives, there is typically a combination of a priori reasoning and deference to science and empirical evidence; but in some cases the latter is more to the fore (e.g. work on biological individuality within philosophy of biology, as a science-first approach), while in other cases a priori reasoning is more to the fore (e.g. the appeal to biological examples by metaphysicians interested in identity, persistence etc. for their own metaphysical sake, as a metaphysics-first approach). And these approaches will be associated with different stances: naturalistic stances, such as the MoS stance, are in play to the extent that contact with and inferences from the science predominates; the armchair metaphysics stance is in play to the extent that a priori reasoning predominates. What we should jettison is not the labels 'naturalistic' ('science-first') and 'non-naturalistic' ('metaphysics-first'), but the idea that there is a hard and fast distinction between forms of metaphysics of science that fall into one category and forms that fall into the other.

5.3 Theories Or the World?

Some have drawn a somewhat different distinction between the project of appealing to science to solve metaphysical problems, and the project of exploring the metaphysical concepts, commitments and assumptions of our current scientific theories—the focus here is on theories, not the world[43] (Stanford 2017; Waters 2017[44]). I am sceptical of this distinction. I think it is fairly clear that, at least most of the time, when metaphysicians of science examine the metaphysical concepts and assumptions of a scientific theory, what they are really interested in is the world, not the theory for its own sake. Thus when, to use two of Stanford's examples, philosophers inquire into whether quantum theory is indeterministic, or whether evolutionary theory treats natural selection and drift as forces akin to Newtonian forces, what they are really interested in (*pace* Stanford *ibid.*) is whether *the world* is indeterministic, and whether natural selection and drift *really are* forces. The reason such inquiries can appear to be in the first instance about theories, not the world, is because of the (often unstated) assumption, expressing of course the naturalist commitment, that the best way of finding out about parts or aspects of the world is by determining what our best current scientific theories say or imply about those parts or aspects of the world. I think it is very clear, for example, that when philosophers argue that evolutionary theory treats species as individuals, what they are nearly always trying to show is that species *really are* individuals. Or that when they argue that evolutionary theory treats functions as selected effects, what they are trying to show is that biological functions *really are* selected effects. We can of course define a project like the one Stanford is describing if we wish, that of exploring the metaphysical commitments or content of a scientific theory while remaining neutral on whether the world is the way the theory says it is,[45] but I claim that in fact very little metaphysics of science is like this.

---

[43] This is one way of interpreting Triviño's 'metaphysics in science' (2022, 2023).

[44] Waters (*ibid.*) also distinguishes between an approach to metaphysics that emphasises the *results* of science and one that seeks to emulate the *methods* of science.

[45] van Fraassen, for one, views this type of inquiry—exploration of how scientific theories represent, and speculation about what the world would have to be like if certain theories were true—as a legitimate form metaphysics can take (2004a, 180; 2007, 379–81). Ladyman (2011, 94) calls this 'quasi-metaphysics'. But the focus is on interpreting theories, not gaining metaphysical knowledge about the world.

5.4 The Critique of the Methodology of Armchair Metaphysics

Central to recent defenses of naturalised metaphysics has been a critique of the methodology of orthodox armchair metaphysics, in particular its reliance on intuitions about metaphysical matters. Ladyman and Ross (2007) are careful to point out that their objection to armchair metaphysics is not that its deliverances are meaningless (as the logical positivists held), but that its methodology is deficient. This methodology, they claim, characteristically involves a kind of cost-benefit analysis of the virtues and drawbacks of certain metaphysical hypotheses, where consistency with intuitions is accorded a central role in this process of theory-evaluation. To the extent that results from contemporary science are relevant to answering the sorts of metaphysical questions they are interested in—for instance when it comes to questions about material composition, or the existence of individuals, or the nature of modality, or space and time—these results are generally ignored.[46] Ladyman and Ross (2007) and Humphreys (2013) argue for instance that one thing science has demonstrated conclusively is that the world is not scale-invariant, yet traditional metaphysicians continue to theorise as if it is—assuming that, for example, intuitions about the behaviour of entities at the macro-level are applicable to phenomena at the micro-level.

Some of the questions this form of metaphysics addresses are, they claim, genuine questions worth investigating: the problem is with the attempt to answer them from the armchair, by consulting intuitions and common sense (perhaps along with appeal to non-empirical, theoretical criteria such as plausibility, coherence, elegance, explanatory power etc.[47]), rather than by consulting the relevant science. And part of

---

[46] Thus several theorists (e.g. Humphreys 2013; Ritchie 2010) have argued that the Humean supervenience thesis defended by Lewis appears to have been refuted by quantum mechanics, yet continues to be debated a priori by philosophers as if it were a live option.

[47] Defenders of traditional metaphysics such as Paul (2012) claim that metaphysics and science employ a similar methodology inasmuch as, given varying degrees of underdetermination, they both appeal to these theoretical virtues in theory-choice, especially explanatory power. Naturalists have responded that the use of IBE in science is very different from its use in metaphysics: in science it is always subordinate to empirical adequacy and empirical strength, whereas in metaphysics it is entirely divorced from empirical evidence (see Ladyman 2012; Ney 2012).

Ladyman and Ross' critique (*ibid.*) is the claim that mainstream metaphysics, to the extent that it does consult science, seeks to 'domesticate' it, that is, render it and its results compatible with our inherited commonsense conceptual scheme.

Other questions it addresses—such as those concerning universals and nominalism, bundle vs. substance-attribute theories, etc.—are, they argue, too distant from any potential evidential contact with the world to be answerable by us.[48] Thus, even if they are meaningful questions, they should be ignored, in accordance with Ladyman and Ross' principle that 'no hypothesis that the approximately consensual current scientific picture declares to be beyond our capacity to investigate should be taken seriously' (*ibid.*, 29). Ladyman (2017) calls this verificationism about knowledge, as opposed to about meaning or truth.

This critique, and related ones, of the methodology of armchair metaphysics has been the focus of much recent work on naturalised metaphysics, both for and against.[49] But this debate need not detain us here, since, as should have been clear from our previous discussions, I shall be following Chakravartty in *assuming* that metaphysics ought to be naturalistic in the broad sense of based on or informed by science, and that armchair metaphysics that does not make sufficient contact with science or empirical evidence should be rejected. Chakravartty's point is that *even if* this 'norm of naturalised metaphysics' is assumed, this leaves open a great many questions concerning precisely what constraints the norm imposes on metaphysical theorising and speculation, e.g. what counts as 'sufficient' contact with science? Indeed what counts as 'contact' with science? Thus I will not aim in Part I to *defend* naturalised metaphysics; my arguments will be directed to those who accept the naturalist norm, and accept the general critique of traditional, armchair metaphysics that naturalists have advanced. I acknowledge that there are those who do not

[48] Bennett (2009) defends an 'epistemicist' position with respect to metaphysical disputes such as these, according to which they are not just verbal disputes, and there may be a fact of the matter about which position is correct, but given our evidential situation, there is little justification for taking one position rather than another.

[49] See Ladyman and Ross (*ibid.*), Ladyman (2017), and Humphreys (2013) who endorse the critique; for responses to the critique see Dorr (2010) and Guay and Pradeu (2020), who, while sympathetic to the program of naturalised metaphysics, argue that there are several potential targets of the naturalists' critique, and these have tended to be confused.

accept these things, and these questions have been and continue to be debated. But there are enough interesting questions to raise *within* the broad naturalistic approach or framework to keep us well occupied, for instance we shall see that one need not accept Ladyman and Ross' specific, strong version of naturalised metaphysics in order to count as a naturalist.[50] The tenability of that framework itself is something I shall be presupposing.

6 Naturalised Metaphysics of Biology?

With a few exceptions (Slater 2017; Waters 2017; Triviño 2022, 2023) there has been relatively little work done explicitly addressing the question of whether and how naturalised metaphysics may apply to the metaphysics of biology. This has seemingly been for two, diametrically opposed reasons. The first is that it has just seemed *obvious* that contemporary metaphysics of biology is and must be naturalistic. Since David Hull founded the modern discipline of philosophy of biology in the 1970s it has been axiomatic within that discipline that philosophical reflections on biology must be informed by a deep and serious engagement with the content of biological science, and one aspect of this has been the near-universal acceptance of the idea that to the extent that metaphysical questions arise about biological matters—e.g. are species real? Are species individuals or natural kinds? Is there design and teleology in the natural world? Is there a fact of the matter about the level at which natural selection acts? Is natural selection a force? Is reductionism in biology correct?—their answers will very largely depend on the details of our best contemporary biological theories and explanations. Even if the answers to such questions cannot be straightforwardly 'read off' the science, as the fact that metaphysics is for the most part underdetermined by science means that philosophy has an important role to play in answering them (Humphreys 2013, 72), such answers will be grounded in and informed

[50] See, for example, Morganti and Tahko's more moderate version of naturalised metaphysics (2017).

by the science in just the way the naturalist about metaphysics requires.[51] Non-naturalistic, armchair metaphysics of biology would not today be taken seriously by anyone.[52] Thus Slater (2017, 56) speculates that the reason there has been very little discussion of the extent to which the species-as-individuals (SAI) thesis may exemplify naturalised metaphysics is that it just seems obvious that it does so exemplify it, given the way the arguments for it appeal to current theory and practice in biological systematics (this judgement, he argues, may be premature[53]).

The second reason for the neglect of naturalised metaphysics of biology is that discussions of naturalised metaphysics have been heavily physics-focused. This is partly no doubt due to the contingent fact that many of those arguing for naturalizing metaphysics are more interested in or knowledgeable about physics than other sciences, so have tended to draw their examples from this science. But the idea that for metaphysics to be naturalistic it is must be based on not just any science but fundamental physics, has been explicitly articulated and defended (Ney 2012; see Guay and Pradeu *ibid.*). Ladyman and Ross (*ibid.*), for instance, argue that to be a legitimate piece of naturalised metaphysics, a claim or theory must show how to unify two (or more) claims in contemporary science, at least one of which is taken from fundamental physics (others may be from the special sciences)—their 'Principle of Naturalistic Closure' (*ibid.*, 44).[54] It follows that the metaphysical questions about biology I mentioned above—since they clearly do not satisfy this requirement—are not legitimate metaphysical questions. Far from it being almost trivial that

[51] In my (2017) I suggest that debates concerning the ontological status of species are governed by what I call the 'Principle of Theory Dependence', which (following Hull 1976, 180; 1977, 92–93) states: 'Ontology is theory-dependent in the following sense: The ontological status and nature of entities, processes etc. of which science gives us our best theories is determined by, and relative to, the best scientific theories in the relevant domain.'

[52] See Triviño (2022, 2023), discussed in the previous section, on the different forms that science-informed metaphysics of biology can take.

[53] I discuss the relation between SAI and naturalistic metaphysics in Chap. 4.

[54] This requirement is too strong to be plausible even if we do think that naturalised metaphysics must be based solely on fundamental physics. The claim, for example, that determinism is false because one of our best-confirmed contemporary scientific theories, quantum mechanics, is indeterministic, is surely an uncontroversial specimen of naturalistic metaphysics, yet it does not appear to connect two (or more) specific hypotheses. See Melnyk (2013, 82–3). Another counterintuitive consequence of their requirement is that metaphysics no longer includes ontology, since ontological claims to the effect that certain entities exist, do not involve *unifying* claims at all.

metaphysics of biology is or should be naturalistic, there can be no naturalised metaphysics of biology at all, only naturalised metaphysics of physics. This might, incidentally, be at odds with Ladyman and Ross' view—part of their 'rainforest realism'—that '[t]he ontological commitments of the special sciences should be taken as metaphysically on a par with those of physics. For example, there are atoms, cells, organisms...' (Ladyman 2017, 152). We have to assume that they do not regard these sorts of ontological commitments as *metaphysical* hypotheses, despite the presence of the words 'ontological' and 'metaphysically' in the above sentence; at least, not until they have been related to claims in fundamental physics.

Each of these reasons for neglecting or rejecting naturalised metaphysics of biology is unpersuasive. With respect to the triviality claim, while nearly everyone in the philosophy of biology at least pays lip-service to the idea that metaphysics of biology to be at all credible must be based on or sensitive to our best contemporary science, there is considerable disagreement in practice concerning what exactly this means, i.e. the precise extent to which, and manner in which, metaphysical claims can be derived from biological claims and theories, and the naturalistic credentials of particular hypotheses in metaphysics of biology. We have encountered such disagreement above when discussing scientific ontology and epistemic stances: a fairly strict naturalist may only wish to advance metaphysical claims about the existence and nature of entities explicitly mentioned and theorised about in science (species, clades, units of selection, genes, etc.), while a less-strict naturalist will be willing to commit to the existence and speculate about the nature of implicit entities of biology (natural kinds, essences, functions etc.). Still others may be willing to go further, and posit and theorise about entities (universals, sui generis teleology, Aristotelian substances etc.) that don't even appear to be *implicit* entities of biology. Thus, while all may profess their allegiance to a broadly naturalistic approach, the question of how direct the 'metaphysical inferences' from biology must be for a piece of metaphysics to count as genuinely naturalistic will be one on which there can be legitimate disagreement, and thus the issue of what 'naturalised metaphysics of biology' involves, or should involve, is, far from being trivial, a controversial and interesting question worth pursuing.

With respect to the claim that metaphysics must be based on physics, I, and many others, find very implausible the claim that there are no genuinely metaphysical questions—none that should be taken seriously, anyway—that are distinctively biological in nature or origin, and that the only science that does or should matter for the naturalised metaphysician is physics. The question of the ontological status of species for instance—whether species are individuals, sets, natural kinds, or something else—has been extensively discussed, and is standardly taken to be a paradigmatically metaphysical question. Yet it concerns biological entities and it has nothing to do with fundamental physics (Guay and Pradeu *ibid.*).[55] It seems that the reason questions such as this have not been taken by orthodox naturalised metaphysicians to be relevant to scientific metaphysics (though they are by philosophers of biology) is that they are considered to be insufficiently general or fundamental (Waters 2017). Metaphysics is concerned with the most general or fundamental features of the *whole of* reality, and the science that is thought to be best placed to inform us about this is fundamental physics (see Ladyman and Ross *ibid.*, 37, Maudlin 2007, 1). But this is a mistake. As Waters argues, metaphysical questions concerning biology, no less than those concerning physics, 'can be interpreted as a … quest to identify fundamentals. The fundamentals sought in these inquiries are the fundamentals of the living world (rather than of the universe at large)' (*ibid.*, 86).[56] If metaphysics is concerned with the most general and fundamental aspects of being, then only an unjustified physics-bias could lead one to deny that investigation

[55] In fact in a later paper Ladyman (2017, 150) seems to accept that precisely this question—about the ontological status of species—provides a counterexample to the Principle of Naturalistic Closure.

Similar claims could be made about metaphysical questions that arise in the philosophy of social science. Surely the question of whether genders or races exist, and if so what their ontological status is, is paradigmatically metaphysical in nature, even though this has nothing to do with fundamental physics, or possibly indeed with anything ontologically 'fundamental' at all, as Barnes argues in an important paper (2014). What naturalised metaphysics in such cases would look like is a difficult question. Presumably the relevant sciences would be social sciences.

[56] See also Guay and Pradeu (*ibid.*) who similarly take issue with the exclusive physics-focus of some advocates of naturalised metaphysics, and argue that metaphysics of biology is a particularly clear example of scientific metaphysics.

into the most general and fundamental aspects of being in the living world[57] counts as metaphysical inquiry in the full sense of the term.[58]

Ladyman and Ross connect their physics-bias with their 'primacy of physics' principle, which states that special-science hypotheses that conflict with established claims of fundamental physics should for that reason be rejected, but the reverse does not hold.[59] (Note that this claim is distinct from the thesis of *physicalism*, which, at least in its standard forms, they reject.) Even if we accept this view however, it's not clear that it follows that there can be no metaphysics of biology, that naturalised metaphysics can only take its cue from fundamental physics.

Notoriously, some philosophers of science such as Smart (1963) once held that biology was not a genuine science, since genuine sciences traffic in universal truths and laws that hold throughout all of space and time; biology deals only with contingent truths and generalizations about the nature of systems localized to (as far as we know) one planet. It is thus at best an applied science like engineering. Smart's view is widely rejected today, but there are echoes of it in the attempt to restrict naturalised metaphysics to the universal and fundamental theories and laws of physics and their implications, and deny the status of metaphysics to inquiry into general features or ontological structures of being in the biological world.[60]

[57] Or, for that matter, the social world; see footnote above.

[58] Waters (*ibid.*) offers an account of generality—his 'generality across scales' thesis—according to which metaphysical claims about the living world may even have a *greater* degree of generality than claims about fundamental physics.

[59] See Melnyk (2013) for a critique of their arguments for this view.

[60] Ladyman (2012) does suggest that one form the recent revival of metaphysics has taken has been work on metaphysical issues in the special sciences such as biology—metaphysical issues around functions, natural selection, natural kinds, essences etc. Ladyman does seem here to regard this as a legitimate form of naturalised metaphysics; though near the end of this paper he wonders if it really counts as theoretical science rather than metaphysics.

7 Stances, Values, and the Limits of Rational Argument

The upshot of the epistemic stance approach to understanding debates about scientific ontology is that questions of scientific ontology have much more to do with values and attitudes, and much less to do with objective assessments of the evidence, than has standardly been thought. Beliefs and commitments concerning scientific ontology, and arguments for them, are downstream from the adoption of value-laden stances that specify which types of belief and commitment are V-permissible and which V-prohibited. This should open the door to a less dogmatic and rigid, more pluralistic, and more tolerant conception of these debates and the rival positions therein. In Part I I aim to make this case for several important debates in the metaphysics of biology. But first we need to address a particular worry about values and rational discourse.

In presenting this material one objection I have sometimes encountered is that I am presupposing something that is at least debatable, viz. that rational arguments about which values we ought to hold are impossible or at least extremely difficult. Thus the inference from 'stances are value-laden' to 'rational debate about which stance to accept is difficult or impossible' is too quick.

In the debate about stances, the idea that it is hard or impossible to rationally debate or defend values has often been unquestioned, and has indeed been seen as a powerful argument against construing philosophical positions as values-based stances. The worry is that we would end up with a kind of metaphilosophical emotivism, according to which philosophical opinion is taken to be merely the subjective expression of values (Ho 2007; Jauernig 2007; Ladyman 2004). Van Fraassen has responded to this criticism precisely by denying that it is not possible to rationally debate value-choice: 'After all, we have learned something since those heady but simplistic days of *Language, Truth and Logic*… we know very well how to defeat the simplistic philosophies that make values just a matter of subjective preference, dismissably relative' (see also van Fraassen 2004b, 9, 17). '[W]e are members of a highly politicised open society in which ethical and ideological differences are precisely what are most up for debate. We need not look far to see that rational discourse is possible on matters that touch our values, attitudes and commitments' (2002, 62; see also 2007, 375–8).

In my (2018a) I considered this response, and suggested that while it may be possible to (a) rationally evaluate which stance to adopt on the basis of which values, and (b) rationally evaluate which values we ought to hold given certain other values we hold—in particular, which less-fundamental or less-basic values follow from or cohere with more-fundamental or more-basic values—nonetheless, when it comes to the justification of fundamental or foundational values themselves, it is likely that rational argument reaches its limits. We do not need to embrace the outmoded positivist conception of all value judgements as subjective expressions of feeling hermetically sealed off from fact, reason and argument in order to accept that with respect to certain basic values we are sometimes at bedrock.

I've suggested, following Chakravartty, Wylie and others, that the central disagreement between partisans of the realist and empiricist stances derives from the fact that they ascribe different value to explanatory power and minimising epistemic risk, and that this is a special case of the more general question, highlighted by James, of how to balance the competing aims or values of believing the truth and avoiding error. It is plausible that these qualify as relatively basic or fundamental (epistemic) value judgements, and thus we should expect them to be relatively impervious to rational argument. We can perhaps best make this case by asking the rhetorical question, just what would a rational argument between someone who thought that, granting both are important and desirable, believing the truth was more important (more to be valued) than avoiding error, and someone who thought that avoiding error was more important (more to be valued) than believing the truth, look like? How would one side in the 'debate' hope to rationally persuade the other?' Compare: some philosophers think truth is intrinsically valuable, while others think it is only instrumentally valuable. How could this dispute be rationally decided? These seem like basic value judgements, with respect to which argument gives out, and a form of philosophical quietism is probably appropriate.[61] If we move from epistemic to non-epistemic values, it is well-known that equality and liberty, though both are valued by most people in liberal, Western societies, can conflict in practice, and various political philosophies can (in part) be defined in terms of how they advocate striking the balance between them. But how would a rational argument between someone who held that equality was more important than liberty, and someone who held that liberty was more important than equality, proceed? Here, surely, we reach bedrock. Given the postulate that equality (liberty) is the most basic societal value or aim, one that, in the event of a conflict, trumps all others, we can elaborate a political philosophy on that basis, but that postulate is to be treated as a basic axiom to be accepted without argument, or as self-evident. The competing values that characterise rival stances are for the most part like this, I claim.[62]

This does not mean that whether, say, one does or should hold that certain explicit or implicit entities of science exist has nothing to do with evidence or rational argument. Seeing beliefs about ontology as being situated within, or accepted against the background of, epistemic stances, and seeing epistemic stances as being a function of values, does not amount to the radically (and implausibly) relativist/constructivist view that one may adopt any position on matters of scientific ontology that fits with one's values or inclinations, and that any view is as good as any other. As should become clear in the course of Part I, the relativism, pluralism and voluntarism that van Fraassen, Chakravartty and I endorse is limited. Evidence, reasoning, and epistemic justification still have an important role to play. As the pragmatic encroachment literature in epistemology (see e.g. Kim and McGrath 2019), and the recent literature on values in science (see e.g. Douglas 2000, 2013; Elliott 2011; Elliott and McKaughan 2014) have amply shown, epistemic rationality may be

[61] The dispute between those who think life (or human life) is intrinsically valuable and those who think it is merely instrumentally valuable, may be similar.

[62] Serebrinksy (2024) accepts that stance-based disagreement is 'deep' disagreement in the sense that there is a lack of common ground between the disputants and arguments presuppose the stance being defended. But she suggests that stance-based arguments can nonetheless play a useful dialectical role, in clarifying the intuitions of the rival disputants, which I am happy to concede. My claim is not that all stance-based disputes are pointless and should cease forthwith, but that we should recognise the inevitable limits to the effectiveness of rational argument across stance-divides.

constrained by values and non-epistemic factors without being swallowed up by them. There is a difference between standards of evidence and the legitimacy of forms of argument being sensitive to non-epistemic factors such as values, and simply believing whatever we deem is useful or corresponds to our values.

I turn now to consider several issues in the metaphysics of biology. I will not for the most part be endorsing any particular first-order views in these debates. I have my sympathies that no doubt will emerge at times. But what I am arguing for is, firstly, that interpreting these debates in terms of the epistemic-stance approach to scientific ontology is illuminating and fruitful, and secondly, that such an interpretation has quite striking meta-level implications. I'd suggest that when it comes to the epistemic-stance framing of issues in scientific ontology, the proof of the pudding will be in the eating: if this approach helps us to understand, and provide a coherent and unified meta-philosophical interpretation of, a number of actual, current debates in scientific ontology, as I argue it does, it will have proved its worth.

References

Achinstein, P. 2002. Is there a valid experimental argument for scientific realism? *Journal of Philosophy* 99: 470–495.

———. 2010. *Evidence, explanation, and realism*. New York: Oxford University Press.

Arenhart, J. R. B., & Arroyo, R. W. 2021. The spectrum of metametaphysics: Mapping the state of art in scientific metaphysics. *Veritas* (Porto Alegre), 66(1), 1–21.

Armstrong, D.M. 1997. *A World of States of Affairs*. Cambridge: Cambridge University Press.

Barnes, E. 2014. Going Beyond the Fundamental: Feminism in Contemporary Metaphysics. *Proceedings of the Aristotelian Society* 114 (3): 335–351.

Bhakthavatsalam, S. 2023. Revisiting stance voluntarism: in defense of an active stance pluralism. *Synthese* 202, 184.

Baumann, P. 2011. Empiricism, Stances and the Problem of Voluntarism. *Synthese*, 178 (1). 27–36.

Bennett, K. 2009. Composition, Colocation, and Metaontology. D. Chalmers, D. Manley and R. Wasserman (eds.), *Metametaphysics: New Essays on the Foundations of Ontology*. Oxford: Clarendon Press. 38–76.
Berenstain, N. and Ladyman, J. 2012. Ontic Structural Realism and Modality. Landry, E. & Rickles, D. (eds.), *Structural Realism: Structure, Object, and Causality*. Springer.
Bird, A. 2007. *Nature's Metaphysics: Laws and Properties*. Oxford University Press.
———. 2022. *Knowing Science*. Oxford University Press.
Boucher, S.C. 2017. Gould on species, metaphysics, and macroevolution: a critical appraisal. *Studies in the History and Philosophy of the Biological and Biomedical Sciences*, 62: 25–34.
———. 2018a. Stances and epistemology: values, pragmatics and rationality. *Metaphilosophy* 49 (4): 521–547.
———. 2018b. What is the relation between a philosophical stance and its associated beliefs? *dialectica* 72 (4): 509–524.
Browning, D. (ed.). 1965. *Philosophers of process*. New York, Random House.
Bryant, A. 2021. A Thousand Flowers on the Road to Epistemic Anarchy: Comments on Chakravartty's *Scientific Ontology. Dialogue* 60 (1): 1–13.
Chakravartty, A. 2007. *A Metaphysics for Scientific Realism: Knowing the Unobservable.* Cambridge: Cambridge University Press. Kindle edition.
———. 2011. A Puzzle about Voluntarism about Rational Epistemic Stances. *Synthese*, 178 (1). 37–48.
———. 2013. On the Prospects of Naturalised Metaphysics. Don Ross, James Ladyman & Harold Kincaid (eds.), *Scientific Metaphysics*. Oxford University Press. 27–50.
———. 2017. *Scientific Ontology: Integrating Naturalised Metaphysics and Voluntarist Epistemology.* Oxford University Press.
———. 2024. Resolving Debates about Scientific Realism: The Challenge from Stances. *Philosophy of Science* 91, 1306–1315.
Chalmers, S., D. Manley and R. Wasserman (eds.). 2009. *Metametaphysics: New Essays on the Foundations of Ontology*. Oxford: Clarendon Press.
Clarke, E. 2016. A levels-of-selection approach to evolutionary individuality. *Biology and Philosophy* 31: 893–911.
Della Rocca, M. 2020. *The Parmenidean Ascent*. Oxford University Press.
Devitt, M. 2011. Natural Kinds and Biological Realisms. In J.K Campbell, M. O'Rourke, & M.H. Slater (eds.), *Carving Nature at its Joints: Natural Kinds in Metaphysics and Science*. Cambridge, MA: MIT Press. 155–174.

Dorr, C. 2010. Review of *Every Thing Must Go: Metaphysics Naturalized* by James Ladyman and Don Ross. *Notre Dame Philosophical Reviews*, June 16. https://ndpr.nd.edu/reviews/every-thing-must-go-metaphysics-naturalized/

Douglas, H. 2000. Inductive Risk and Values in Science. *Philosophy of Science* 67 (4): 559–579.

———. 2013. The value of cognitive values. *Philosophy of Science* 80. 796–806.

Dupré, J. 2021. *The Metaphysics of Biology*. Cambridge: Cambridge University Press.

Dupré, J. and Nicholson, D.J. (eds.). 2018. *Everything Flows: Towards a Processual Philosophy of Biology.* Oxford University Press.

Dupré, J. & Meincke, A.S. (eds.). 2020. *Biological Identity: Perspectives from Metaphysics and the Philosophy of Biology.* Routledge.

Elder, J. 2019. Defending stance voluntarism. *Philosophical Studies*, 176(11), 3019–3039.

Elliott, K.C. 2011. Direct and indirect roles for values in science. *Philosophy of Science* 78 (2): 303–324.

Elliott, K.C. & McKaughan, D.J. 2014. Nonepistemic values and the multiple goals of science. *Philosophy of Science* 81 (1): 1–21.

Ellis, B. 2001. *Scientific Essentialism*. Cambridge University Press.

Epstein, B. 2015. *The Ant Trap: Rebuilding the Foundations of the Social Sciences.* Oxford: Oxford University Press.

Eronen, M.I. 2019. Robust realism for the life sciences. *Synthese* 196 (6): 2341–2354.

Fitzpatrick, S. 2013. Doing Away with the No Miracles Argument. In D. Dieks and V. Karakostas (eds.), *Recent Progress in Philosophy of Science: Perspectives and Foundational Problems*. Springer.

French, S. and McKenzie, K. 2012. Thinking Outside the Toolbox: Towards a More Productive Engagement Between Metaphysics and Philosophy of Physics. *European Journal of Analytic Philosophy* 8, no. 1: 42–59.

French, S. and McKenzie, K. 2015. Rethinking Outside the Toolbox: Reflecting Again on the Relationship Between Philosophy of Science and Metaphysics. In T. Bigaj and C. Wüthrich (eds.), *Metaphysics in Contemporary Physics: Poznan Studies in the Philosophy of the Sciences and the Humanities*, vol. 104. Leiden: Brill: 25–54.

Guay, A. and Pradeu, T. 2020. Right out of the box: How to situate metaphysics of science in relation to other metaphysical approaches. *Synthese* 197: 1847–1866.

Henschen, T. 2021. Subatomic Particles, Epistemic Stances, and Kantian Antinomies. *Journal for General Philosophy of Science* Vol. 55: 247–26.

Ho, D. 2007. Farewell to Empiricism. B. Monton (ed.). *Images of Empiricism: Essays on Science and Stances, with a Reply from Bas C. van Fraassen*. Oxford: Oxford University Press. 319–333.

Hull, D. 1976. Are Species Really Individuals? *Systematic Zoology*, 25. 174–191.

———. 1977. The Ontological Status of Species as Evolutionary Units. R. Butts and J. Hintikka (eds.), *Foundational Problems in the Special Sciences*. Dordrecht, Holland: D. Reidel Publishing Company. 91–102.

Humphreys, P. 2013. Scientific Ontology and Speculative Ontology. Don Ross, James Ladyman & Harold Kincaid (eds.), *Scientific Metaphysics*. Oxford University Press. 51–78.

Jaksland, R. 2022. A trilemma for naturalized metaphysics. *Ratio* 36: 1–10.

Jauernig, A. 2007. Must Empiricism Be a Stance, and Could it Be One? How to Be an Empiricist and a Philosopher at the Same Time. B. Monton (ed.). *Images of Empiricism: Essays on Science and Stances, with a Reply from Bas C. van Fraassen*. Oxford: Oxford University Press. 271–318.

Kim, B. and McGrath, M. (eds.). 2019. *Pragmatic Encroachment in Epistemology*. Routledge.

Kitcher, P. 2023. *What's the use of philosophy?* Oxford University Press.

Kusch, M. 2020. Stances, voluntarism, relativism. In Finkelde, D. & Livingston, P. (eds.). *Idealism, relativism and realism: New essays on objectivity beyond the analytic-continental divide*. De Gruyter: 131–153.

Ladyman, J. 2004. Empiricism Versus Metaphysics. *Philosophical Studies*, 121 (2).133–145.

———. 2011. The Scientistic Stance: The Empirical and Materialist Stances Reconciled. *Synthese*, 178 (1). 87–98.

———. 2012. Science, Metaphysics and Method. *Philosophical Studies* 160: 31–51.

———. 2017. An apology for naturalised metaphysics. Slater, M.H. and Yudell, Z. (eds.). *Metaphysics and the Philosophy of Science: New Essays*. Oxford University Press. 141–161.

———. 2018. Scientific realism again. *Spontaneous Generations: A Journal for the History and Philosophy of Science*, Vol. 9, No. 1. 99–107.

Ladyman, J. and D. Ross, with D. Spurrett and J. Collier. 2007. *Every Thing Must Go: Metaphysics Naturalized*. Oxford: Oxford University Press.

Lipton, P. 2004. Epistemic Options. *Philosophical Studies*, 121 (2). 147–158.

Maddy, P. 2007. *Second Philosophy: A Naturalistic Method*. Oxford University Press.

Magnus, P.D. 2012. *Scientific Enquiry and Natural Kinds: From Planets to Mallards*. Palgrave Macmillan.
Maudlin, T. 2007. *The Metaphysics within Physics*. Oxford University Press.
Melnyk, A. 2013. Can Metaphysics Be Naturalized? And If So, How? Ross, D., Ladyman, J. & Kincaid, H. (eds.), *Scientific Metaphysics*. Oxford University Press.
Morganti, M. & Tahko, T.E. 2017. Moderately Naturalistic Metaphysics. *Synthese* 194 (7): 2557–2580.
Mumford, S. & Tugby, M. (eds.). 2013. *Metaphysics and Science*. Oxford University Press.
Ney, A. 2012. Neo-positivist metaphysics. *Philosophical Studies* 160: 53–78.
Paul, L.A. 2012. Metaphysics as modelling: the handmaiden's tale. *Philosophical Studies*, 160 (1): 1–29.
Pincock, C. 2024. Defending a Realist Stance. *International Studies in the Philosophy of Science*, 37 (1–2): 1–15.
Psillos, S. 2011. Choosing the realist framework. *Synthese* 180 (2): 301–316, 338–339.
———. 2021. Scientific Ontology: Fact or Stance? *Dialogue* 60 (1): 15–31.
Quine, W.V.O. 1953a. On What There Is. Quine, W.V.O. *From a Logical Point Of View*. Cambridge, MA: Harvard University Press. 1–19.
———. 1953b. Two Dogmas of Empiricism. Quine, W.V.O. *From a Logical Point of View*. Harvard University Press. 20–46.
Rief, S. & Scott, A. 2025. A tale of two sociologies: empiricism versus 'metaphysics' in social research and the early sociology of knowledge. In: Fran Collyer (ed.) *Research Handbook on the Sociology of Knowledge* (pp. 102–121). Cheltenham and Northampton, MA: Edward Elgar.
Ritchie, J. 2010. Naturalized Metaphysics. *International Journal of Philosophical Studies* 18 (5): 673–685.
Russell, B. 1956. Logical Atomism. In Russell, B. *Logic and Knowledge*. New York: George Allen and Unwin. 321–43.
Schaffer, J. 2009. On what grounds what. D. Chalmers, D. Manley and R. Wasserman (eds.), *Metametaphysics: New Essays on the Foundations of Ontology*. Oxford: Clarendon Press. 347–383.
———. 2010. Monism: The Priority of the Whole. *Philosophical Review*, Vol. 119, No. 1: 31–76.
Serebrinsky, D. 2024. Epistemic Stances, Arguments and Intuitions. *Journal for General Philosophy of Science* Vol. 55: 79–94.
Silva, B.M.E. 2021. What's voluntary in stance voluntarism? *Manuscrito* 44 (2): 52–88.

Slater, M. 2017. Naturalized Metaphysics and the Contention over the Ontological Status of Species. Slater, M.H. and Yudell, Z. (eds.). *Metaphysics and the Philosophy of Science: New Essays*. Oxford University Press. 55–80.

Smart, J. 1963. *Philosophy and Scientific Realism*. Routledge & Kegan Paul.

Stanford, P. K. 2017. So Long And Thanks For All The Fish: Metaphysics and the Philosophy of Science. In Slater, M. and Yudell, Z. (eds.). *Metaphysics and the Philosophy of Science: New Essays*. Oxford University Press. 127–140.

Steup, M. 2011. Empiricism, metaphysics, and voluntarism. *Synthese* 178: 19–26.

Strevens, M. 2018. Explanation and reality: comment on Chakravartty. *Metascience* vol. 27: 371–378.

Surovell, J. R. 2019. Stance Empiricism and Epistemic Reason. *Synthese* 196, 709–733.

Teller, P. 2011. Learning to live with voluntarism. *Synthese* 178(1): 49–66.

Triviño, V. 2022. Towards a characterization of metaphysics of biology: metaphysics *for* and metaphysics *in* biology. *Synthese* 200, 1–21.

———. 2023. Exploring the interactions between metaphysics and science: lessons from the metaphysics of biology. *Philosophica*, 31(1), 51–69.

van Fraassen, B.C. 1980. *The Scientific Image*. Oxford: Clarendon Press.

———. 2004a. Replies to Discussion on The Empirical Stance. *Philosophical Studies*, 121 (2). 171–192.

———. 2004b. Reply to Chakravartty, Jauernig, and McMullin, unpublished typescript of a talk delivered at a symposium on *The Empirical Stance* at the Pacific APA in Pasadena, CA.

———. 2007. From a View of Science to a New Empiricism. B. Monton (ed.). *Images of Empiricism: Essays on Science and Stances, with a Reply from Bas C. van Fraassen*. Oxford: Oxford University Press. 337–383.

———. 2009. The perils of Perrin, in the hands of philosophers. *Philosophical Studies* 143: 5–24.

Waters, C.K. 2017. No general structure. In Slater, M. and Yudell, Z. (eds.). *Metaphysics and the Philosophy of Science: New Essays*. Oxford University Press. 81–107.

Wilson, M. 2008. Beware of the Blob: Cautions for Would-Be Metaphysicians. D. Zimmerman (ed.). *Oxford Studies in Metaphysics Vol. 4*. Oxford University Press. 275–320.

Wolff, J. 2019. Naturalistic quietism or scientific realism? *Synthese* 196: 485–498.

3

The Units of Selection

1 Introduction

Recall the general picture of epistemic stances I set out in the previous chapter. In the context of naturalised metaphysics, the adoption of an epistemic stance licenses, or discourages, certain inferences from the relevant science to certain beliefs concerning scientific ontology. That is, it licenses or discourages[1] certain *metaphysical inferences.* The realist stance licenses inferences (i.e. it *allows*, but does not *compel* such inferences) to the existence of unobservable entities, forces, and processes, while the empiricist stance licenses inferences only to observables (in the sense of V-Prohibitions). The realist stance comes in stronger and weaker versions, where the stronger the stance, the more extravagant or adventurous the metaphysical inferences it treats as legitimate. We can distinguish broadly between a standard scientific realism (SSR) stance, which licenses belief in the explicit subject matter of science but demurs when it comes

[1] I say 'discourages' rather than 'prohibits', since, as I explained in the previous chapter, for a voluntarist like van Fraassen, it is not irrational *per se*, even for an empiricist, to believe in unobservables. But part of what is involved in being an empiricist is that one tends to regard belief in unobservables as not worth the candle.

S. C. Boucher, *Empiricism, Naturalism and the Metaphysics of Biology*,
https://doi.org/10.1007/978-3-032-00555-7_3

to the entities that are part of the implicit subject matter of science (e.g. essences, causal powers, dispositions etc.), and the metaphysics of science (MoS) stance, which is willing to embrace both the explicit and the implicit entities of science (but demurs when it comes to 'purely' metaphysical entities too far removed from science).

In this chapter I will apply this epistemic stance framework to the units of selection problem in the philosophy of biology, specifically the question of realism vs. pluralism about the units of selection. This will involve interpreting the latter as an instance of the debate in general philosophy of science concerning scientific realism and anti-realism, which is somewhat unorthodox. While certainly, as we shall see, there has been much discussion among philosophers of biology of realism and anti-realism regarding the units of selection, the mainstream debate about scientific realism within *general* philosophy of science has been overwhelmingly preoccupied with the propriety of belief in the theoretical entities postulated by the hard sciences, physics and chemistry: electrons, protons, quarks etc. To the extent that biology has been considered in this context, the relevant entities have tended to be explicit theoretical entities of biology such as cells, chromosomes, DNA molecules, microbes and the like thought of along the lines of the unobservable entities of physics and chemistry. Thus Eronen offers mitochondria, cell membranes, pollen, the *Eschericia coli* bacterium and amacrine cells in the retina as examples of biological entities with respect to which one may be a realist (2019). These are all explicit biological entities so would only require the SSR stance.

But, firstly, the life sciences postulate and theorise about unobservable entities that can be thought of as explicit scientific entities but do not fit the hard science paradigm, such as species, higher taxa and units and levels of selection. And secondly, biology also provides reason, arguably, to postulate *implicit* entities, such as natural kinds, essences, and teleology. Committing to these requires at least the MoS stance. None of these things can be observed directly; they are theoretical posits of, or possibly implications of, biological theory, thus they are all potential subject matter for the scientific realism debate.[2] They are the subject of Chapters 3, 4, 5, and 6 of this book.

2 Realism vs. Pluralism About the Units of Selection

The issue of realism vs. conventionalist pluralism about the units of selection is (as Sober notes (2011, 162)) one of the most important, yet unresolved, philosophical questions about the nature of selection.[3] There are different forms of pluralism about the units of selection: some, e.g. 'unit pluralism', which accepts that as a matter of objective fact selection operates at multiple levels, are compatible with realism. Others, such as 'model pluralism', which holds that particular selective episodes can be legitimately modelled in term of different units or agents of selection, and there is no fact of the matter concerning the level at which selection is acting, are anti-realist, or conventionalist views. My focus in this chapter is on conventionalist pluralism about the units of selection, and that is what I shall mean by 'pluralism'.

Realism is usually taken to be the view that for the selective episode in question, there is an objective fact of the matter concerning the level(s) at which selection is acting. Realism comes in different forms. If realism is correct, there is always a determinate fact of the matter concerning the

[2] I thus see myself as engaged in a similar project to that of Derek Turner, in his (2007). Turner suggests that the scientific realism debate has been overly preoccupied by the hard experimental sciences, and has neglected important realism-related issues in historical sciences such as archaeology and palaeontology.

[3] For a recent, rather technical discussion see Bourrat (2021). See also Shanahan (1996) and Wilson (2007). I am not in this chapter discussing the causalism vs. statisticalism issue about the units of selection (see Sober 1984 for the classic statement of the causalist view, Walsh et al. (2002) and Matthen and Ariew (2002) for classic defences of the statisticalist view, and Pence (2021) for a recent summary of this now quite extensive controversy). The reason is that while this issue is not unrelated to the realism vs. conventionalism issue, the two issues are distinct, and it is not necessary to take a position on the causalism vs. statisticalism issue in order to take a position on the realism vs. conventionalism issue, as evidenced by the fact that several prominent discussions of the latter do not discuss the former, at least not directly, e.g. Okasha (2006), Bourrat (2021), Sober (2011), and Boucher (2023).

entities (or combination of entities) subject to selection in any given case, but this leaves open the question of *which* entities are being selected. Darwin held that it is always (or nearly always) the organism; Dawkins (circa 1976) held that it is always the gene; multi-level selectionists (e.g. Gould 2002; Sober and Wilson 1998) hold that selection acts at multiple levels. We can pass over this issue; for us all that matters is the realist thesis that there is always a fact of the matter about the level(s) at which selection is acting, and the (as I shall suggest) concomitant ontological claim that units of selection exist. *Which* entities are units of selection is an empirical scientific question we need not take a stand on.

3 Realism vs. Pluralism About the Units of Selection as an Ontological Issue

Realism vs. conventionalist pluralism about the units of selection is often understood in epistemic terms, with the central question being whether there is, so to speak, 'one true story' about the units of selection (monism), or more than one equally correct description of selection processes (pluralism). Even when expressed in more metaphysical terms, the issue is seen to be not so much an ontological matter, but rather the question of whether there is a 'fact of the matter' about the level(s) at which selection is operating (that is how I expressed it above). I am interpreting it as at least in part an ontological question—the question of whether certain entities, i.e. units/targets/levels/vehicles/interactors/agents of selection, exist.[4] This is a somewhat unorthodox but, I claim, quite natural interpretation of what is at issue in this dispute.

[4] From now on I shall mainly use the term 'units' of selection, to abbreviate the longer list. There may be subtle differences between these concepts, but we can set this issue aside for the purposes of the discussion. Even if not quite synonymous, they clearly all belong to the same family of notions, such that ontological commitment to some will tend to involve ontological commitment to the others (though this may not hold in every case). Suárez and Lloyd (2023) argue that 'unit of selection' is ambiguous; it 'refers to at least three non-co-extensional functional concepts: interactor, replicator/reproducer/reconstitutor, and manifestor of adaptation/type-1 agent' (1). If that is correct, there may be different realism questions concerning these different entities; one could be a realist about replicators but not interactors, for instance. Their arguments for the polysemous character of 'unit of selection' are not, in my view, persuasive. Even if it is useful to distinguish between

And indeed this is seemingly how the issue has been understood, either explicitly or implicitly, in at least some of the literature. In their influential (1988) paper, Sterelny and Kitcher endorse the position that there are typically several possible ways of representing or modelling a selection process, such that no one way is more correct than the others. 'There is no privileged way to segment the causal chain and isolate the (really) real causal story…there are alternative, maximally adequate representations of the causal structure of the selection process' (*ibid.*, 358). They note that what they are arguing for is a kind of 'instrumentalism', according to which certain putative entities, *'targets of selection', 'do not exist'* (*ibid.*, 359; emphasis added; see also Waters 1991, 2005). According to Kitcher et al. (1990, 159), '…asking about the real unit of selection is an exercise in muddled metaphysics.' We may ask questions about the level(s) in the biological hierarchy from the standpoint of which it is theoretically or explanatorily useful to model selection processes, but the problem of which levels are 'real' loci of evolutionarily relevant causation is in some sense a pseudo-problem (*ibid.*, 161).

Kitcher (2004, 12) asserts that those who hold that there is a genuine metaphysical issue about which units of selection are 'real' are 'held captive by a picture'. Once we free ourselves of the mistaken conception that there are real entities in nature called 'units of selection', we will see that 'the ways in which the causal chain [of a selection process] is segmented is entirely up to us … All that matters is that the models we construct be adequate to the phenomena, generating the right predictions and doing so in a way that fits the causal process that produces the outcomes.' Elsewhere Kitcher (2003, xii) writes that: 'there's a constant danger of overinterpreting Darwin's metaphor, so that there's an alleged causal fact about the world concerning the exact place at which natural selection 'acts'. Better, I believe, to see that selection is just our metaphorical way of approaching the complex facts of birth, mating and death; we can organise those facts in various Darwinian ways, and our choices among styles of organisation should be thoroughly pragmatic' (see also Kitcher 2008). As Okasha notes, in Kitcher's view, 'realism about the levels of

---

these concepts, it doesn't follow that they all pick out *units of selection* of different kinds, or that 'unit of selection' is not univocal; see my (2024).

selection is the result of inflationary metaphysics' (Okasha 2006, loc. 1677).

Thus the (epistemic) pragmatist-pluralist position appears to be closely connected to an anti-realist or instrumentalist view (as the name 'conventionalist pluralism' indicates) according to which 'certain putative entities'—units of selection—'do not exist'.

In order to be more precise about the content of realism and pluralism about the units of selection, we can adopt an idiom employed by Boghossian (1990). Consider some putative selective context C with respect to which one may be a realist or a pluralist. The realist holds there will be sentences of the form 'U was the (or a) unit of selection in C' that are such that the predicate 'unit of selection' picks out a property that something in C possesses, and thus the sentence possesses truth conditions that obtain. The pluralist by contrast holds that in all sentences of the form 'U was the (or a) unit of selection in C', the predicate 'unit of selection' expresses a property that nothing in C possesses, as a matter of objective fact, and all such sentences thus possesses truth conditions that do not obtain, and so are false, strictly speaking, though perhaps some are 'useful fictions': we may take an instrumentalist attitude towards them.

This error-theory-instrumentalist interpretation of the model-pluralist view as the view that units of selection talk is systematically false, albeit potentially useful, and nothing answers to the concept 'unit of selection'—is the most natural interpretation of that view. The alternative is to interpret model pluralism as a kind of nonfactualism. While this has the advantage of enabling pluralists to deny that sentences that quantify over units of selection are systematically false (they rather come out as neither true nor false), it comes at the significant cost of having to deny what seems obviously true—that units-of-selection discourse is in the business of describing the world, that its sentences possess truth conditions, and its characteristic predicates denote properties (see Boucher 2025).

Thus realism about the units of selection can be understood as the view that the predicate 'unit of selection' expresses a property that some entities (e.g. genes, organisms, groups) possess; therefore we do, and should, quantify over units of selection, and be ontologically committed to them. Pluralism is the view that the predicate 'unit of selection' expresses a property that nothing possesses. We should not be ontologically committed to units of selection. We should be error-theorists about units-of-selection talk. This can then, as in other domains, be combined either with a recommendation of eliminativism (we stop using units-of-selection talk), or of instrumentalist retention (we keep using it for pragmatic purposes).

4 Local Pluralism?

I have thus-far been interpreting conventionalist pluralism as a global position, the view that there is never a fact of the matter about the units of selection. But some theorists have defended *local pluralism*—the view that realism is true of some episodes (global pluralism is false) and pluralism is true of others. On the epistemic stance interpretation, this would involve an unusual combination of the realist stance for some selective regimes, and the empiricist stance for others. And indeed when we look closely at the view, we find, it seems to me, that it is of doubtful coherence (see Boucher 2020). In Sterelny and Griffiths' version of local pluralism (1999), when selection acts on organisms, or superorganisms, there are facts concerning the level at which selection acts, but when it 'acts' in population-structured trait-group evolution, such facts drop out of the picture, and we may adopt whatever perspective we find most useful in modelling the process. But this is hard to make sense of. We may express this in the idiom we adopted above. When, in a case C of organism selection, we utter 'the organism was the (or a) unit of selection in C', the predicate 'unit of selection' picks out a property that something in C (i.e. organisms) possesses, and our utterance possesses truth conditions that obtain. But when, in a case of population-structured evolution P, we utter a sentence of the form 'U was the (or a) unit of selection in P', 'unit of selection' expresses a property that nothing in P possesses, as a matter of objective fact, and all such utterances thus possesses truth conditions that do not obtain, and so are false, strictly speaking, though perhaps some are 'useful fictions'. This combination of realism for some selection contexts with an error-theoretic interpretation of *all* units of selection talk with respect to other selection contexts may not be logically contradictory, but it is, to say the least, a strangely disunified view about the metaphysics of evolution (Sober 2011, 167). It suggests that local pluralism is fundamentally unstable. It would be like adopting the realist stance for some entities of physics, say, electrons, while adopting the empiricist stance for other entities in physics, say, protons and neutrons, suspending judgement about their existence. Thus I will set local pluralism aside, and assume that our options are global realism or global pluralism.

5 Epistemic Stances and the Units of Selection

We saw that the SSR version of the realist stance licenses inferences to the existence of the explicit entities of science, while the MoS version also licenses inferences to the existence of the implicit entities of science. I will take it that the entities the existence of which is at issue in the realism vs pluralism debate about the units of selection are part of the explicit (but not directly observable) subject matter of biology, although of course philosophers have contributed as much to debates about the units of selection as biologists. Thus, realism about the units of selection may be an expression or manifestation of either the SSR stance or the MoS stance. If one adopts either stance, one treats as potentially legitimate belief in the existence of things such as units of selection. When Dawkins subtitled the first edition of his book on the extended phenotype 'the gene as the unit of selection', he was referring to units of selection as part of the explicit subject matter of evolutionary biology (Dawkins 1982). Realists would argue that he was thereby ontologically committing himself to units of selection, much like the fact that, as Platonists about numbers argue, mathematicians are (whether they realise it or not) ontologically committing themselves to numbers when they assert things like 'there are an infinite number of primes'. Pluralists will be quick to point out that such apparent commitment can be revoked upon further analysis, as indeed Dawkins does himself in the book, when he claims that the gene's eye view and the orthodox individualist view are factually equivalent. On my interpretation it follows from Dawkins' model pluralism that in fact there are no such things as units of selection after all. In any event, it is certainly not just philosophers offering metaphysical interpretations of science who talk and theorise about units of selection, interactors, vehicles and the like. The nature of units of selection is, it seems to me, a scientific question (though of course one that philosophers may contribute to answering) in a way that the nature of, say, natural kinds, causation or laws is not. However *realism* about units of selection as an ontological claim is best regarded a philosophical thesis. Thus a 'localist' version of the SSR stance (as I mentioned in Chap. 2), on which the existence of certain explicit scientific entities is settled within science by standard empirical means, whether it's applicable anywhere (which I doubt), will not be applicable here. The realist about the units of selection may, and should, as we shall see, help herself to ideas, evidence and concepts from biology, but realism itself is a philosophical claim, and arguing for it requires distinctively philosophical ideas and arguments. Compare again Platonism about numbers. It may be that Platonism is supported by the fact that mathematicians endorse certain claims (e.g. 'there is a prime number greater than 5') but Platonism itself is a philosophical (ontological), not a mathematical, thesis, and arguing for it is a philosophical, not a mathematical business.

I have suggested that units of selection are among the explicit entities of science, but if I am wrong about this, and they should in fact be thought of as part of the implicit subject matter of science, that will not affect the main part of my argument; it will just mean I am wrong in supposing that one who adopts the SSR stance may commit to them; a stronger stance would be needed. Below I shall mainly be referring to the *realist stance* in general in any case, without distinguishing stronger and weaker stances within it.

Below I will consider an argument for believing in units of selection. This argument, and the beliefs it supports, are, I maintain, available to, and only to, those who have, implicitly or explicitly, adopted the realist stance. Yet whether the argument succeeds is not settled by the adoption of a stance. The argument must be judged on its own merits. And it can be challenged on its own terms. But I'll nonetheless suggest that we can fruitfully interpret the disagreement between those who do and those who don't accept this and similar arguments as deriving in part from disagreement at the higher, stance level.

We can equally interpret conventionalist pluralism as an expression or manifestation of the empiricist stance. When Kitcher states that 'selection is just our metaphorical way of approaching the complex facts of birth, mating and death', the claim is naturally interpreted in instrumentalist terms, in keeping with the empiricist stance. The role the 'metaphor' of selection plays is, he suggests, to allow us to make accurate predictions with regard to the observable facts of 'birth, mating and death'. We shouldn't 'overinterpret' this metaphor in a way that reifies units of selection. What matters in the end, is that our models are 'adequate to the phenomena, generating the right predictions'. Anything further is unjustified inflationary metaphysics.

6 Pluralism About the Units of Selection, Epistemic Relativism, and Stance Relativism

I am suggesting conventionalist pluralism about the units of selection can usefully be understood as an expression of the empiricist stance. But there is an important difference between conventionalist pluralism about the units of selection and Chakravartty's paradigm examples of empiricist-stance positions, such as van Fraassen's (1980) constructive empiricism (CE). The latter take an *agnostic* attitude towards the entities in question, suspending judgement about, for example, the existence of electrons. They do not, typically, deny that they exist (Chakravartty 2017, 50). This is important in the context of Chakravartty's defence of stance relativism, or stance pluralism. The latter would be untenable, he suggests, if it involved *epistemic* relativism, the view that there are rationally sanctioned contradictions. If one stance had it that Fs exist, and another stance had it that Fs did not exist, then the stances would conflict in terms of the truth-claims they endorsed—they would endorse incompatible claims—and thus relativism with respect to these stances would entail epistemic relativism. That result, Chakravartty argues, can be avoided once we see that the empiricist is not in fact *denying* any claim endorsed by the realist; she is rather *suspending judgement* about those claims. There is no logical contradiction between believing P and suspending judgement about P. The stances thus are not making incompatible claims: one is merely committed to more than the other. It ventures opinions on matters on which the other is silent. Thus relativism about these stances does not entail epistemic relativism.

Conventionalist pluralism about the units of selection as I am interpreting it, on the other hand, *is* in the business of denying the existence of the relevant entities. This is explicit in Sterelny and Kitcher: 'certain entities do not exist'. One could imagine a suspend-judgment version of anti-realism about the units of selection. The view would be that we ought to commit ourselves only to the existence of the observable entities, facts and events of the evolutionary process—(Kitcher's facts about living and dying)—while remaining agnostic regarding the existence of the unobservable entities postulated by biologists to explain this process, such as units of selection. But this is not in fact the form that conventionalist pluralism has taken. I am not aware of such a view being explicitly formulated or defended. Conventionalist pluralism has rather, as the name suggests, taken the form of error-theory-instrumentalism. It has, either explicitly or implicitly, involved the view that units of selection do not exist.

Such a view still appears to be an expression of the empiricist stance however. As I argued in Chap. 2, empiricism can take 'atheist' as well as 'agnostic' forms. Chakravartty's suggestion that the empiricist stance must take the suspend-judgement form is unjustified. Let us assume then, that in the case at hand, the realist stance characteristically issues in the claim that units of selection exist, and the empiricist stance issues in the claim that they do not exist. What does this mean for stance relativism and epistemic relativism? There are three choices. First, we could deny that in such a case stance relativism would in fact entail epistemic relativism. If however Chakravartty is right that in such a case stance relativism would entail epistemic relativism, we have two choices. We could embrace both stance relativism and epistemic relativism. Or we could reject stance relativism along with epistemic relativism.

The third option is the standard view, or would be if theorists interpreted the debate in terms of stances. It is thought that one must adopt either realism or pluralism (local pluralism aside), and relativism at the higher level is rarely if ever proposed as a serious option. I however follow Chakravartty, van Fraassen and others in the 'stance' literature who have advocated versions of pluralism and relativism about stances. So the choice is between options one and two. I agree with Chakravartty that epistemic relativism, or truth relativism, is to be avoided if possible, which leaves us with option one: denying that stance relativism entails epistemic relativism.

When we said above that in the case at hand, the empiricist stance issues in the claim that the relevant entities do not exist, we spoke too loosely. 'Issued in' is not to be taken too strictly. To adopt a stance is not yet to believe certain claims (Boucher 2018). One may adopt a kind of relativism or pluralism about rival realist and empiricist stances with respect to the metaphysics of evolution and selection without *thereby* holding that 'there are units of selection' and 'there are not units of selection' are somehow both true, or epistemically justified. All that adopting, say, the empiricist stance in this context commits one to is approaching this issue from the standpoint of empiricist goals, values, attitudes and commitments. One may of course *go on* to assert, from this standpoint, that units of selection do not exist, which claim may contradict certain factual claims believed by partisans of the realist stance. But that is a further claim that adoption of the stance does not *rationally compel* one to make (just as, as we have seen, adopting the realist stance doesn't rationally compel one to believe in the existence of any particular theoretical entity). The key point is that may be the case that (1) stance choice is a function of values and is thus relativistic; and (2) the belief that there are Xs finds a natural home within the framework of one stance, while the belief there are not Xs finds a natural home within the framework of a different stance, *without it being the case that whether or not there are Xs depends on one's values, or is relativistic.* This is essentially the same point as that which Chakravartty makes regarding voluntarism at the stance level and voluntarism at the level of belief (2017, 219–220), and which I discussed in the Introduction and Chap. 2. One may adopt a stance, he notes, and this may partly be a matter of choice, based on one's values.

This stance then carves out the space of the types of beliefs that are considered admissible. Thus the prior voluntaristic adoption of the stance does indirectly influence what is believed downstream, but those beliefs can and should still be sensitive to evidence in the standard way, and we need not hold that they are or should be believed merely on the basis of the will, or one's values. Stance voluntarism doesn't entail implausibly-strong belief-voluntarism. Equally stance relativism doesn't entail belief-relativism.[5]

7 An Argument for Realism About the Units of Selection[6]

So far the discussion been fairly abstract. I wish now to consider a concrete example of an argument for realism about the units of selection. I will set out the argument in some detail. If the idea that the units of selection debate can be understood as an instance of the scientific realism debate (shaped by epistemic stances) is to be taken seriously, we need to go beyond vague, intuitive or hand-waving arguments. We need to know what a careful, detailed argument for including units of selection in our ontology might look like, the sense in which it is an expression of the realist stance, and consider how those who embrace the empiricist stance in this domain might try to respond (just as the epistemic-stance interpretation of the more typical scientific realism debate has been developed while keeping in mind detailed arguments for realism and anti-realism in the more typical domains). Given our focus in Part I, it will also be worth considering the argument as a specimen of naturalised metaphysics (of biology). It will give us an idea of what naturalised metaphysics of biology looks like in practice.

[5] Kusch (2020) argues that stance voluntarism does involve a form of epistemic relativism, but agrees with van Fraassen that it is a 'benign' rather than a 'debilitating' form.

[6] This section draws on my (2023).

I will first set out the argument, then consider ways those who adopt the empiricist stance might reply.[7]

The argument makes use of the following principle:

> Units of selection exist (should be included on our ontology) if there is sometimes irreducible selection at a level in nature.

The realist takes this principle as a premise, then proceeds to (a) identify sufficient conditions for selection acting at a level where the process is not re-describable in terms of selection acting at some lower-level, and (b) show that these conditions are satisfied in nature. Given the above principle, it follows that units of selection exist (should be included on our ontology). I will now explain how the realist might go about doing (a) and (b).

7.1 Fitness in MLS2 and Conditions for Natural Selection

First, we need to distinguish two forms of multi-level selection: 'multi-level selection 1' (MLS1) and 'multi-level selection 2' (MLS2). In collective selection of the MLS1 variety the focus is on changes in the prevalence of types of particles in the metapopulation over time, with collectives acting as a context or environment affecting the fitnesses of particles. 'Collective' selection of this type leads to the evolution of properties of particles (such as altruism), not collectives, and while it is collectives (as well as particles) that are the bearers of fitness (of a kind), it is the particles that reproduce and pass on their traits. The fitness of a collective in MLS1 (fitness$_1$) is defined as the number of particles it contributes to the next generation, or the average fitness of the particles composing the collective. In multi-level selection type 2 (MLS2), the focus is rather on changes in the prevalence of types of collectives over time. Collectives, as

[7] In my (2023) I endorse the argument (or part of it, at least). Here my primary aim is not to endorse it (though it will be clear that I do), but just to present it (as well as possible replies to it, which I shall also not endorse or reject), and consider the sense in which it can be understood as, firstly, an instance of naturalised metaphysics, and, secondly, as an expression of the realist stance.

well as particles, reproduce, and pass on their traits to offspring collectives, i.e. there is collective heritability. In MLS2 collective fitness (fitness$_2$) is defined in terms of number of offspring collectives, and can vary independently of the fitness of the particles that compose the collectives, and of the fitness$_1$ of the collectives (on the difference between MLS1 and MLS2 see e.g. Okasha 2006; Arnold and Fristrup 1982; Sober 1984; Damuth and Heisler 1988; see Jeler 2020 for a recent discussion). Take two collectives, X and Y, containing equal numbers of particles. X may contribute more particles to the next generation than Y, but Y may contribute more collectives to the next generation. In that case X would have the higher fitness$_1$, but Y would have the higher fitness$_2$ (Okasha 2006, loc. 729).

It might seem that MLS2 processes, as Okasha describes them, give us the sufficient conditions we require (sufficient conditions for irreducible selection acting at a level): there is evolution by natural selection at level L if there is heritable variation in fitness$_2$ among entities at L. After all, unlike in MLS1, in MLS2 our focus is on the evolution of collective-level characters through differential higher-level fitness that is irreducible. As Oksaha writes, '…in MLS2, group selection affects the evolution of a group character, not an individual character; so there is no way that an 'individualist' re-description is possible. This ties in with the point that pluralism is only a theoretical option in relation to MLS1, for only there is it possible to describe the overall change in both multi-level and single-level terms' (*ibid.*, loc. 2505). (MLS2 processes (as the name implies) are usually discussed as examples of group, species and higher-level selection, but the concept easily generalises to any level in the hierarchy at which entities reproduce, giving rise to other entities of the same kind. MLS2 processes operate at the multicellular organism-level, for example (Michod 1999, 2005). I return to this point below.[8])

In fact Okasha's own discussion (2006, ch. 3) shows why MLS2 processes on their own do not provide us with sufficient conditions for irreducible selection acting at a particular level. This is because of what he

[8] This means, incidentally, that the name 'multi-level selection 2' is misleading. One could hold that MLS2 processes occur only at the level of the individual organism (or below). On such a view there would be no 'higher-level' selection (at the group, or species level) at all, as multi-selectionists hold, but still there would be 'MLS2' processes at lower levels.

calls ‘cross-level by-products’ in MLS2. Consider some level L. In some cases differential success of entities at level L, where this is understood in terms of $fitness_2$ differences, is entirely a by-product of fitness differences, and selection, at a lower-level. Thus, Okasha suggests, for selection at a level there needs to be a causal, not just statistical connection between character and fitness at the appropriate level. Fitness-character covariance can be result of a causal relation between character and fitness, or can be a result of (by-product of) lower-level fitness.

Take the classic case of the evolution of running speed in zebras (Williams 1966). Fast herds of zebras may give rise to more groups than slow herds (assuming the herds can reproduce, which may be questionable (Godfrey-Smith 2009)), and thus have higher $fitness_2$. There is covariance of zebra group character (group fleetness) and zebra group fitness, but this is a by-product of fitness and selection at the level of individual zebras; it is a merely statistical connection/correlation: the group character does not cause the group fitness. Species sorting is another example of this (Vrba and Gould 1986).[9] This is an MLS2 process. The differential fitness of species that is a product of lower-level selection processes can be $fitness_2$: some species do better than others in the sense of leaving more offspring-species; but, although this kind of fitness is not defined in terms of lower-level fitness, this can be (mere) species sorting rather than species selection if it results from selection on individuals. If species A has higher $fitness_2$ than species B merely because the members of A have higher average fitness than the members of B—that is, because species A has higher $fitness_1$ than species B—the $fitness_2$ differences are a cross-level by-product, and we are dealing with mere species-sorting, not selection. There will be some (aggregate) species-level trait that is correlated with success or failure characterised in terms of $fitness_2$, but in species sorting the species-level character does not cause the $fitness_2$ of the species; the latter is caused rather by the fitness of the individual members of the

[9] More precisely, ‘sorting’ refers to differential species proliferation, with selection being one possible explanation for this. Sorting without selection is ‘mere’ sorting. (Vrba calls this ‘effect macroevolution’, though it may happen in standard microevolution.) From now on by ‘sorting’ I shall mean ‘mere sorting’.

species; see Fig. 3.1. (While mere sorting has been discussed primarily at the species-level, it easily generalises to any collective level. From now on by 'sorting' I shall mean sorting at any level.)

The point is that although fitness$_2$ is defined independently of fitness$_1$, the two can be aligned: fitness$_2$ is defined independently of fitness$_1$, but

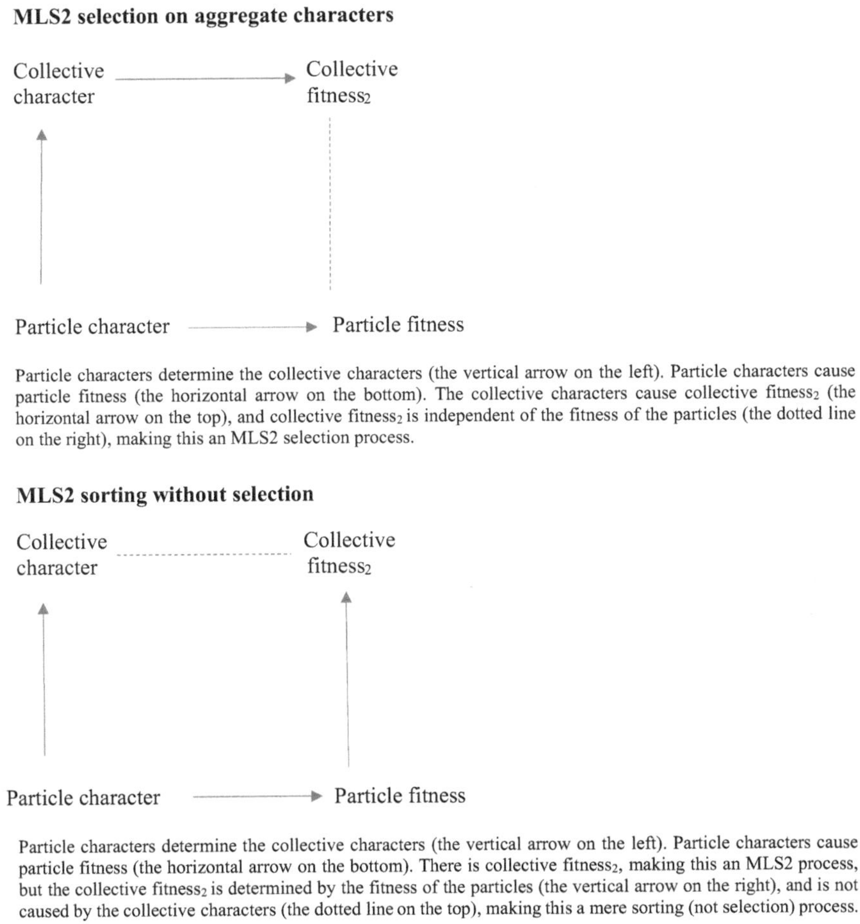

Fig. 3.1 MLS2 selection on aggregate characters, MLS2 sorting without selection, MLS2 selection on emergent characters, and MLS1 selection. Horizontal and diagonal arrows indicate causal relations, vertical arrows indicate relations of determination or constitution. Dotted lines indicate the absence of causal relations (if horizontal) or the absence of relations of determination or constitution (if vertical)

MLS2 selection on emergent characters

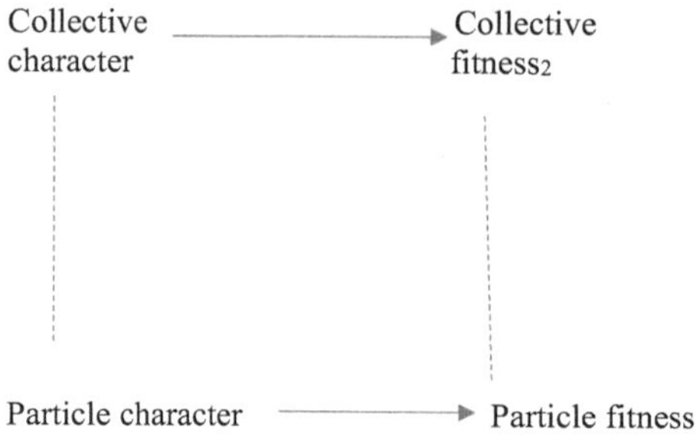

Particle characters cause particle fitness (the horizontal arrow on the bottom). The collective characters are not determined by the particle characters (the dotted line on the left). Collective fitness_2 is independent of the fitness of the particles (the dotted line on the right). Collective fitness_2 is caused by the collective characters (the arrow on the top), making this an MLS2 selection process.

MLS1 selection

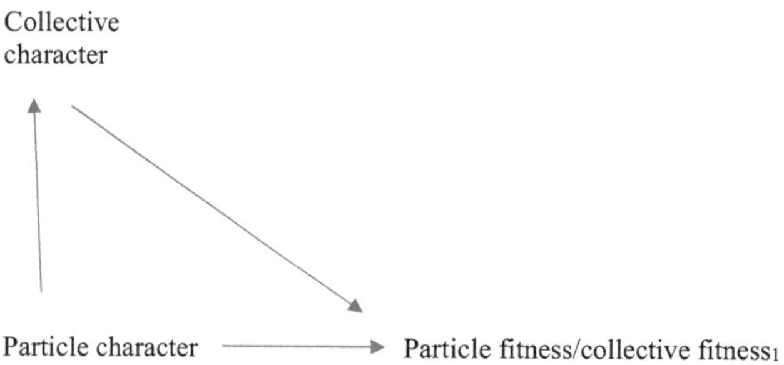

Particle characters determine the collective characters (the vertical arrow on the left). Particle characters and collective characters both cause particle fitness and thereby cause collective fitness_1 (the horizontal arrow on the bottom and the diagonal arrow).

Fig. 3.1 (continued)

can still be entirely the result of fitness_1, and lower-level selection. This suggests, as Okasha notes, that fitness_2 differences are not a sufficient condition for selection at L: it also needs to be the case that the differences are not a by-product of lower-level processes—there must be irreducible causal processes at the higher-level where this goes beyond just fitness_2 differences: there needs to be the causal connection between character and fitness.

This, then, suggests generalised sufficient conditions for selection at a level—

> There is selection at level L if there is heritable variation in fitness$_2$ among entities at L, and the character-fitness covariance at L is causal.[10]

Thus it isn't quite right to suggest, as Okasha does, that in MLS2, 'there is no way that [a lower-level] … re-description is possible' because in MLS2, collective selection affects the evolution of a collective, not a particle character (*ibid.*, loc. 2505). If the fitness$_2$ differences at the collective level are just a by-product of particle-level selection, then there is no collective-level *selection* at all, and presumably a particle-level re-description in such cases, at least with respect to the relevant *causes*, is not only possible but mandatory. In cases of species sorting, for instance, the differences in fitness$_2$ among species, while not themselves being strictly reducible to fitness$_1$ differences (as by definition fitness$_1$ and fitness$_2$ are characterised differently), are nonetheless entirely a by-product of selection on individuals, so that the species 'selection' must be re-described as an individual-level selection process, albeit one that may influence collectively characterised traits, and have macroevolutionary effects. Here, the collective (species) selection is spurious.

As noted above, MLS2 processes are standardly thought to apply to 'higher-level selection', or at least levels where collectives produce other collectives, and I have not always been able to avoid talking in these terms, but this is an artefact of the history of the concept, and is not (or should not be) built in to the very concept of MLS2 selection. What matters for MLS2 is that the fitness of entities at a level is understood in terms of the generation of more entities of that type (at that level) whether or not these are collective entities. (This contrasts with MLS1: since the fitness$_1$ of an entity concerns the fitness of the particles within it, it only applies to collective levels.) Thus the above sufficient condition equally

---

[10] This is a version of the 'heritable variation in fitness' account of the conditions for evolution by natural selection. This account, in one form or another, has been standard since Lewontin (1970). There are a number of questions that may be asked about its status and applicability, and how precisely it should be formulated—see Godfrey-Smith (2007, 2009). But it is widely enough accepted, at least in broad outline, that I can take it for granted here. I am interested in whether, given we accept this account, it can be combined with Okasha's analysis to yield plausible general conditions for selection at a level.

applies to non-collective levels. If genes are not collectives in the relevant sense, they may still possess fitness$_2$, understood as their propensity to replicate, i.e. produce more copies of things like themselves. (Of course, in such a case the fitness of the entities will be trivially irreducible, since there is nothing it can be reduced to.)[11]

7.2 Emergent Fitness and Emergent Characters

Some, such as Vrba (1984, 1989), have maintained that, at least in the case of species selection, my proposed sufficient conditions are not in fact sufficient. For species selection to take place, there also need to be emergent (not just aggregate) species-level characters.[12] Whether this is required for species selection, or whether emergent species fitness is sufficient, has been the subject of an ongoing debate. Lloyd (1988), along with Okasha (2006), Gould (2002) and Grantham (1995), deny that emergent traits are necessary, and I think they are correct. Take two species of horses, one small-bodied and one large-bodied. Large-bodied horses may have a competitive advantage over small-bodied horses at the organismic level, such that (a) within each species larger body size is selected for, and (b) the larger-bodied species contributes more individuals to subsequent generations that does the small-bodied species (it has higher fitness$_1$); yet the small-bodied species may speciate more rapidly than the large-bodied species (it has higher fitness$_2$). Unlike in the zebra case, the aggregate species character (being a small species) *is* the cause of the fitness$_2$ of the species (see Fig. 3.1). This can give rise, under the right conditions, to a cladogenetically-driven trend of decrease in body size in the horse clade (Grantham 1995). In such a case individual selection and species selection are pushing in opposite directions (a sufficient, but not

[11] So note that Okasha's definition of particle fitness: 'the number of offspring particles it leaves' (loc. 717) is, aside from the words 'particle' and collective', identical to his definition of collective fitness$_2$: 'the number of offspring collectives it leaves' (loc. 720). There is one type of fitness here: the number of offspring of that type produced, which may apply to collectives, or particles that are not themselves collectives. This means that we need to qualify Okasha's claim, discussed below, that MLS2 processes evolve out of MLS1 processes. This is not true of all MLS2 processes, but only those involving collectives.

[12] See Grantham (1995, 2007) and Turner (2011) on different accounts of what it is to be an 'emergent' property. We can set aside this issue, as emergent characters are not in fact required for selection at a level.

a necessary condition for higher-level selection). Whether this has actually happened is of course an empirical question, but it is seemingly coherent as a species-selection scenario. Yet average body size is plainly an aggregate, not an emergent trait of horse species. Species selection happens when species fitness is caused by species characters, and the species fitness is not reducible to individual (lower-level) fitness (i.e. the evolution of the character is not explainable in terms of selection at lower-levels), whether the species character is emergent or aggregate (which is not to deny that cases that involve emergent species characters are the paradigm cases of species selection).

Again, this point generalises to selection at any level. The generalised version of Vrba's requirement would be that for selection to take place at level L, there must be emergent characters at L. But that is too strong.[13] For there to be genuine selection at level L, it is not necessary that there be emergent characters at L, only that there be emergent, irreducible fitness defined at L. The criterion for selection acting at level L is that the fitness of entities at level L is caused by level-L characters, and this fitness is not reducible to the fitness of entities at level L-n (i.e. the evolution of the character not explainable in terms of selection at lower levels). For this irreducible (emergent) level L characters are not required.

It is true that at levels other than the species-level—the organism level, for instance—there may not be many non-emergent traits that cause irreducible fitness, in which case the differences between Vrba's view and the one I favour would not show up at that level. When there is irreducible fitness at the organism-level, it is likely to be caused by emergent organism traits. Organisms are highly cohesive, integrated, and tightly organised entities, such than many of their traits are good candidates to be emergent characters—'more than the sum of the parts'. Organisms are 'manifestors of adaptation' (Suárez and Lloyd 2023) or 'type-1 agents' (Suárez and Lloyd 2023; Okasha 2018). The cases where the two views disagree, i.e. cases of aggregate traits at a level causing irreducible fitness differences at that level, may be found predominantly at the species, and

[13] There is also the problem that emergent characters at L are presumably typically a *product* of a selection process at L, i.e. L-level adaptations. But in that case, they cannot also be a *precondition* for such a process (Okasha 2006, loc. 1463).

perhaps the deme, level. Thus it is likely that the falsity of conventionalist pluralism could be established even if we accepted Vrba's requirement: whether her more stringent conditions are ever met in the case of species—and this is an open question—it is very plausible that they are met at other levels, such as the organism-level, and all we require to establish P1 is some case or other of irreducible selection at a level. It is nonetheless preferable to work with the less restrictive view, not only because it is more plausible in itself, but because, for the purposes of the argument, firstly, if it turned out that no characters at any level were emergent (and this metaphysical claim, however implausible, can't be entirely ruled out in advance), the argument would fail, if it relied on Vrba's criteria. And secondly, we can thereby sidestep the vexed question of how to define 'emergent character' (Turner 2011).

7.3 MLS2 Processes in Nature

Recall our sufficient conditions:

> There is selection at level L if there is heritable variation in fitness$_2$ among entities at L, and the character-fitness covariance at level L is causal.

If these conditions are satisfied, there is collective selection which is not reducible to particle selection (more generally, there is irreducible selection at a level), and thus realism is unavoidable with respect to it, and pluralism is ruled out. But are the conditions ever satisfied? Species selection is the paradigm case of MLS2 (Damuth and Heisler 1988), but, while there are theorists who believe that our two conditions are often met in the case of species, and thus that species selection is a real and important phenomenon, this is a controversial, and seemingly a minority, position. So it would be unwise to tie the argument against conventionalist pluralism too closely to the empirical case for species selection. But there is no need to. We have an MLS2 process whenever there are heritable fitness$_2$ differences and this may occur at any level in the hierarchy. Two, perhaps more plausible, examples of processes satisfying our two conditions are the evolution of superorganisms (e.g. ant colonies) and, most obviously, individual organisms.

As Sterelny notes (1996), the most persuasive example of genuine, irreducible group selection has always been the evolution of organism-like colonies in the eusocial insects—ant colonies, termite colonies, etc.—classic 'superorganisms'. There are two reasons, he suggests, for taking superorganism evolution to be irreducible group selection, with respect to which a particle-level description is unavailable. Firstly, superorganisms display a degree of cohesion, integration, and functional specialisation that approaches that of individual organisms, and thus have almost as strong a claim to be considered objective vehicles/interactors as do individual organisms. Secondly and relatedly—and this is the key point for our purposes—we have reason to think the fitness of superorganisms varies independently of the fitness of their component organisms—in our (but not Sterelny's) terms, it is $fitness_2$, in which the fitness of the nest or hive has been completely decoupled from the fitness of its individual members: the hive or nest has $fitness_2$ in its own right, not as a by-product of its $fitness_1$. Thus, if he is right, superorganism evolution is an example of our sufficient conditions being satisfied in nature. It follows that conventionalist pluralism is false.

And finally and most convincingly, the work of Michod (1999, 2005) and Michod and Roze (1999) suggests (a) that our two conditions are the appropriate criteria for genuine selection to occur at the level of *individual multicellular organisms* (indeed, for it to be meaningful to speak of such organisms existing as evolutionary individuals at all), and (b) that the conditions are often satisfied at the individual organism level (see Okasha *ibid.* loc. 3066 for discussion).[14] If they are right about (b)—and

[14] Sterelny and Griffiths also defend realism (and reject pluralism) with respect to evolution at the level of individual organisms (1999, 171), thus rejecting global, or what they call 'radical' pluralism, but their reasons don't have to do with our two conditions; indeed, they seem to accept the theoretical *possibility* of a lower-level (cell-level) redescription of the relevant processes. This redescription, while possible, is inappropriate, they claim, as individual organisms are integrated, cohesive and organised enough to count as objective interactors. Applying our two conditions however, allows us to see that the cell-level description is not even a possible option in cases of genuine individual selection. Sterelny (1996) also ties the interactor-status of individual organisms, and the rejection of pluralism at their level, primarily to their cohesiveness, although see p. 581, where he hints at organisms possessing $fitness_2$.

I take their claim to be relatively uncontroversial—there is often selection at the level of multicellular organisms that is not reducible to lower-level (e.g. cell) selection.

Given the principle we started with—i.e. units of selection exist (should be included on our ontology) if there is sometimes irreducible selection at a level in nature—it follows that (we should believe that) units of selection objectively exist (they should be included in our ontology).

8 The Argument as an Example of Naturalised Metaphysics of Biology

Before considering possible responses to the above argument, it is worth pausing to reflect on the sense in which it counts as a piece of naturalised metaphysics. The conclusion—realism about the units of selection—is clearly a metaphysical (ontological) claim. It is not a scientific claim, and few biologists *qua* biologists would have reason to take much interest in it. Yet the argument for it draws on claims, theories, concepts and distinctions in biology, albeit some of these are relatively conceptual and thus philosophers (such as Okasha, whose work I have drawn on) as well as biologists have contributed to their formulation and defence. The distinction between MLS1 and MLS2, the issue of cross-level by-products, and the subsequent formulation of sufficient conditions for selection at a level, are based on contemporary evolutionary theory. The debate concerning whether emergent characters or just emergent fitness is required for MLS2 selection is also a debate within evolutionary biology, though admittedly a partly-conceptual one (Gould says that was one of the most important questions that the theory of species selection needed to address (2002)). And most obviously, the argument relies on empirical and theoretical work by scientists such as Michod that (the argument claims) establishes (or at least strongly suggests) that the sufficient conditions for selection at a level are satisfied in nature. The inferences from these claims, theories, concepts and distinctions in biology to realism about the units of selection are metaphysical inferences in Chakravartty's sense.[15] Yet they are not immediate or direct; realism about the units of selection cannot be just 'read off' the science. The argument also involves distinctively philosophical reasoning, most obviously in its use of the claim that units of selection exist (should be included on our ontology) if there is sometimes irreducible selection at a level in nature. This is a classical specimen of realist ontological reasoning. It is not part of or in any way based on science. It expresses the realist metaphysical impulse, the realist commitment to explanation of the observable in terms of the unobservable. And it is the part of the argument to which empiricist anti-realists will most strongly object, as we shall see in the next section.

9 Empiricist Responses

To summarise the argument for realism about the units of selection I have presented:

1. There is irreducible selection at level L if there is heritable variation in fitness_2 among entities at L, and the character-fitness covariance at L is causal.
2. The sufficient conditions specified in 1 are sometimes satisfied in nature.
3. There is sometimes irreducible selection at a level in nature (from 1 and 2).
4. Units of selection exist (should be included on our ontology) if there is sometimes irreducible selection at a level in nature.
5. Thus units of selection objectively exist (should be included on our ontology).

The partisan of the empiricist stance must reject the argument. How could she do so? There are seemingly two options. Firstly, she could accept the argument for irreducible selection at a level, but deny that this

[15] The surely very plausible claim that this argument counts as a piece of naturalised metaphysics is further evidence against the idea that I discussed and critiqued in Chap. 2, that physics is the only science on which naturalised metaphysics may be based.

has the ontological implication that units of selection exist. That is, she could block the argument by denying 4. The argument may show that there is sometimes 'selection at a level' that can't be reinterpreted in terms of selection at lower levels. But 'selection at a level' need not be understood in a way that requires us to admit special entities, *units of selection* into our ontology. The idea here is that the step to ontological commitment to units of selection in the argument is fundamentally an *explanatory* one, and it is part of what it is to be an empiricist that one rejects such explanatory inferences (they are V-prohibited for the empiricist). We have no reason to accept 4, the empiricist may claim, unless we have adopted a stance that values the explanatory power that admitting such entities into our ontology potentially affords. Units of selection are explanatory posits the empiricist thinks we can do without, even if the argument to 3 is sound. Perhaps all the argument to 3 shows is that for certain selective episodes only certain types of selection models are adequate to the phenomena. On this view it is a mistake to interpret (as I did above) the units of selection issue as an ontological issue. Or, more precisely, such an interpretation is not neutral but presupposes the realist stance with its explanatory demands and ontological-explanatory projects.

In other words, the empiricist could accept monism[16] without (what she sees as) inflationary metaphysics. This can take either the error-instrumentalist form, according to which units of selection do not exist, or the agnostic form, according to which we should suspend judgement about whether they do.

While monism without inflationary metaphysics is a possible empiricist response it is unlikely to be attractive to the conventionalist pluralist, precisely because it involves abandoning the pluralism which is the core of her position. The conventionalist pluralist cannot accept the argument for irreducible selection at a level (the argument to 3). She must reject 1 or 2. The pluralistic gene selectionist (who holds that while multiple representations of selection processes are generally available and permissible, the genic perspective is always available, and thus should be privileged as providing the most general account of evolution (Sterelny and Kitcher

[16] This could be hierarchical or multi-selection monism; the point is it is a rejection of conventionalist pluralism.

1988)) could, for example, respond that even in cases of MLS2 selection where the character-fitness covariance at the relevant level is causal, the evolutionary change can be measured in genic terms, such that the genic perspective is available and permissible even in such cases. This would involve rejecting 1. But the conventionalist-pluralist, insofar as she is an empiricist, would also note that *even if* the argument to 3 went through, she would block the argument to 5 by rejecting 4 as presupposing the realist stance with its explanatory demands.

So the two responses—monism without inflationary metaphysics, and conventionalist pluralism—can be understood as different ways of denying that units of selection exist. According to the epistemic-stance interpretation of the dispute I have outlined in the chapter, the disagreement between the realist and the empiricist (of either sort) has its ultimate source in the different value each of them attributes to, on the one hand, the potential explanatory power afforded by belief in the explicit but theoretical entities of science (such as units of selection), and on the other, the need to minimise the epistemic risk involved in committing to the existence of such entities.

So, like other disputes over scientific ontology, the realism vs pluralism dispute over the units of selection reflects a deeper disagreement at the level of values; in particular, diverging conceptions of the value or disvalue of ontological commitment to the theoretical entities of science. And once this is recognised, the kind of tolerant pluralism, relativism, or voluntarism about stance choice that van Fraassen, Chakravartty and others have advocated in other domains starts to look like the view we ought to adopt with respect to the stances that give rise to rival views on the metaphysics of selection. Which is not to say, of course, that one ought to step back from the dispute and declare every view and every stance equally good, refusing to take sides. As van Fraassen and others have made clear, pluralism or relativism about stance choice is compatible with passionate partisanship and commitment to one's favoured stance and theory. But it does require that we recognise that rival stances on scientific ontology are likely to be equally rational, and that one's choice of stance is therefore more a reflection of values and attitudes than an apprehension of the objective truth.

References

Arnold, J. and Fristrup, K. 1982. The Theory of Evolution by Natural Selection: A Hierarchical Expansion. *Paleobiology* 8 (2): 113–129

Boghossian, P. 1990. The Status of Content. *Philosophical Review* 99: 157–184

Boucher, S. C. 2024. Review of Javier Suárez and Elisabeth A. Lloyd, *Units of Selection*. *Metascience* 33: 423–426

———. 2023. An Argument for Global Realism about the Units of Selection. *Biology and Philosophy* 38 (5): 1–22

———. 2020. Pluralism, Realism and the Units of Selection. *South African Journal of Philosophy*, 39 (1): 47–62

———. 2018. What is the relation between a philosophical stance and its associated beliefs? *dialectica*, 72 (4): 509–524

———. 2025. Functionalism and Structuralism in Biology: Stances or Explanatory Strategies? *Synthese*, 206 (1): 1–23.

Bourrat, P. 2021. *Facts, conventions, and the levels of selection*. Cambridge University Press

Chakravartty, A. 2017. *Scientific Ontology: Integrating Naturalised Metaphysics and Voluntarist Epistemology*. Oxford University Press

Dawkins, R. 1976. *The Selfish Gene*. Oxford: Oxford University Press

———. 1982. *The Extended Phenotype, 1st ed.* Oxford: Oxford University Press

Damuth, J. and I. L. Heisler. 1988. Alternative Formulations of Multi-level Selection. *Biology and Philosophy*, 3, 407–30

Eronen, M.I. 2019. Robust realism for the life sciences. *Synthese* 196 (6): 2341–2354.

Godfrey-Smith, P. 2007. Conditions for Evolution by Natural Selection. *Journal of Philosophy* 104: 489–516.

———. 2009. *Darwinian Populations and Natural Selection*. Oxford: Oxford University Press

Gould, S. J. 2002. *The Structure of Evolutionary Theory*. Cambridge, MA: Harvard University Press

Grantham, T. 1995. Hierarchical Approaches to Macroevolution: Recent Work on Species Selection and the "Effect Hypothesis". *Annual Review of Ecology and Systematics* 26, 301–321

———. 2007. Is Macroevolution more that successive rounds of microevolution? *Palaeontology* 50 (1): 75–85

Jeler, C. 2020. Explanatory goals and explanatory means in multilevel selection theory. *History and Philosophy of the Life Sciences* 42: 36

Kitcher, P. 2003. *In Mendel's Mirror: Philosophical Reflections on Biology*. New York, NY: Oxford University Press

———. 2004. Evolutionary Theory and the Social Uses of Biology. *Biology and Philosophy*, 19. 1–15

———. 2008. Carnap and the Caterpillar. *Philosophical Topics*, 36 (1). 111–127

Kitcher, P., K. Sterelny, and C.K. Waters. 1990. The Illusory Riches of Sober's Monism. *Journal of Philosophy*, 87 (3). 158–161

Kusch, M. 2020. Stances, voluntarism, relativism. In Finkelde, D. & Livingston, P. (eds.). *Idealism, relativism and realism: New essays on objectivity beyond the analytic-continental divide*. De Gruyter: 131–153.

Lewontin, R.C. 1970. The Units of Selection. *Annual Review of Ecology and Systematics*, 1. 1–14

Lloyd, E. A. 1988. *The Structure and Confirmation of Evolutionary Theory*. New York: Greenwood

Matthen, M. and Ariew, A. 2002. Two ways of thinking about fitness and natural selection. *Journal of Philosophy* 99(2): 55–83.

Michod, R. E. 1999. *Darwinian Dynamics: Evolutionary Transitions in Fitness and Individuality*. Princeton NJ: Princeton University Press

———. 2005. On the transfer of fitness from the cell to the multicellular organism. *Biology and Philosophy*, 20(5), 967–987.

Michod, R. E. and Roze, D. 1999. Cooperation and Conflict in the Evolution of Individuality III. In C. L. Nehaniv (ed.) *Mathematical and Computational Biology: Computational Morphogenesis, Hierarchical Complexity, and Digital Evolution*, vol. 26, American Mathematical Society, Providence, Rhode Island, 47–92.

Okasha, S. 2006. *Evolution and the Levels of Selection*. Oxford: Clarendon Press. Kindle edition

———. 2018. *Agents and Goals in Evolution*. Oxford University Press

Pence, C. 2021. *The Causal Structure of Natural Selection*. Cambridge University Press

Shanahan, T. 1996. Realism and Antirealism in Evolutionary Biology. In Cohen, R.S., Hilpinen, R., Renzong, Q. (eds). *Realism and Anti-Realism in the Philosophy of Science. Boston Studies in the Philosophy of Science, vol 169*. Springer, Dordrecht. 449–466

Sober, E. 1984. *The Nature of Selection: Evolutionary Theory in Philosophical Focus*. Chicago: University of Chicago Press

———. 2011. *Did Darwin Write the Origin Backwards? Philosophical Essays on Darwin's Theory*. New York: Prometheus Books. Kindle Edition.

Sober, E. and D. S. Wilson. 1998. *Unto Others: The Evolution and Psychology of Unselfish Behavior.* Cambridge, MA: Harvard University Press
Sterelny, K. 1996. The Return of the Group. *Philosophy of Science*, 63 (4). 562–584
Sterelny, K. and P. Griffiths. 1999. *Sex and Death: An Introduction to Philosophy of Biology.* Chicago: The University of Chicago Press
Sterelny, K. and P. Kitcher. 1988. The Return of the Gene. *Journal of Philosophy*, 85 (7). 335–358
Suárez, J. and Lloyd, E.A. 2023. *Units of Selection.* Cambridge: Cambridge University Press
Turner, D. 2007. *Making Prehistory: Historical Science and the Scientific Realism Debate.* Cambridge: Cambridge University Press
———. 2011. *Paleontology: A Philosophical Introduction.* Cambridge: Cambridge University Press
van Fraassen, B.C. 1980. *The Scientific Image.* Oxford: Clarendon Press
Vrba, E. 1984. What is Species Selection? *Systematic Zoology* 33, 318–28
———. 1989. Levels of Selection and Sorting with Special Reference to the Species Level. *Oxford Surveys in Evolutionary Biology* 6, 111–68
Vrba, E. and Gould, S. J. 1986. The Hierarchical Expansion of Sorting and Selection: Sorting and Selection Cannot be Equated. *Paleobiology* 12 (2): 217–228
Walsh, D. M., Lewens, T. and Ariew, A. 2002. The trials of life: Natural selection and random drift. *Philosophy of Science* 69(3): 429–446.
Waters, C.K. 1991. Tempered Realism about the Forces of Selection. *Philosophy of Science*, 58. 553–573
———. 2005. Why Genic and Multilevel Selection Theories are Here to Stay. *Philosophy of Science*, 72 (2). 311–333
Williams, G. C. 1966. *Adaptation and Natural Selection: A Critique of Some Current Evolutionary Thought.* Princeton, NJ: Princeton University Press
Wilson, R.A. 2007. Levels of Selection. In Matthen, M. & Stephens, C. (eds.). *Handbook of the Philosophy of Science, Volume 3: Philosophy of Biology.* Cambridge University Press. 155–176.

4

Biological Kinds I: Species and Natural Kinds

1 Introduction

Recall the distinction between the explicit and the implicit entities of science. The units of selection debate concerned only explicit entities. In this chapter and the next we shall be concerned both with explicit entities of biology (systematics in particular) such as species and higher taxa, and with implicit entities such as natural kinds, essences, and species-individuals.

I shall be spending more time on the topic of the implicit entities, as it is here that most of the interesting philosophical work has been done in recent years. Or, more precisely, to the extent that questions about the explicit entities—species and higher taxa—have been discussed in the philosophical literature, it has largely been under the aspect of, and in relation to, questions concerning the nature and existence of the implicit entities (natural kinds and the like) of which they are thought to be instances (e.g. Slater 2013).

It is nonetheless important to discuss the explicit entities separately, as there are specific questions concerning realism and anti-realism with regard to them and, as we shall see, it is possible, from the standpoint of

S. C. Boucher, *Empiricism, Naturalism and the Metaphysics of Biology*,
https://doi.org/10.1007/978-3-032-00555-7_4

the standard scientific realism (SSR) stance, to be a realist about explicit entities such as species without committing to implicit entities such as natural kinds.

2 Explicit Entities: Species and Higher Taxa

In this section I discuss the realist stance and the empiricist stance in relation to ontological questions concerning certain explicit (but unobservable) entities in biology, i.e. entities whose nature it is the business for biological science to investigate: species, higher taxa, and monophyletic clades. The entities at issue in this section are explicit entities of science, so can be accepted by partisans of either the SSR stance or the metaphysics of science (MoS) stance (or for that matter the armchair metaphysics stance). I shall assume that these entities are not observable, though of course their members/parts are. This is clearly the case for higher taxa and clades. The claim that species are not observable is I take it relatively uncontroversial. Some defenders of the species-as-individuals thesis argue that species are in-principle observable; if we were large enough and long-lived enough, we'd be able to observe them as we observe individual organisms. But even they would accept that in practice we cannot observe a species in its entirety.

One who adopts the SSR stance may, if the arguments in favour of the view(s) are judged to be cogent, accept realism about species and/or higher taxa and/or monophyletic clades. But she would stop there; she would not go on to engage in deeper metaphysical explanation and speculation that could issue in ontological commitment to entities such as natural kinds and essences, that are arguably tacitly implicated in scientific theorising about species and biological kinds, but the nature of which it is not among the tasks of science to investigate directly. One may be a species realist *and* think that species are natural kinds; this would require the MoS stance. Or, one could be a species realist while taking no interest whatsoever in the question of the ontological status of species; this would require only the SSR stance.

Though I don't know of such a view being defended, one could even imagine a version of naturalistic 'localism' (as discussed in Chap. 2) in this domain, according to which there are convincing arguments and evidence for believing in explicit entities such as species that are internal to science, such that philosophy has nothing further to add. Certainly there are biologists who have claimed that there are firm scientific grounds for thinking species are objectively real (in a way that higher taxa typically are thought not to be; Eldredge and Cracraft 1980), and others such as cladists who commit, again largely on empirical or theoretical scientific grounds, to the reality of monophyletic clades. These would be roughly analogous then to the Perrin experiments and arguments putatively establishing the reality of atoms, and the localist defender of the SSR stance can point to the absence of comparable scientific evidence or argument for the existence of natural kinds or essences, as grounds for rejecting the existence of the latter, a rejection that could take the 'agnostic' or 'atheist' form.

2.1 Species, Higher Taxa and the Realist Stance

If one adopts the realist stance, one may consider arguments for the view that species or higher taxa are objectively real. As explicit but unobservable entities of science they can potentially be admitted into one's ontology as explanatory posits.

It is customary to distinguish realism about species *taxa* from realism regarding the species *category*. The thought is that one could hold that species taxa, or what are *called* 'species taxa', exist, but the species *category* does not: the particular taxa are real, but there is nothing significant they all have in common, and thus no general species category that they and only they belong to, in virtue of which they count as *species*. We shall see that Ereshefsky (1992) defends a version of this view. In my view this distinction (between category and taxa) comes to less than is sometimes supposed. After all, if there is no species category, then there are no species taxa: the taxa that have been supposed to be species taxa might exist but they are not *species* taxa. If there is no species category, then nothing is a species. Contrapositively, if there are species taxa, the species category must be real. So realism (and anti-realism) about the species category and species taxa go together. Compare: if the category 'witch' is not real, then there are no witch-people. The people who have been *called* 'witches' exist, but they are not witches. And if there are witch-people, then there is a witch category. Realism (and anti-realism) about the witch category and witch-people go together.

A number of arguments have been offered for species realism, including the fact that some biological theories and explanations indispensably quantify over species, so given the Quinean criterion of ontological commitment (that we should be ontologically committed to all and only the entities quantification over which is indispensable to the truth of our best-confirmed scientific theories), we should believe that species exist (Quine 1953); the fact that anthropological evidence seems to show that many different kinds of human societies and cultures identify the same species taxa in nature (Atran 1999), the best explanation of which is that species exist (here the connection to the explanatory demands characteristic of the realist stance are particularly evident); the fact that species realism follows from, or is supported by, certain well-supported macro-evolutionary theses, such as Punctuated Equilibrium (Gould and Eldredge 1972; Gould 2002); and the fact (sometimes connected to the previous point) that species have a certain ontological status—they are concrete, cohesive, spatiotemporally bounded individuals, and thus are real, objective, discrete objects[1] (Eldredge and Cracraft 1980).[2] This last view we will return to in the section on implicit entities.

Realism about higher taxa is rarer. Many theorists have held that species are objectively real, whereas higher taxa are artificial constructs. If biologists themselves tend to take this view, as many do, it is hard to regard quantification over higher taxa as indispensable to our best biological theories, and the Quinean argument can't get a grip. The most promising path to realism about higher taxa is the cladist route. Cladists hold, firstly, that all (and only) biological taxa are monophyletic (consist

[1] Currie (2016) has recently offered a novel argument for species-category realism: the 'argument from indifference'. In their work paleobiologists are free to be indifferent to the question of the correct species concept, since their work in identifying and theorising about particular species does not depend on which concept is correct. This supports realism about the species category, he claims, as 'species' as a category is explanatorily useful for paleobiology whatever precise specification is given of the nature of the category. See also Hazelwood (2021).

[2] The species-as-individuals thesis may be sufficient for species realism, but it is seemingly not necessary for it; one may be a species realist while holding that species are natural kinds (Boucher 2022).

of an ancestor and all and only its descendants), and, secondly, that all (and only) monophyletic taxa are real groups or kinds (or real individuals).[3] Higher-taxa realism obviously follows from the conjunction of these two theses, though, firstly, both are controversial,[4] and secondly, they will only certify the reality of taxa insofar as they are monophyletic; there is no support to be found here for realism about traditional Linnaean higher taxa that are paraphyletic or polyphyletic (indeed the cladist explicitly denies that such taxa are real).

Of course, if one holds that (some or all) higher taxa are natural kinds (whether or not they are monophyletic[5]), and that natural kinds are real, then higher-taxa realism will follow. But since this higher-taxa realism comes via realism about natural kinds, we can hold off discussion of such views till Sect. 3, when we introduce the implicit entities.

Whatever the merits of these various arguments for species- and higher-taxa realism, the key point for us is that unless one has adopted the realist stance one has no reason to consider such arguments, as they presuppose that belief in the reality of these entities is V-permissible. If ontologically committing to species is V-permissible for one, one may, should one be convinced by the relevant arguments, adopt species realism as explanatory of biological (and perhaps folk) practice.

The Stance-Dependence of Arguments for Species Realism

But is this correct? It might be wondered why I am claiming that certain arguments for certain types of realism are only be available to those who have adopted the realist stance. Consider the arguments for species realism above. Why can't anybody, whatever stance they have adopted (or indeed if they have adopted no stance), accept the Quinean criterion of ontological commitment, then proceed to consider whether it applies to species? Why can't anybody, whatever stance they have adopted (or if

[3] Monophyletic clades are themselves among the explicit entities of biology.

[4] I criticise the second, metaphysical claim (that all (and only) monophyletic taxa are real groups or kinds (or real individuals)), while endorsing the first claim, in my (2022).

[5] Boyd (1999, 2010) has defended the view that at least some higher taxa are natural kinds, on his 'homeostatic property cluster' account of natural kinds, which I discuss below.

they have adopted no stance), accept the theory of punctuated equilibrium—an acceptance which itself is surely not realist-stance-dependent—and accept that the theory entails (or strongly supports) species realism? Even if we concede that once one has accepted species realism, to that extent one has adopted, or should adopt, the realist stance, at least implicitly, why can not the adoption of the stance be the *consequence*, not the *precondition*, of considering and accepting as compelling these sorts of arguments for species realism? Why, in other words, cannot one be *persuaded* that species realism is true purely on the strength of the arguments and objective evidence in its favour, without first having to adopt the realist *stance*?

My reply is that the realist stance with its commitment to the value of explanatory power is always lurking in the background of these arguments, such that, on closer inspection, the arguments will only be regarded as compelling if the stance has already been adopted (at least implicitly). So, for instance, the empiricist of the van Fraassian variety can accept Quine's criterion of ontological commitment, but withhold assent (in empiricist fashion) from the truth of claims referring to species, statements whose truth (we can concede) would require the existence of species, just as the empiricist withholds assent from (suspends judgement about) the truth of claims referring to electrons, statements whose truth would require the existence of electrons.[6] And the *reason* she withholds assent from these claims is precisely that she rejects the value the realist stance places on the putative explanatory power that commitment to these unobservable entities potentially brings. Thus the Quinean argument for species realism will only be compelling to one who has already adopted the realist stance with its characteristic values and attitudes. In including as a premise of the argument the assumption that (we should believe that) statements quantifying over species are true it in a sense begs

[6] When, in his (1953) Quine mentions this argument for species realism, he imagines the empiricist replying by accepting the truth of claims referring to species such as 'some zoological species are cross-fertile', and attempting to block the inference to species realism (and, for that matter, abstract entities as Quine supposes species to be), by offering a meaning-preserving paraphrase of the claim that does without quantification over the problematic entities (*ibid.*, 13). This is a possible empiricist strategy, but one associated with the forms of empiricism that van Fraassen's CE was intended to be an improvement on, and one that is unlikely to be viable in all cases. See Daly (2010, 89–98) for a useful discussion of the various problems faced by the Quinean idea of paraphrase.

the question against a certain type of empiricist. That is why I claim that such arguments are downstream from stance choice.

Or consider the inference from punctuated equilibrium. The idea is that in treating species as Darwinian individuals that experience a relatively rapid cladogenetic birth, remain relatively stable throughout their lifespan, and then go extinct, punctuated equilibrium removes much of the arbitrariness involved in delineating species that is associated with the traditional gradualist anagenetic approach to speciation, making species realism more defensible:

> If population A changes so extensively by anagenesis that we feel impelled to provide the resulting population with a new Linnaean name (as species B), then where should we place the breakpoint between A and B? Any boundary must be arbitrary ... Punctuated equilibrium took a radically different approach ... Eldredge and I argued that the vast majority of species originate by splitting, and that the standard tempo of speciation, when expressed in geological time, features origin in a geological moment followed by long persistence in stasis. Thus, the classic and endlessly-fretting 'species problem in paleontology' disappears because species act as well-defined Darwinian individuals, not as arbitrary subdivisions of a continuum. (Gould 2002, 775–6. See also Sterelny and Griffiths 1999, 182; Stamos 2003, 220–31)

The truth of punctuated equilibrium may remove *some* (though not all, as Gould admits (*ibid.*)) of the 'fuzziness' of species boundaries; but it certainly does not thereby logically entail that species are real. There is a 'gap' between punctuated equilibrium and species realism, and this gap can only be filled by presupposing the realist stance. If one is inclined to accept the latter, inclined that is to accept into one's ontology entities like species as explanatory posits, the inference from punctuated equilibrium to species realism can seem compelling. To put it another way, if one is antecedently inclined to accept species realism (because one accepts the realist stance) one can find in punctuated equilibrium (and associated views like species selection) a picture or model of species, their agency and boundaries, that buttresses that view. If one is not so inclined, there are certainly ways of accepting many or all of the explicit claims of punctuated equilibrium without having to endorse species realism.[7]

Eronen (2019, 2347) argues, in his defence of 'robust entity realism' for the life sciences, that when we have multiple independent lines of evidence in support of the reality of some unobservable entity in science, it 'borders on irrationality' for anybody, whatever their general commitments, to withhold assent from believing in it, even if such withholding is technically permissible on the empiricist-voluntarist view. But this, again, overlooks the fact that, as I have argued, questions of evidence are downstream from stance choice. If we have adopted the realist stance with its characteristic V-permissions, we may consider evidence, such as Eronen presents, for the reality of unobservable entities, and at that point the evidence may be compelling. But according to the voluntarist, it is not irrational to adopt the empiricist stance, on which such beliefs are V-prohibited, such that questions of evidence in favour of them just can't get a look-in. To declare such a position 'irrational', as Eronen does, merely begs the question against the voluntarist. Eronen also begs the question against the empiricist in that the type of argument he recommends for justifying realism about certain entities is a species of inference to the best explanation, as he admits (*ibid.*, 2349), albeit a local one; an argument form that is internal to the realist stance, and one the empiricist rejects.

Similarly, Strevens (2018) suggests that realist commitment is justified when the entity in question plays a range of explanatory roles that would not easily be satisfied by any replacement entity, and implies that this criterion is stance-independent. But as Chakravartty points out in reply (2018), it is not: it presupposes the realist stance with its valuing of explanation.

We encountered just this issue in the previous chapter. The argument for realism about the units of selection I presented there might seem at first blush to be available to anybody, and not to presuppose any particular stance, even if those who accept it *thereby* adopt (or require) the realist stance. But I suggested that on closer inspection, it involves a crucial premise whose justification appeals fundamentally to the values and

[7] I explain how one might do this in Boucher (MS).

attitudes typical of the realist stance. That argument, too, presupposed, rather than merely supported, the realist stance.[8]

In response to the question of whether someone may not accept species realism, say, purely on the strength of the arguments in its favour, quite apart from stance choice, the answer is that while this is not feasible inasmuch as the realist stance is presupposed by such arguments, still, whether one who has accepted the realist stance ought to accept any *particular* form of realism, such as species realism, *will* be a matter, in the end, of whether the arguments and evidence for that view are thought to be convincing.[9] One who has adopted the realist stance is not thereby *compelled* to accept species realism. To reiterate a key theme of Part I, the fact that argument and evidence are downstream from stance choice doesn't make them unimportant.

2.2 Species, Higher Taxa, Anti-Realism and the Empiricist Stance

In this section I shall discuss anti-realism about species and higher taxa, and consider its relationship to the empiricist stance. Recall that the empiricist stance does not licence belief in the explicit but unobservable entities of science, such as species and higher taxa, but that anti-realism about such entities need not involve the empiricist stance.

I shall focus here mainly on species, since, as we've seen, even scientific realists often reject realism with respect to higher taxa and clades. But scientific realists tend to be species realists. Anti-realism about species tends to take one of two forms: nominalism, and pluralism.

---

[8] As I noted in the chapter 2, Psillos (2011) suggests that arguments for realist *theses* presuppose the realist *framework*.

[9] The realist stance is, if you like, one premise in any argument for species realism (to the extent that it make sense to call non-propositions premises), but it is not the *only* premise.

Nominalism

Nominalism—the view that only concrete, particular things exist—is closely associated with empiricism. More precisely, on the (plausible) assumption that no abstract entities are observable, nominalism is necessary for empiricism; but since not all concrete particulars are observable (electrons, Lewisian possible dragons, etc.), nominalism is not sufficient for empiricism.

Whether nominalism is consistent with species realism depends on the metaphysical status of species.[10] One traditional view was that species were sets, i.e. abstract entities, and thus if species exist some abstract entities exist, and nominalism is false (Quine 1953). But any view on which species are natural (i.e. real) kinds also conflicts with nominalism, whether natural kinds are thought of as sets, or properties (as Soames 2007 argues), or as sui generis entities. According to nominalism, only concrete, individual things exist; nature creates particular things, not kinds; all 'kinds' are imposed on the world by us, as it were, and don't exist independently of us (Kornblith 1993). But, as I shall discuss below, a popular contemporary view of the metaphysics of species is that species are not natural kinds (or sets or classes) but concrete individuals. Such a view is consistent with nominalism, as it holds that species are concrete particulars just as individual organisms are. So on this view, nominalists may be species realists which, of course, requires the realist stance.

The main *motivation* (or argument) for the species-as-individuals (SAI) thesis was not the truth of nominalism, and defenders of the SAI thesis can hold, and have held, that there are natural kinds elsewhere in biology (or other sciences)—Hull (1988) holds that replicators, interactors and lineages are natural kinds for instance—but it remains the case that SAI theory taken on its own is consistent with nominalism. Whether a nominalism that recognises species-individuals is consistent with *empiricism* is questionable however, as species-individuals are presumably not directly observable.

[10] This reinforces a theme of this chapter, that philosophical discussion concerning explicit entities of science such as species tends to quickly embroil us in deeper questions of metaphysics involving implicit entities such as natural kinds, essences and species-individuals, and hence why the greater part of this chapter shall focus on the latter questions.

I shall discuss nominalism, and closely-related versions of conventionalism, again in Sect. 3.

Pluralism

Pluralism about species concepts—the view that there are multiple legitimate species concepts (e.g. the biological species concept, the phylogenetic species concept, the phenetic species concept, the evolutionary species concept, the ecological species concept, and many others) that cross-classify organisms—has become the orthodoxy in philosophical discussions about species (some classic defences include Kitcher (1984), and Dupré (1981, 1999)). Several theorists who defend pluralism argue that it is compatible with species realism, that is, they seek to defend pluralism from the standpoint of the realist stance. (In this they differ from the pluralists about the units of selection we discussed in the previous section, who are not typically realists.) Two exceptions are Stanford (1995) and Ereshefsky (1992).

Stanford follows pluralists such as Kitcher in suggesting that what counts as a species will be relative to the interests and goals of certain groups of biologists. If certain legitimate epistemic goals are well served by demarcating organisms in terms of a particular concept of what a species is, then that concept is *ipso facto* legitimate, and the groups it picks out count as species. Like other pluralists, Stanford does not advocate an 'anything goes' view; there are objective constraints on legitimate species concepts, in that they must be based on real and biologically interesting properties of organisms. And neither is his view an across-the-board anti-realism: the biological properties of organisms in terms of which species are delineated exist objectively and mind-independently.

So far this is common ground among pluralists; where Stanford departs from many of them is in claiming that if we think through the logic of the pluralist position, anti-realism about species will be seen to follow. This is because epistemic interests and goals change over time, without the natural world itself changing. '[W]hich divisions are biologically interesting changes dramatically over time without any corresponding change in the organismic world' (Stanford *ibid.*, 82). As interests and goals change, what counts as a species will change: At time t, a certain set of interests and goals will determine that certain groups of organisms count as species; at time t+n, a different set of interests and goals will determine that different groups of organisms count as species. Given that pluralists such as Kitcher hold that 'the legitimate interests of biologists *constitute* those divisions of organisms recognized as species' (*ibid.*, 83) (so we are not entitled to claim that the earlier divisions of organisms into species were incorrect, so long as the interests and purposes they subserved were legitimate), it follows that *what species there are* changes over time without any change in the natural world. And it follows from this that species do not enjoy mind-independent existence. They are constructed rather than discovered.[11]

The typical, and most natural, interpretation of Stanford's position takes it to be the view that *there are no* species in an objective and mind-independent sense. Currie however interprets his view as a weaker *agnosticism* about the reality of species (2016). He suggests Stanford holds that the 'biological world is subject to many possible carvings, but we cannot establish whether any particular carving is the right one, and so we should remain agnostic about the general species category: one specification might, after all, be correct, but we can never know' (*ibid.*, 4). Whether or not this is accurate as a reading of Stanford (I'd suggest it is not, but won't argue for that here), Currie has here identified a form that anti-realist empiricism with respect to species can take, one that is kin to van Fraassen's constructive empiricist view that we ought to remain agnostic (suspend judgment) regarding the existence of unobservable scientific entities such as electrons and quarks. While this is a possible empiricist view about species, it is not typically the form that anti-realism about species has taken (as we saw that it wasn't in the case of the units of selection).

Ereshefsky (1992) offers a somewhat different argument for anti-realism about species, one we touched on earlier. Given pluralism about species concepts, and the resultant heterogeneity of the types of things

[11] See Crane and Sandler (2011) for a similar argument; they don't say explicitly that they are rejecting species realism; they claim rather that species are not natural kinds, and that species concepts do not carve at the joints.

that go by the name of 'species', it follows, he argues, that the species *category* does not exist, that is, that there is nothing interesting that all the groups of organisms that have been called species have in common, such that it makes sense to group them all together under the general heading 'species'. It might be that the particular *groups* or base taxa that have been *called* species exist *qua* groups or taxa; but nothing is achieved by insisting on calling them species (it may be however legitimate to call them 'phylospecies', 'ecospecies' etc.; these more specific categories plausibly *are* real and important). We should eliminate the general species category. It does no explanatory or theoretical work.

There has been much discussion about both Stanford and Ereshefsky's views about species, that we need not go into here.[12] Although, as we shall see, both thinkers, in other respects, appear to be sympathetic to the empiricist stance, their pluralist positions on species and arguments for them are not obviously expressions of the empiricist stance.[13] A view is not an expression of the empiricist stance just in virtue of being (or implying) an anti-realist view of the relevant entities. It must also be associated with or motivated by empiricist values and attitudes. That doesn't appear to be the case here: neither Ereshefsky nor Stanford is suggesting that species realism is unacceptably metaphysical, that it involves too great a degree of epistemic risk, or that it is a response to the (merely optional) realist demand for explanation of the observable in terms of the unobservable.

In general, unlike realist views, anti-realist views do not require any particular stance. One can be an anti-realist about species from the standpoint of the empiricist stance, the SSR stance, the MoS stance, or for that matter the armchair metaphysics stance. Anti-realism with respect to any kind of scientific entity, especially of the suspend-judgement type, is after all the 'default' position, in the absence of positive reasons to accept realism. It is true that the *basis* for one's anti-realism may differ depending upon one's stance. For the empiricist, belief in species is V-prohibited, so in a sense the particular *arguments* for the view don't come into it. For

[12] See e.g. Brigandt (2003), Currie (*ibid.*).

[13] Unless, that is we accept Currie's interpretation of Stanford's view given above.

those who accept the other stances, belief in species is V-permissible, so they need to consider whether the stance-internal arguments for the view are persuasive. But what about arguments *for anti-realism*, such as the ones from pluralism above (or the one from vagueness I discuss below)? Do they presuppose any particular stance? In one sense they don't, as the view being defended is V-permissible on any stance. But in another sense, it might be thought that they do presuppose a view on which belief in species is V-permissible, since offering arguments of this type *against* the existence of certain entities presupposes that the existence of the entities is, so to speak, up for debate; something to be rationally decided, by means of argument and evidence, something belief in which is not V-prohibited. For the empiricist, belief in species is V-prohibited, so the argument *from* pluralism *to* anti-realism is, in a sense, beside the point. That is, it may be that both arguments for species realism and arguments against it must take place *within* the realist stance: they are both downstream from stance choice. If that is right, while anti-realism *as such* (as a *view*) does not require any particular stance, certain *arguments* for anti-realism might.

Two Forms of Naturalism

Stanford's and Ereshefsky's views are motivated by a form of naturalism, that is, by the need to take seriously the role that species concepts play in actual biological practice. As I noted in the Introduction, this is a somewhat different (though not entirely unrelated) form of naturalism from the one we have been largely working with in Part I, which has to do with the directness of metaphysical inferences from science to ontology, and with the extent to which our metaphysics is answerable to empirical evidence. One potentially violates naturalism in this sense when one posits or ontologically commits to some metaphysical entity that is 'distant' from science in the sense of enjoying a low or non-existence degree of empirical vulnerability, and/or pursues one's ontological inquiries in a largely a priori way. The sense of naturalism that Ereshefsky and Stanford (and other pluralists) are trying to honour has to do rather with taking a 'science-first' or 'deferring to science' approach when it comes to offering philosophical accounts of important scientific (or science-related) concepts. Think of this as a type of naturalism concerning scientific *concepts*—I shall refer to it as 'conceptual naturalism'—and the former type as concerning scientific *ontology*.

One violates conceptual naturalism when one fashions (and applies) a philosophical account or definition of a scientific concept (or concept of relevance to science) without paying close attention to the way in which the concept, or related concepts, theories and explanations, are actually employed in science; this would be a 'philosophy-first' approach, and it involves philosophy 'dictating to' or 'legislating about' science and its use of concepts in a way that naturalists regard as illegitimate. One example (that we shall return to in Chap. 10) is reductionism in genetics. Philosophers who defend an account of theory-reduction according to which Mendelian genetics is not reducible to molecular genetics have (as I would put it) been accused of violating conceptual naturalism. It is widely accepted by the relevant scientific community that the former does reduce (in some sense) to, or is explained by, the latter, so this is (other things being equal) a basic datum that, according to conceptual naturalism, ought to be respected by any philosophical account of reduction. A science-first account of reduction would seek to preserve the judgements of the scientists, if at all possible. As Stent puts it, 'What geneticist could take seriously any explication of "reductionism" which leads to the conclusion that molecular genetics does *not* amount to a successful reduction of classical genetics?' (Stent quoted in Sterelny and Griffiths 1999, 143). Another example (that I shall return to below) concerns philosophical accounts of physicalism and the physical. If such an account has the implication that certain respectable entities, forces or processes referred to physics are non-physical, and so violate physicalism, this is a significant objection to, if not an outright *reductio* of, said account, given conceptual naturalism.

The initial shift from monism to pluralism in the debate about species was justified in similar terms: biologists find different, incompatible views of species useful for their different, seemingly equally legitimate epistemic projects and purposes; for philosophers (or biologists for that matter) to insist that at most one of these species concepts is correct—that species concepts that do useful or even indispensable work in promoting legitimate epistemic ends within science are nonetheless *wrong*—was rejected primarily because it was seen as an unacceptable violation of conceptual naturalism. As Stanford puts it, monism 'hobble(s) significant investigations in biology by limiting our conceptual and taxonomic resources …' (*ibid.*, 77). We shall see this idea at work later in this chapter when we discuss natural kinds. Defenders of currently popular 'epistemic' accounts of natural kinds argue, in ways not dissimilar to pluralists about species, that conceptual naturalism requires that we recognise as genuine natural kinds whatever categories scientists find useful in satisfying their legitimate epistemic goals, even if these will be metaphysically quite heterogeneous. Trying to shoehorn all natural kinds in science into a particular metaphysical mould is a failed strategy.

What is interesting is the way in which, if Ereshefsky and Stanford are correct, naturalistically-inspired pluralism can lead to anti-realism. We shall also see this when we consider epistemic accounts of natural kinds below. I shall argue that such views entail that natural kinds do not enjoy mind-independent existence. Thus naturalism can cut both ways. A traditionally Quinean naturalised metaphysician with respect to species will reason in the following flat-footed way: we should base our ontology on science; scientists indispensably quantify over species; so we should believe that species exist. Stanford and Ereshefsky on the other hand, point out that close attention to the diverse roles played by different species concepts in biology makes pluralism compulsory; and like it or not, pluralism brings with it anti-realism. One may be a naturalist without embracing species realism.

The Vagueness Argument

The final point to make before moving on to the implicit entities of systematics is that aside from the argument from pluralism, the other main argument for anti-realism about species is based on the vagueness of species boundaries. Given evolution, on any species concept species change into new species gradually, such that there are no sharp boundaries separating one species from its immediate descendant; it follows (according to this argument) that the precise place where we draw the line separating species diachronically is arbitrary, and thus species are not objectively real, but just arbitrarily and artificially demarcated/constructed segments of what is in reality a smooth continuum. This argument seems to have tempted Darwin himself (1859; see Ereshefsky 2011).[14] This argument is independent of both pluralism about species concepts and nominalism, and is consistent with the denial of both. As with the arguments from pluralism above, the vagueness argument is not obviously motivated by the empiricist stance, and may in a sense presuppose the realist stance, as it assumes that realism about species is in play as a live option (is V-permissible). It is clearly a specimen of naturalised metaphysics, as it appeals (or claims to appeal) to our best current scientific understanding of the nature of species and the evolutionary process, in arguing for the (negative) ontological thesis that species do not exist.

In Sect. 2 we have been examining the role of epistemic stances with respect to ontological questions concerning explicit (but unobservable) entities in biology: species, higher taxa, and monophyletic clades. I have suggested that while realism with respect to these entities presupposes the realist stance (in the form of either the SSR or the MoS stance), anti-realism with respect to them does not *in itself* presuppose any particular stance. This does not rule out however, that in some cases it may be motivated by, say, the empiricist stance. I suggested further that not only do arguments for realism with respect to these entities presuppose the realist stance, but, paradoxically, some arguments for *anti-realism* with respect to them may also presuppose the realist stance, inasmuch as they assume that the specific realist thesis in question is on the table as a genuine option, even if it is ultimately rejected. For those who reject the realist stance, such realist theses are not even live options, so *arguments* against them are beside the point. We now turn our attention from explicit entities such as species and higher-taxa to the implicit scientific entities of which they are plausibly instances, such as natural kinds.

---

[14] We shall see that Ellis (2001) argues this vagueness means species cannot be natural kinds.

3 Implicit Entities: Natural Kinds

In Sect. 3 we shall consider debates surrounding the existence and nature of implicit entities that are arguably implicated in scientific work on species and higher taxa, in particular (but not restricted to) natural kinds.[15] The first period of sustained discussion of these questions began in the early 1970s, and concerned the metaphysical status of species: whether species are, in metaphysical terms, natural kinds, sets, classes, or concrete individuals. This spawned a very large literature in philosophy and biology that began to peter out, or perhaps more accurately, became subsumed in a new and more general debate in the philosophy of science concerning natural kinds in biology and elsewhere, a debate that has flourished and itself generated a formidable literature in the last couple of decades. We shall begin (in Sect. 3.1) by looking briefly at the former debate, in particular the species-as-individuals (SAI) thesis alluded to above, then (in Sect. 3.2) move on to look in some detail at the more general recent debate concerning natural kinds. Our discussion here shall be somewhat intricate, involving multiple intersecting taxonomies. This is, unfortunately, unavoidable, if we are to do justice to the various positions and arguments that have emerged in a sprawling and at times confusing literature. I hope to bring a degree of order and coherence to this debate, in any case.

The question of essences in biology will be touched on at times in Sects. 3.1 and 3.2, but in Chap. 5 we shall look much more closely at this topic. Our focus throughout Chaps. 4 and 5 shall be on the question of ontology, i.e. which views in these debates are committed to the existence of the implicit entities in question, and on the epistemic stances that underlie and motivate the different views and arguments on these topics.

[15] We shall be concerned primarily with the natural kinds of systematics: species and higher taxa. Of course there are natural kinds in other areas of biology, but I shall not be focusing on those.

3.1 Species as Individuals

The species-as-individuals (SAI) thesis was first proposed by Ghiselin (1974) and Hull (1976, 1978). They argued that contemporary evolutionary biology teaches us that species are not natural kinds, or classes, as had traditionally been supposed, but historical individuals—'chunks of the genealogical nexus'. Their main arguments were as follows:

- Species evolve over time, unlike classes which, as abstract entities, cannot change or evolve.
- Species taxa, unlike natural kinds, are not defined by intrinsic, microstructural essences: sets of intrinsic, microstructural properties that are necessary and sufficient for an organism to belong to them. There is no genetic property, for instance, possessed by all and only members of a particular species.
- Species are spatiotemporally bounded. Again, this makes them like concrete individuals but unlike natural kinds or classes. Just like an individual organism, once a species goes out of existence, it cannot reappear later on. No matter how similar a newly evolved species may be to it, if it is not historically connected to the original species in the right way, it is a different species. Moreover, there is in principle no limit to the amount of evolutionary change a species can undergo while remaining the same species, so long as it doesn't split into two species cladogenetically. This is also true of organisms. What makes me at age two and me at 90 the same individual is causal connection and spatiotemporal continuity, not similarity. (The spatiotemporal boundedness of species can be expressed by saying that similarity is neither necessary nor sufficient for two organisms to be conspecific, either synchronically or diachronically.)
- Often, a key criterion of individuality is that the parts of an individual *causally interact* in certain characteristic ways.[16] Individuals possess *internal organisation and structure*. Corporations, clubs and other social organisations are not spatially united or continuous, but they are con-

[16] This may not be true of all individuals. Elementary particles with no internal structure would be an exception.

sidered to be individuals because they act cohesively, exhibiting internal organisation and structure. Their parts causally interact in the right sort of way. They are *functionally integrated*. So species can be individuals even if their parts (organisms) aren't spatially contiguous, as long as they have internal organisation, and their parts interact causally. They are cohesive (largely due to gene flow).

- Species names are *proper names* that pick out their referents directly, rather than via any mediating description. The names of natural kinds abbreviate descriptions, e.g. 'gold' means 'the chemical element with atomic no. 79'. But *Homo sapiens* names a species *directly*.
- Species do not figure in laws of nature. In this they are like individuals but unlike natural kinds. There are laws of nature that refer to natural kinds, but not individual objects. Arguably there are no laws about tigers, or *Homo sapiens*.

Are the entities the existence of which is in dispute in the debate on the ontological status of species part of the explicit or implicit subject matter of science? Natural kinds are clearly part of the implicit subject matter of science, thus those that hold that species are natural kinds are committed to (at least) the MoS stance. SAIers hold that species are concrete individuals. These *species-individuals* are quite unlike cats, tables, and cricket bats, in that they are not directly observable, and are not part of our common-sense ontology. SAI is indeed notoriously ontologically *revisionary* in its claim that species are individuals; common-sense has it that they are collectivities, not individuals. And species-individuals *as such* are certainly not part of the explicit subject matter of science. What scientific theories say about the role species play in various evolutionary and ecological processes may *imply or suggest* that they are concrete individuals (as SAIers argue), but this is not explicit in the science; it is the job of the metaphysician of science to draw it out. The scientist *qua* scientist need not take a stand on it. SAI has always been understood to be part of metaphysics (albeit of the naturalised variety), not science. Similarly Platonists about numbers argue that it is an implication of statements mathematicians make and accept (e.g. 'there are an infinite number of primes') that numbers exist. But this implication is not explicit in the mathematics; it is the job of the philosopher of mathematics to draw it out. The mathematician *qua* mathematician need not take a stand on it.

Also, while the category 'individual' is not in itself metaphysically obscure, SAIers have had to develop metaphysically sophisticated definitions of 'individual' in order to show why species may be, despite appearances and common-sense, individuals (for example, by pointing out that we classify some entities, such as nations and corporations, as individuals despite their parts being spatially scattered or physically discontinuous). That is, a category can be part of common-sense, yet its articulation and definition may call for sophisticated philosophical analysis.

Thus I shall interpret SAI as requiring (at least) the MoS stance, inasmuch as the entities it is committed to—species-individuals—are—at least under that description—part of the implicit subject matter of science. Thus both SAI and the species-as-natural-kinds view require (at least) this stance, such that the debate about the ontological status of species is, usually, a dispute *within* this stance. I think that will fit with most peoples' intuitions.

To what extent, then, did the SAI thesis, and the arguments offered for it, exemplify the naturalistic approach to metaphysics? The form of naturalism most relevant here is the conceptual variety. While in principle one could, from the standpoint of the empiricist or SSR stance, object to species-individuals as being posits too far removed from science or empirical inquiry to be naturalistically respectable, in practice it is conceptual naturalism that some have claimed SAI has failed to be true to.

The arguments for SAI proceeded in two stages. In the first, an account of both natural kinds and concrete individuals was offered (or presupposed), drawing on metaphysical tradition (in the armchair, i.e. non-naturalistic mode). Natural kinds were construed as spatiotemporally unrestricted sets or classes of entities defined by an intrinsic essence, and featuring in laws of nature. Individuals were construed partly negatively—as entities that are *not* spatiotemporally unrestricted sets or classes of entities defined by an intrinsic essence, and featuring in laws of nature—and partly positively, as entities that (often) exhibit internal organisation and structure, act as cohesive wholes, change over time, have parts that causally interact in certain characteristic ways, are functionally integrated, and are individuated by causal connections and spatiotemporal relations. The second step involved looking closely at contemporary biological science, to determine whether, given what it seems to be telling us about the nature of species, species ought to be construed as natural kinds or individuals.

The second step was unimpeachably naturalistic. Unlike Kripke and Putnam, and some of the previous generation of would-be philosophers of biology, the prominent early defenders of the SAI thesis, Hull and Ghiselin, had a firm grasp of contemporary biological science, and took what the science was saying very seriously. Expressed in our terms, they urged that we endorse as compelling certain *metaphysical inferences* from various claims, theories and assumptions in (especially evolutionary) biology to the conclusion that species are metaphysical individuals, or, to put it in terms of scientific ontology, that species-individuals exist.[17] For instance, the claims in the arguments for SAI above that species evolve, that there are no intrinsic properties that are such that possessing them is necessary and sufficient for an organism to belong to a given species, that species are cohesive due to gene flow, that they are spatiotemporally bounded, that their parts causally interact, that they exhibit internal organisation and structure, and that they do not figure in laws of nature, were all taken from or based on contemporary evolutionary theory. Hull also appealed to cladogenetic and allopatric models of speciation, partly in order to strengthen the analogy between the births of species and the births of organisms (1977, 153), while Gould argued (2002) that SAI followed from or was associated with macroevolutionary theories such as punctuated equilibrium and species selection.

The naturalistic character of (the second part of) the SAIers' approach—in the sense of responsive to science—was thought to be evident in the fact that not only did they argue that it follows from central assumptions of evolutionary theory (or the relevant parts of it[18]) that species are individuals, but that to the extent that this conclusion conflicted with common sense—which assumes that species are collectivities—we ought to reject common sense (Hull 1976, 1977, 1978, 1981; see Slater 2017, 65).

[17] As Hull admitted, these inferences are not necessarily immediate or direct; the SAI thesis could not, he recognised, be 'read off' the science, but required philosophical articulation and reasoning.

[18] *Which* parts of science a properly naturalised metaphysics ought to defer to, especially when scientists disagree with one another, is not a straightforward question; see Slater (2017).

The first step of the argument for SAI, however, came to be regarded by many as not in the spirit of conceptual naturalism. The suggestion that the account of the nature of natural kinds to be found in the philosophical tradition—one developed without close attention to natural-kind-involving scientific practice (especially in biology and the social sciences)—was the correct one, and the implication that putative biological kinds (species) were not natural kinds at all, had, for those who pressed this objection, too much of the flavour of philosophy dictating to, and passing judgement on, science and its putative kinds.[19] I discuss this, and subsequent developments, in the following section.

3.2 Natural Kinds: The Recent Debate

In this section I consider the recent debate over the nature of natural kinds.[20] The debate has concerned natural kinds in general, not just in biology, but as we shall see, the views on offer all have implications for the nature of biological kinds. While it may be possible in principle to defend one of the accounts of natural kinds we shall consider while holding that there are not natural kinds in biology so understood, in fact nearly every view we will look at is intended to capture the natural kinds of biology, in particular species and biological taxa. Indeed, one of the central *motivations* for many of the views has been the need to develop an account of

---

[19] Slater (2017) offers an insightful criticism of the view that SAI can be justified by appeal to a principle of naturalised metaphysics, but he does not discuss this particular sense in which arguments for it violated conceptual naturalism. Slater's claim is that while the general case against traditional essentialism was a 'naturalist success story', it is less clear that we can say the same thing for the case *for* SAI theory. The problem is that the latter has the unacceptable metaphysical implication that there can be vague objects: objects for which it is indeterminate whether or not they exist. So we have metaphysical reasons to reject it despite its apparent naturalist credentials. Furthermore, any naturalist attempt to argue that in such a case the fit with science trumps the problematic metaphysics faces several objections. In short, he argues that there is no straightforward, defensible conception of naturalised metaphysics on which we can justify the claim that it is illegitimate to reject a theory with this sort of unacceptable metaphysical consequence, even if in other respects in meshes well with parts of science.

[20] As Crane (2021) notes, there are two traditions or approaches to theorising about natural kinds, the philosophy of science approach and the philosophy of language approach, and there may be certain tensions between the two. I focus here on the philosophy of science approach (since I am interested in scientific ontology, not reference and semantics), though some of the thinkers I refer to, such as Putnam and Kripke, have contributed to both traditions.

natural kinds that fits the biological world better than more traditional accounts. Thus, in considering the debate about natural kinds in general, we are considering the question of the nature of biological kinds, in particular species and biological taxa.

The subject of natural kinds is often introduced as concerning kinds that exist objectively and mind-independently; natural kinds are kinds that are 'made by nature' rather than being imposed on nature by us. But such a characterisation of the topic, while perhaps appropriate in the past, is no longer, since it begs the question against mind-dependent views of natural kinds. One who holds that, in some sense, there are natural kinds (one who rejects outright eliminativism) but holds they are mind- or language-dependent (as a number of people do, as we will see), does not thereby believe an analytic falsehood. The strong-realist, mind-independent view of natural kinds might be correct, but should not be built into the very concept of natural kinds. While it is coherent to hold that natural kinds are mind-dependent, any account which claims to be an account of *natural* kinds still needs to do justice to the 'natural' part: the idea of 'natural kinds' is still being contrasted with fully 'conventional kinds': even one who holds that all natural kinds in biology are mind-dependent will regard the categories *mammal*, *chordate*, or *arthropod*, as 'natural' in a way that categories such as 'fish that humans enjoy eating' or 'green-coloured living things' are not. If the distinction is not to be cashed out in terms of mind-independence, some other means of doing this is needed. Typically this will be by way of epistemic criteria such as scientific-theoretical significance, usefulness for induction, generalisation or prediction, as we shall see.

Metaphysical, Mixed, and Epistemic Accounts

On the metaphysical account, natural kinds are characterised in metaphysical terms. The traditional essentialist view of natural kinds as sets of entities united by possession of an essence—a set of intrinsic properties individually necessary and jointly sufficient for membership in the kind—was a classic example of a metaphysical account. Epistemic accounts on the other hand, understand natural kinds in terms of human epistemic practices, in particular the practices of induction, projection, explanation, prediction and generalisation. These need not be mutually exclusive. Even traditional metaphysical accounts tended to accept that natural kinds played an important epistemic role, and one could of course define natural kinds as kinds or categories that (a) satisfy certain metaphysical criteria, *and* (b) play a certain role or roles in our epistemic practices. In practice, accounts of natural kinds have tended to emphasise either the metaphysical or the epistemic dimension at the expense of the other. But those that do emphasise each more-or-less equally I shall call 'mixed' views. As we shall see, the views I am calling mixed views tend to be more liberal or permissive with respect to the metaphysical component of their view than is the case for exclusively or predominantly metaphysical accounts.[21]

In recent years largely or exclusively epistemic accounts have gained in popularity. This is partly due to the idea that while the question of the metaphysical status or nature of natural kinds is controversial and uncertain, what we do know, and what there is widespread agreement on, is that whatever the correct metaphysical characterisation of natural kinds is, they are kinds that permit projection, explanation, inference, generalisation etc. This latter epistemic role, especially in the sciences, is thought by many to be what is fundamental about natural kinds, with metaphysical questions being secondary and subordinate to this.

Buttressing this view is the idea that only largely or entirely epistemic accounts can satisfy conceptual naturalism; our account of natural kinds should capture (that is, should classify *as* natural kinds) the kinds and categories that play the epistemic role of natural kinds in the sciences, that is, the kinds and categories that are implicitly treated as natural kinds in the sciences. Whether the account should classify as natural kinds *only* the kinds and categories that play the epistemic role of natural kinds in the sciences, or whether we should allow for non-scientific natural kinds, is more controversial, but it is widely accepted the account should capture *all* plausible scientific kinds. The idea is to start with the science, i.e.

[21] See Massimi (2022, Part II) for an interesting and complex account of natural kinds that is a 'mixed' account in that it emphasises metaphysical and epistemic considerations (though she argues that the former are downstream from the latter), but foregrounds *phenomena*, and *inferences*, to a greater extent than other accounts in the literature.

start with the kinds that scientists find useful in helping them achieve their epistemic aims, and craft an account of natural kinds that will vindicate that all or most of these scientific 'kinds' really are natural kinds, rather than starting with a philosophical account of natural kinds and using it to pass judgements on the status of the putative kinds of science.

A classic example of an account that, according to many, violated this requirement was the traditional account of natural kinds mentioned above: natural kinds as spatio-temporally unrestricted sets of entities united by possession of an intrinsic essence that feature in laws of nature (what Millikan calls 'eternal kinds' (1999)). This view notoriously failed to fit the biological world. Those who revived essentialism in the 1970s such as Kripke and Putnam failed to notice this, assuming that species no less than chemical kinds could be understood as natural kinds in this sense, defined by microstructural essences. But as we saw in the previous section, those who were the first to notice the mismatch (due to possessing a better grasp of biological science), such as Hull and Ghiselin, did not thereby reject that account of natural kinds; they accepted the account, and drew the conclusion rather that species were not natural kinds at all, but concrete individuals. Those of their critics who have had the most lasting influence in the debate charged them with violating conceptual naturalism. Species manifestly play the epistemic role in biology that natural kinds characteristically play, allowing for induction, projection, inference, generalisation etc. If there is an account of natural kinds according to which they fail to qualify as natural kinds, then so much the worse for that account (Haber 2016).[22] Thus alternative, more permissive accounts of natural kinds were developed, such as the homeostatic

[22] See Ellis (2001) for an opposing view. The vagueness of species boundaries implies, he argues, that species are not natural kinds: 'If evolution occurs in the gradual way that Darwin supposed, or if small changes in genetic constitution can be brought about artificially, then the distinctions between adjacent species—living, dead or yet to be created—must ultimately be arbitrary' (Ellis *ibid.*, 169). Ellis, in the vein of defenders of the SAI thesis, offers a metaphysical account of natural kinds (one that among other things requires natural kinds to be discontinuous and have sharp boundaries) on which species fail to count as natural kinds. He is thus vulnerable to the charge of violating conceptual naturalism.

Devitt (2008, 2021, 2023) defends an unusual view on which natural kinds are characterised in (relatively) traditional metaphysical (essentialist) terms yet biological taxa do count as natural kinds so understood. I shall discuss this view in the following chapter on essentialism.

property cluster (HPC) account, due initially to Richard Boyd (1991, 1999).[23]

Boyd suggests that natural kinds are characterised by sets of properties, both intrinsic and extrinsic, that account for the suitability of the kinds for induction and generalisation. The clustering is not accidental or a brute fact but a reflection of the nature of the world: the causal structure of the world is such that certain clusters of properties fit naturally together; they stand in a relation of *causal homeostasis*. The presence of one property is indicative of the others, supporting inductive inferences from the presence one to the presence of others in the cluster. The cluster is stable in virtue of underlying homeostatic mechanisms that operate to ensure this stability is maintained, and the integrity of the kind is kept intact in the face of external perturbations in the environment. In the paradigm case the members of a species reliably share a number of properties, and this is reliably reproduced over the generations, due to the operation of several causal processes: shared ancestry, developmental mechanisms, reproduction and ecological interactions. In the standard version of the view no one of the properties within the cluster (and no particular *set* of them) is *necessary* for membership of the kind. More precisely, possession of *one or other* minimal cluster of the relevant properties is necessary for membership of the kind—there is a kind of *disjunctive* necessity—but there is no individual property, or cluster of properties, that is such that possession of *it* is necessary for membership of the kind (Wilson et al. 2007).[24] But something is a member of the kind if it possesses a *sufficient number* of these properties, where what counts as a sufficient number will depend on the kind and properties in question. In these respects the view is weaker than the traditional essentialist account of natural kinds.[25]

We are focusing in this chapter on the kinds of systematics, but the HPC account has been applied to other biological kinds. For instance

---

[23] We shall see in the following chapter that as well as (and related to) more liberal accounts of natural kinds, more liberal versions of essences and essentialism, relaxing some of the traditional requirements on essences, better suited (it is claimed) to the biological world, have also been developed.

[24] The view is in the 'family resemblance' tradition associated with Wittgenstein.

[25] See also Kornblith (1993) and R. Wilson (1999, 2005). For thorough defences of the HPC view see Wilson et al. (2007), and Brigandt (2009).

Williams (2011) suggests that diseases do not count as natural kinds on the traditional essentialist account; hence, given there are compelling reasons for treating them as natural kinds, we ought to reject essentialism, and endorse a more permissive account of natural kinds which can vindicate the idea that diseases are natural kinds. The account that can best do this, he argues, is the HPC view.

Note that the HPC account is what I am calling a mixed account; while its advocates (e.g. Boyd, Brigandt) tend to accept that natural kinds are, fundamentally, kinds that play certain characteristic epistemic roles, they also offer a quite specific metaphysical account of natural kinds in terms of property clustering, homeostatic causal mechanisms, and so on.

So, perhaps not surprisingly, the HPC account has in turn been accused of being too restrictive and thus violating conceptual naturalism (Ereshefsky and Reydon 2015, 2023; Khalidi 2013a; Slater 2013, 2015[26]; Chakravartty 2007; Magnus 2014; Papineau 2022). The claim is that some, perhaps many, plausible natural kinds in science are not HPC kinds. Thus Ereshefsky and Reydon argue that while the HPC view 'aims to be a naturalistic account …', it 'is not naturalistic enough' (*ibid.*, 978). Notice that this concerns *conceptual* naturalism: it is quite different from accusing the HPC account of violating naturalism through positing metaphysical items too far removed from science, and too epistemically risky, to be naturalistically respectable. There is no suggestion here that HPC theory is guilty of *that.* Ereshefsky and Reydon (*ibid.*) give the examples of functional kinds (kinds characterised in functional terms that are multiply-realisable by entities with widely varying properties and structures), and 'heterostatic' kinds (kinds whose members display persistent differences, such as stable polymorphisms in most biological taxa), as non-HPC kinds, while a number of critics have noted that the natural kinds of fundamental physics, such as bosons and quarks, are defined by clusters of co-occurring properties where these do not apparently cluster on the basis of underlying causal mechanisms (e.g. Chakravartty 2007). While there is some debate about whether the HPC account can capture

[26] While Slater rejects HPC theory for being inapplicable to certain scientific kinds such as species taxa, he regards some would-be naturalistic criticisms of HPC theory—for instance Ereshefsky's complaint that it conflicts with the principle of monophyly for biological classification—as misguided (2018, 70–71).

some of these apparent counterexamples—Brigandt (2009) argues that functional kinds and heterostatic kinds can be understood as HPC kinds, for instance—it has become increasingly clear, many think, that even if some, perhaps many, natural kinds in science may be HPC kinds, others are not, so HPC theory fails as a general account of the nature of natural kinds.

Rather than overrule the science, as it were, and reject the putative scientific kinds that don't fit the HPC model, conceptual naturalism requires, the critics argue, that we reject HPC theory, at least in its standard version. A moderate version of this response recommends more metaphysically liberal 'mixed' views. So Slater (2013, 2015) rejects standard HPC theory but develops a modified version of it that, he argues, better satisfies the naturalist constraint. This has been called the 'bare (or 'stable') property cluster (BPC)' account. On this view the epistemic utility of natural kinds for induction and explanation derives from and is underwritten by the tendency of properties to stably cluster together. But there is no single, unified metaphysical account of what gives rise to this clustering; in particular, it is not always the result of *causal mechanisms*, as HPC theory would have it.[27] This is not to say that there is not, in any particular case, anything that grounds the clustering; it is just to say that what grounds the clustering varies from kind to kind, and defies any simple, uniform account. Khalidi (2013a, b) defends a similar view. He agrees with HPC theory that it is insufficient when characterising natural kinds to refer simply to properties clustering together; for property clusters to constitute real natural kinds there must be some metaphysical ground for the clustering. And he agrees with HPC theory that this grounding must be, in some sense, causal (making his view stronger than Slater's; see Brigandt 2020). But, like Slater and others, he argues that HPC theory is too restrictive in its claim that this *always* takes the form of homeostatis-generating causal mechanisms.[28] Natural kinds are defined in terms of clusters of properties, where this clustering derives from or is

[27] See Chakravartty (2007) for a similar account.

[28] He gives the example of the kind *larva:* there are no central causal mechanism underlying the processes that reliably generate instances of the kind across the generations. It is thus not an HPC kind (2013a).

a reflection of the objective causal structure of the world that science aims to uncover; but that's all we can say as a general claim.

However even seemingly maximally metaphysically-neutral mixed accounts such as Slater's have been criticised as insufficiently naturalistic, as forcing us to rule out plausible examples of scientific kinds. The BPC account accepts that natural kinds are defined by clusters of properties that are such that the presence of one of the properties is indicative of the presence of the others, such that the presence of one gives strong inductive grounds for inferring the presence of the others. But Martínez (2020) argues that some natural kinds are what he calls 'synergistic kinds': these are kinds in which the above condition is violated, since the relevant properties must be jointly instantiated in order for the inductive potential of the kind to be realised. Examples of synergistic kinds include those characterised by epistatic interactions such as *fruit-fly wing* (*ibid.*, 1943–1944; see Valero 2023 for discussion).

Another, perhaps more basic worry is that in focusing exclusively on kinds whose members *share* clusters of properties, the BPC view, like the HPC view it aims to improve upon, uncritically adopts the traditional assumption about natural kinds that they are fundamentally sets of *similar* entities, sharing a large number of properties, making them suitable for induction and prediction (Magnus calls this 'similarity fetishism' (2011, 862–3, 2012)). Yet, as we saw above, many natural kinds in science appear to comprise entities that *differ* in persistent ways, i.e. that only share a small number of properties (Papineau 2022). Importantly for us, some key counterexamples we noted above come from biology, such as historically-defined biological taxa and functionally-defined biological kinds. With respect to the former, many regard monophyletic clades as natural kinds. While the members of such clades do typically share a number of intrinsic properties in virtue of common ancestry and phylogenetic inertia (Griffiths 1999), they also often differ substantially from one another, especially in large clades such as *Tetrapoda*, due to adaptive divergence, drift and other processes, making them ill-suited on the whole, *qua* phylogenetic groups, for induction and prediction (Chakravartty 2007). With respect to functionally-defined kinds, ecological kinds such as *predator* and *parasite*[29] are kinds whose members may share some common adaptive features related to functional or ecological role but also differ in many ways (Boucher 2022; Wilson et al. 2007; Griffiths 1994; Papineau 2022). Chakravartty (2007) invokes functional kinds as a counterexample to Mill's view of natural kinds, yet such kinds also do not fit easily within the BPC view he advocates. Indeed he allows that some natural kinds may comprise members that share only a single property, which on the face of it contradicts his 'sociability' view, since the latter surely requires at least a few properties to robustly *cluster* together in order for a natural kind to be constituted.

Thus, in moving beyond HPC theory, it may not be enough to reject the causal-mechanism requirement to satisfy conceptual naturalism; we may need to reject the more basic shared-property-cluster requirement too. In doing so, we seem to abandon any prospect of providing metaphysical criteria for natural kind-hood, even weak and easy-to-satisfy ones, that is, we abandon any prospect of providing necessary and sufficient—or even just necessary—metaphysical conditions for something to count as a natural kind. If that is correct, the higher-order kind picked out by 'natural kind' is not an essentialist-metaphysical kind. If one wanted to pursue the analogy between particular natural kinds and the kind 'natural kind', one could either (a) claim that 'natural kind' names an essentialist-epistemic kind, in the sense that satisfying certain epistemic criteria, or performing certain epistemic roles, is necessary and sufficient for being a natural kind, i.e. counts as so to speak the epistemic essence of the natural kind kind (Dupré 2002, 48); or (b) one could claim that 'natural kind' picks out a higher-order metaphysical-cluster kind, i.e. there is a set of higher-order metaphysical properties a majority of which most natural kind tend to satisfy (such as being defined by an essence), but none of which is necessary to being a natural kind. I will not pursue these particular suggestions further as I am wary of pushing the analogy between particular natural kinds and higher-order kinds such as the kind 'natural kind' too far; these seem to me to be very different things (see Dupré *ibid.*).

[29] And some functional kinds with morphological similarities formed by convergent evolution; Papineau's example is *aerial insectivore* (2022).

In any event, some have responded to the apparent repeated failure of even maximally metaphysically-neutral mixed accounts of natural kinds to satisfy conceptual naturalism by rejecting mixed accounts and recommending purely epistemic accounts of natural kinds, as mentioned earlier in this section. According to those who advocate exclusively epistemic accounts, metaphysical, and mixed, accounts of natural kinds are bound to violate conceptual naturalism, in one way or another, by positing metaphysical conditions for natural kind-hood that some plausible natural kinds in science fail to satisfy. Purely epistemic accounts define natural kinds exclusively in terms of epistemic factors, and say little or nothing about the metaphysics of natural kinds. In particular, there is no attempt to give a *metaphysical explanation* of the epistemic role or value of natural kinds. The accounts have the form:

K is a natural kind if and only if K plays epistemic role(s) E in science, and/or satisfies epistemic constraints C.[30]

The views differ over what they say about E and C. E could be understood broadly, for instance, as 'contributes to the legitimate epistemic goals of scientists in a specified context, whatever they may be'; or more narrowly, as e.g. 'contributes to the goals of, say, induction and/or explanation and/or generalisation'. Lemeire (2018) calls views that specify the epistemic tasks for which natural kinds are valuable 'specific' epistemic views, and views that do not so specify these tasks, 'general' epistemic views. All epistemic views stress the domain- or discipline-relatively of natural kinds (Crane 2021, 12185).[31] Each discipline has its own set of kind categories appropriate to and useful for its specific forms of inquiry and modes of investigation: the kinds of biology are not the kinds of chemistry, which are not the kinds of geology, etc. (Magnus 2012).

Perhaps the most natural 'specific' epistemic view would be one that construed natural kinds as those kinds that facilitate inductive inference,

---

[30] For example, Magnus (2014, 474) defends a view according to which natural kinds are 'the categories which are indispensable to successful science for some domain of inquiry'. Similarly Chang (2016) suggests a concept corresponds to a scientific kind if it is 'sufficiently effective in aiding inquiry in the natural sciences' (Chang, quoted in Crane 2021, 12183). Dupré (2002) defends an epistemic account that is more liberal, in that it allows that some folk (non-scientific) kinds count as natural kinds.

[31] Some mixed views do as well, but epistemic accounts are particularly explicit about it.

as this is the epistemic role that natural kinds are most typically associated with. The problem with this suggestion is, firstly, that, as we saw above when discussing the HPC and BPC accounts, some natural kinds, such as historically-defined biological taxa, may not be well-suited to facilitate induction. Secondly, there is a very close connection between the role of facilitating induction and the *metaphysical* conception of natural kinds as sets of entities that share clusters of properties: typically it is this sharing of properties that is thought to *explain* the aptness of natural kinds for inductive inference. Thus 'epistemic' accounts that focus on induction are in danger of opening the door to precisely the sort of metaphysical accounts they were intended to be an advance on.

So consider, for example, the 'bare projectibilism' account (Häggqvist 2005; Valero 2023), the view that projectability (essentially the usefulness of a kind for the purposes of induction) is necessary and sufficient for a category to count as a natural kind. At first sight this may appear to be a pure epistemic account—all that matters for natural kind-hood is projectability, an epistemic notion—and Valero presents it as a move away from the BPC view. But it's not clear that it really differs fundamentally from the latter. As Valero makes clear in responding to the worry that on the bare projectibilism view, there would be no natural kinds if there were no agents to do the projecting, projectability is underwritten by and explained in terms of the shared co-occurrence of properties (*ibid.*). Valero suggests that the view departs from the BPC view in allowing that certain natural kind-constituting property clusters, such as those that define the kinds of fundamental physics, might have no metaphysical ground at all (*ibid.*)—their clustering may just be a brute fact of nature. But inasmuch as these kinds are still defined in terms of robust clustering of properties, we have here, it seems, a variation of (perhaps a more liberal version of) the BPC view, not an alternative to it (Chakravartty's (2007) version of the BPC view allows for brute property clustering). In other words, in requiring shared property clusters, bare projectibilism is still a mixed view, imposing metaphysical requirements that some scientific kinds seem to violate.

In order for an account to really count as a pure epistemic view, it would need to move beyond this central focus on the aptness of natural kinds for induction. This is what Ereshefsky and Reydon (2015) do.

Their account is motivated by the conceptual-naturalistic principle that an account of natural kinds ought to vindicate all and only the kinds referred to by, and useful for, successful science.[32] They argue that a classificatory program in science picks out natural kinds just in case it satisfies certain entirely epistemically-characterised constraints. 'Unlike HPC theory, there is [in our account] no metaphysical overlay such as the requirement that natural kinds involve causal structure or causal structure of a certain sort' (*ibid.*, 985). Crucially, they also reject 'the limiting assumption that all scientific classifications ought to support inductive practices' (*ibid.*).[33] Natural kinds need not comprise similar entities, and thus need not be useful for inductive inferences (although of course many will be).

Epistemic accounts of natural kinds do not just reject the idea that there are necessary and sufficient—or even just necessary—metaphysical conditions for natural kind-hood. As I intimated above, they also seem to reject the idea that we can and should be aiming to provide *metaphysical explanations* for the suitability of natural kinds for the various epistemic roles they perform in science.[34] This is significant since providing such explanations has been considered the crucial role of philosophical accounts of natural kinds under the metaphysics of science paradigm, an assumption shared by metaphysical and mixed accounts. The primary focus of the chapter on natural kinds in Chakravartty's (2007) book, for instance (in which he defends a realist mixed account), is precisely this. So he sees the main challenge facing cluster accounts of natural kinds as

[32] Thus they criticise HPC theory for being too prescriptive about the metaphysics of natural kinds: 'Who is right? HPC Theorists or the scientists? We side with the scientists' (*ibid.*, 970). This is a clear expression of conceptual naturalism.

[33] Valero (2023) does consider and respond to those (such as Spencer 2015 and Magnus 2012) who argue the bare projectability account (and indeed any account that makes projectability necessary for kindhood) is too restrictive because some plausible natural kinds, such as certain elusive chemical elements and polymorphic biological taxa, do not exhibit a high degree of projectibility. Chakravartty (2023) also responds to those such as Ereshefsky and Reydon who question the induction condition on natural kinds. It is not my aim to settle this dispute here.

[34] I am using 'metaphysical explanation' the way it is used in the philosophy of science and the scientific realism debate; there is a different, more loaded sense of 'metaphysical explanation' that is currently in vogue in contemporary analytic metaphysics, where it has to do with so-called *grounding relations*: issues about what entities 'metaphysical ground' or are 'metaphysically prior to' or 'more fundamental than' other entities (Fine 1994, 2001; Schaffer 2009). This is a clear enough example of non-naturalistic armchair metaphysics.

being how to account for the role of cluster kinds in induction and prediction (something that essentialist accounts have no trouble explaining). For advocates of pure epistemic accounts on the other hand, the furnishing of such metaphysical explanations is both impossible and unnecessary. This seems to put them outside of the MoS stance, and brings them closer to the empiricist stance. I shall return to this point in Sect. 3.3.

To summarise this section. The traditional essentialist account of natural kinds was widely rejected for violating conceptual naturalism. A popular, more liberal successor-account was the HPC view; but it in turn was accused by some of violating conceptual naturalism, and thus still-more liberal accounts were developed, including Khalidi's view (that retains the causal requirement) and Slater's BPC view (that does not). But all these accounts arguably violate conceptual naturalism in their basic agreement that all natural kinds are united by shared clusters of properties. Once this requirement is given up, the prospect of offering *any* sort of metaphysical account of natural kinds appears to be hopeless, and thus purely epistemic accounts—or those that aspired to be—entered the scene. But some of these accounts, in their focus on the role of natural kinds in enabling inductive inference, struggled to free themselves from the 'shared properties' picture. The 'purest' epistemic accounts, such as Ereshefsky and Reydon's, jettison the induction requirement, going 'full epistemic': the *only* thing that unites kinds into the higher-order natural kind kind is epistemic factors. I have been careful to avoid taking a position throughout this discussion.[35] As ever, my interest is in the meta-level questions of the metaphilosophical commitments and epistemic stances underlying and motivating these positions and arguments, something I shall return to below.

For now, two points are worth nothing. We have seen that one trajectory in the recent scientific realism debate has been towards a greater degree of metaphysical commitment, as realists such as Ladyman, Chakravartty and Ellis have claimed that realism needs a firm and explicit metaphysical underpinning to be credible and defensible (i.e. the realist should adopt the MoS stance, perhaps in its strong form, not the SSR

[35] And there are obviously a number of subtleties and complexities I have glossed over in my brief summary of the recent debates.

stance). What is interesting is that, as I have discussed in this section, taking conceptual naturalism seriously, at least in the debate about natural kinds, has led (at least many[36]) theorists ineluctably in the opposite direction, towards successively metaphysically *weaker* positions[37]: strong-realist metaphysical accounts of natural kinds (requiring the strong version of the MoS stance) have been replaced by more moderate and permissive mixed accounts (consistent with the weaker version of the MoS stance), which in turn have given way to pure epistemic accounts (expressions, at least sometimes, of the empiricist stance; see below).

Secondly, it is noticeable that this trajectory of the natural kind debate bears a striking resemblance to the debate concerning the proper characterisation of the physical, and physicalism. Various attempts to *metaphysically* characterise the 'physical' have tended to end up denying the status of 'physical' to scientific entities, forces and processes in good standing, a result that is thought by some an unacceptable violation of conceptual naturalism. An attractive and popular response to these failed metaphysical accounts has been to offer purely epistemic accounts of the physical, e.g. an entity, force, process or fact is physical just in case it features in well-confirmed physical theories or explanations (Stoljar 2010). The physical is defined in terms of the entities of physics, rather than the other way around, in other words.[38] Epistemic accounts of natural kinds similarly suggest that natural kinds should, as it were, be understood in terms of scientific kinds, rather than the other way round. Epistemic accounts of natural kinds tend, however, to be less deferential than epistemic accounts of the physical: as Ereshefsky and Reydon note (2015), we probably don't want to say that just any kinds posited in the natural (or social) sciences are *ipso facto* (real) natural kinds. Nineteenth century science postulated human races as natural kinds; it turned out this was wrong. Then there are the historical cases of postulated kinds/substances that turned out to be non-existent, such as phlogiston, caloric and so on.

[36] Of course not *all* theorists have moved in this direction; the strong realist and mixed accounts still have their able proponents. But the *general* trajectory towards weaker, more permissive accounts, has been unmistakable and striking, I claim.

[37] We saw above that in the debate about species realism, Ereshefsky and Stanford argue that conceptual naturalism leads one in the direction of anti-realism (via pluralism).

[38] Though we saw in the Introduction that such an approach may fall foul of Hempel's dilemma.

We can, that is, while honouring conceptual naturalism, still aspire to give *criteria* that putative kinds in science must satisfy to qualify as natural kinds; whereas giving criteria that entities postulated in physics must satisfy to qualify as 'physical' (aside from 'are postulated in physics') has proved elusive.

Realism and Anti-Realism About Natural Kinds

In the previous section I discussed the question of whether natural kinds should be characterised or defined in metaphysical, epistemic, or 'mixed' terms. This debate has been closely connected to another debate, concerning whether natural kinds are mind-dependent or mind-independent entities, that is, the question of realism vs. anti-realism about natural kinds. According to our taxonomy of epistemic stances, it is those who have adopted the MoS stance (or a stronger one) who would be willing to include natural kinds in their ontology, as these form part of the implicit subject-matter of science.

It might be thought that matters are quite simple: all the theories of natural kinds considered in the previous section are committed to the existence of natural kinds (and thus require (at least) the MoS stance); they just offer different accounts of what natural kinds are. On this view only the extreme nominalist/conventionalist view (that few hold) counts as an anti-realist view of natural kinds, and thus as potentially an expression of the empiricist stance. But that simple answer obscures important questions about realism and ontology that have emerged *within* the debates discussed in the previous section. And we will see that it is questionable whether some within this debate are in fact realists about natural kinds. Thus when we look more closely at these questions, we will recognise the need for a more nuanced interpretation of the ontological commitments of the different views, and thus of the particular epistemic stances at work.

I shall distinguish three types of view with respect to realism about natural kinds: nominalism/conventionalism; mind-dependent accounts; and mind-independent accounts. Under the latter two categories I shall consider again metaphysical, mixed, and epistemic accounts, but this time in relation to realism and anti-realism.

Nominalism/Conventionalism

We saw in Sect. 2 that the nominalist rejects the idea of natural kinds. On this view the world has made individual things; it has not, in addition, made kinds of things (Kornblith 1993). More specifically, on this view there is no difference between natural and non-natural partitions of things or divisions between things. All kind-classifications are conventional and pragmatic. We need no theory of natural kinds. We divide the world up into kinds in ways that are useful for us given our interests and purposes, and the world places few if any objective constraints on the ways in which we do this. Thus this view is often termed 'conventionalism' about kinds. Franklin-Hall (2015) suggests that Hacking and Goodman count as conventionalists in this sense. I interpret Giere (2010, 86–87) as holding a similar view in his scepticism about the very idea of objective, theory-independent natural kinds, and his recommendation that we replace this outworn notion with the idea of model-relative theoretical kinds and their empirical counterparts, which correspond to them in rough and interest-relative ways, i.e. kinds within science that are useful enough for our purposes in certain contexts, and relative to certain perspectives, but that in no way correspond to real, objective, perspective-independent joints in nature (in this he follows LaPorte 2004).[39]

Mind-Dependent Kinds

One may be tempted to say that while the nominalist/conventionalist view holds that there are no natural kinds, the mind-dependent view says that there are natural kinds, but they are mind-dependent.[40] But it is best to avoid expressing the difference in that way at the outset, given that, as we shall see, it may turn out that the mind-dependent view is also a species of anti-realism about natural kinds, that it agrees with nominalism that, in a sense, natural kinds as such are not part of the furniture of the

[39] It is possible that Giere should be understood as endorsing something closer to the epistemic or practice-turn account of kinds I discuss below; the line separating this view from conventionalism can be a little fuzzy.

[40] This is how Franklin-Hall (*ibid.*) expresses it.

objective, mind-independent world. Still, there are surely important differences between these two positions. I shall follow Bird (2018) in expressing the difference in this way: the mind-dependent view, unlike the nominalist view, holds that this *is* a difference between natural and non-natural partitions of things or divisions between things. Not all kind-classifications are conventional. Thus Dupré notes that the mind-dependent view 'is not that there are no real divisions in nature between kinds of things … but that what those division are will depend on what the enquiry is' (2002, 41). So we do need a theory of natural kinds, on this view. A further, related difference is that while, as I noted, according to conventionalism/nominalism, the world places few if any objective constraints on the ways in which we divide it up into kinds, according to some (though not all) mind-dependent accounts, the causal structure of the world does impose significant constraints on the correctness of our kind-classifications; it's just that this does not amount to the claim that the world comes pre-packaged into mind-independent and epistemic-context-independent natural kinds sitting there for science to discover. Thus while conventionalist/nominalist views tend to be motivated by a *general* anti-realism, mind-dependent views often combine anti-realism about natural kinds with realism about other things, such as objects, properties and the causal structure of the world (Chakravartty 2023).

Boyd's Version of HPC Theory

On Boyd's view, natural kinds are characterised in partly metaphysical, and partly epistemic terms (it is a mixed view). The fact that they are partly characterised in metaphysical terms might suggest a mind-independent, strong realist, view. We shall indeed see that some who defend HPC theory take this line. But Boyd is very clear that this is not his view. He explicitly rejects the idea that natural kinds are 'practice-independent', seeing natural kinds as arising out of our epistemic practices (constrained by the causal structure of the world), and not enjoying an existence apart from these. Natural kinds are a product of the interaction between our epistemic practices (especially those of scientists) and the mind-independent causal structure of reality.[41] The idea of natural kind categorisations as constrained by the causal structure of the world is what distinguishes Boyd's view from straight-out conventionalism, which recognises no such constraints. The HPCs exist mind-independently, as part of causal structure of the world; but natural kinds do not.[42]

Boyd suggests that the 'naturalness of natural kinds *consists in* their aptness for induction and explanation' (1999, 147; emphasis added). Now one could, perhaps, hold that there may be natural kinds that are *apt for* induction and explanation in a world with no agents to induct or explain (and thus still hold a realist/mind-independent view of natural kinds); but again, that looks a lot like traditional metaphysical views, since unless we want to say that this aptness is some sort of basic feature of the world, we'll seemingly be compelled to ground it in certain metaphysical properties of natural kinds. And this is just not Boyd's view: natural kinds are *not* mind-independent, they arise from the interaction between the epistemic projects of actual human agents in actual epistemic contexts, and the causal structure of the world. With no actual human agents, there would, for Boyd, be no aptness for induction and explanation, and no natural kinds. If there were natural kinds in a world with no agents (due to them still being apt for induction and explanation), they would be practice-independent natural kinds, which Boyd does not accept.

Boyd's view is sometimes called an anti-realist view (Franklin-Hall *ibid.*) and sometimes a weak realist view (Bird 2018). Which of these is correct? We can put the question in terms of scientific ontology: do natural kinds, on this view, belong in our ontology? Bird suggests that Boyd does not think natural kinds belong in our ontology. He is to that extent an anti-realist about natural kinds. On Boyd's view, natural kinds are not part of the mind- and practice-independent furniture of the world. Below

---

[41] Franklin-Hall also defends an explicitly mind-dependent account, the 'Categorical Bottleneck Account' (see her magisterial 2015), according to which natural kinds 'reflect the categories that both ourselves and a large array of scientific inquirers with epistemic aims and cognitive capacities differing from our own would sanction in common, thereby converging on a single set of categories and kinds from multiple, distinct starting positions or points-of-view' (ibid., 940). This account, she argues, while anti-realist, allows for more objectivity and stance-independence than other mind-dependent accounts such as Boyd's and Magnus'.

[42] Khalidi (2013b) also defends a mixed view that abandons the requirement of mind independence.

I shall consider where this places Boyd in terms of Chakravartty's continuum.

Pure Epistemic Views

We saw above the purely epistemic accounts of natural kinds suggest we understand natural kinds as those kinds that play a role in certain types of epistemic projects, and prescind from offering any general metaphysical characterisation of natural kinds. On this view, natural kinds can vary in an almost unlimited number of ways in terms of their underlying metaphysical character and status. What makes them natural kinds is just their epistemic role(s), in particular in scientific inference, classification, and generalisation. This is a generalisation of the idea that *functional* kinds are multiply realisable (see Brigandt 2009). On the epistemic account, the epistemic roles of aiding in and facilitating explanation, generalisation etc. are multiply realisable by kinds of widely varying metaphysical character. Thus just as the kinds 'money' and 'predator' are multiply-realisable, so too is the higher-level kind 'natural kind'.

Whether epistemic views count as realist or anti-realist views depends on whether the epistemic role of a kind is considered to be constitutive of, or evidence for, its naturalness. This issue tends to be overlooked in the literature, but it is crucial. On the evidential interpretation, a natural kind may *be* an natural kind due to mind-independent, metaphysical facts (these facts constitute its naturalness); but our best (perhaps only) evidence for its naturalness is its epistemic utility (Boucher 2022).

I want to argue firstly, that the constitutive interpretation is the preferable one, and the one that defenders of epistemic accounts ought to favour; and secondly, that on such an interpretation, epistemic views are unquestionably anti-realist views.

The reason for favouring the constitutive interpretation is that on the evidential interpretation, the epistemic view does not obviously differ from mixed, or even metaphysical views. Recall that even these views are happy to emphasise the epistemic role played by natural kinds. But they regard this role as being *explained (or underwritten or grounded) by* the metaphysical nature of natural kinds. On some such views (traditional essentialism, HPC theory) there is a unified metaphysical explanation for all natural kinds; on other views (bare property clusters) there is no such unified story to be had (aside from the bare fact of property clustering itself), but all agree that for each natural kind, there is *some* metaphysical explanation, accessible to us, for its playing the epistemic role it does. If pure epistemic views are to be genuinely distinctive, they must, I suggest reject this assumption.[43]

In short, if projectability, say, were simply evidence for naturalness, naturalness would at the end of the day be a metaphysical affair. On the epistemic view, naturalness is an epistemic affair, end of story. Thus Lemeire (2018) begs the question against epistemic views by complaining that even if usefulness for induction and explanation are good epistemic criteria for naturalness, epistemic accounts can't explain why this is so. The use of 'criteria' suggests the evidential interpretation of the epistemic view, which I've argued against; and the demand for a metaphysical explanation for epistemic usefulness is one the epistemic theorist should simply abjure.

The argument for the claim that on the constitutive interpretation, the epistemic view entails anti-realism about natural kinds, is this. What unites natural kinds *as* natural kinds (i.e. what makes natural kinds into a higher-level kind) is, according to the epistemic view, the role they play for us in our epistemic practices. Metaphysically, they are indefinitely heterogeneous.[44] Even the maximally metaphysically-neutral mixed views turned out to be too restrictive. It seems then that on the epistemic view, there is nothing we can say about what metaphysically unites natural kinds into a kind. As Dupré puts it, 'natural kind' does not name a metaphysical (but at best an epistemic or methodological) kind (2002). Thus, so long as we have reasoners, have agents with epistemic aims of

[43] Chang (2022, 121) draws a similar distinction between operational coherence as *evidence* of the reality of an entity and as *constitutive* of its reality, and suggests that for the operational-coherence criterion of reality that he favours to be new and interesting, it is the constitutive interpretation that we should opt for.

[44] The bare property cluster account holds that the *metaphysical explanations* for the various property clusterings are heterogeneous, but it still holds that we can give a simple, unified metaphysical account of the property clusters themselves (e.g. Chakravartty's account (2007) of the 'sociability' of properties), and to that extent they are metaphysically unified into a higher-level kind; pure epistemic views deny even this.

explanation, generalisation etc., we can demarcate the higher-level kind of all natural kinds. But absent these reasoners, absent these epistemic projects, it seems there would not be a natural kind kind. There would be nothing to unite the kinds into the higher-level kind. Thus, natural kinds are mind-dependent.[45]

More precisely, on this view, although particular 'natural kinds' (water, *Canis familiaris*, electrons) might exist mind-independently (or might not), the *category* associated with 'natural kind' would not exist mind-independently (*it* would not pick out a natural kind). This is the analogue of Ereshefsky's 'eliminative pluralism' about species (discussed above), according to which particular 'species' taxa might exist, but given the heterogeneity of what biologists call 'species', the species *category* does not exist (1992); 'species' does not pick out a natural kind. It would be a mistake however to conclude from this that the view in question is a kind of realism about natural kinds after all: a realism about *particular* natural kinds, combined with anti-realism about the natural kind *category*. Just as, as I argued above, for Ereshefsky, if eliminativism about the species category is correct, then nothing is a species—particular 'species', while perhaps being *real taxa (or real groups)*, are not *species*[46]—similarly on the epistemic account of natural kinds I am considering, particular kinds such as water, electrons etc., may (or may not) be real and mind-independent, but if real they are not *natural kinds*, at least in a mind-independent, epistemic-project-independent way, since on this view nothing is a natural kind in a mind-independent, epistemic-project-independent way.[47]

---

[45] Chakravartty (2023) provides a different argument to the conclusion that views such as those I've been discussing are hard to reconcile with natural kind realism: If an epistemic criterion such as inductive success is our only mark of natural kind-hood, we are forced into an ultra-promiscuous (and non-realist) view on which even paradigm cases of mind-dependent kinds (like 'money') count as natural kinds; yet attempts to avoid this result by, say, limiting *natural* kinds to the kinds posited by science, are ad hoc and unmotivated (since we lack any criterion of naturalness aside from inductive success itself).

[46] He allows that more specific categories—'phylospecies', 'ecospecies' etc.—might exist, i.e. might be natural kinds. Some taxa might be phylospecies, but no taxon is a species (1992).

[47] Ereshefsky (2018) argues that we should reject the metaphysical requirement that natural kinds be mind-independent, and replace it with the epistemic requirement that natural kind classifications be defeasible, i.e. empirically refutable. It's not clear whether he thinks this amounts to abandoning realism about natural kinds, but as I am interpreting mind-dependent views, it does.

A *prima facie* disanalogy here—though not, I think, one that undermines the comparison I am drawing—is that whereas Ereshefsky (1992) is an eliminativist about the species category, those defending epistemic accounts of natural kinds do not typically see themselves as eliminativists about natural kinds or the natural kind concept: we can still make use of the concept *natural kind*, they think, and still talk about natural kinds *qua* natural kinds, as long as we recognise this is an epistemic, not a metaphysical category, and natural kinds are not necessarily mind-independent, but are discipline- and project-relative. And indeed Hazelwood (2021) has attempted to rescue the species category from Ereshefsky's eliminativist arguments by employing precisely the epistemic account of natural kinds defended by Ereshefsky and Reydon (2015). Hazelwood construes the latter view in a more realist way than I have however. If the species category picks out a scientific kind in Ereshefsky and Reydon's sense, and more generally in the sense that fits with the 'practice turn' in the philosophy of science, its natural-kind-reality can, he thinks, be vindicated in a robust way (albeit in a rather pragmatist and pluralist manner befitting the practice turn).[48] I am doubtful about this, for the reasons given above: natural kinds, on the epistemic account, are mind-dependent inasmuch as 'natural kind' does not pick out a mind-independent natural kind.

Some who defend epistemic accounts claim that their views count as realist, since they accept that the correctness of natural kind classifications depends in some sense on the objective nature of the world (Crane 2021, 12184). Thus Magnus (2012) argues that natural kinds are what allow for induction and explanation, and, as Lemeire puts it, it is plausible to suppose that 'a category would not support our inductive and explanatory practices if it were not somehow in accordance with the world' (Lemeire 2018, S2919). Similarly Ereshefsky and Reydon (2023) claim that kinds must be 'grounded in' or depend on *some* aspect of the natural or social world—*which* aspect will depend on the epistemic goals of the scientific practice in question—to qualify as natural kinds.[49] This

[48] Thank you to Caleb Hazelwood for useful discussion on this topic.

[49] These gestures towards the way the world constrains natural kinds categories do not make these views mixed views; to count as a mixed view (like Boyd's) an account must incorporate a significant metaphysical component: a substantive claim concerning what precisely it is about certain features of the world that underlies natural kind classifications.

might suggest these theorists are adopting the evidential view: inductive and explanatory usefulness is *evidence* we've latched onto mind-independent natural kind structure; it doesn't constitute the naturalness of natural kinds. But setting this issue aside, in response to the claim that these accounts are realist ones because they accept that the world constrains natural kind divisions, making some taxonomies fail (for not respecting this mind-independent structure), I would object that this isn't sufficient for realism. Boyd's HPC view also acknowledges the constraints the causal structure of the world imposes on natural kind classifications, but doesn't think natural kinds are mind-independent. And recall again Stanford's views on species: species classifications, he says, must respect objective facts about the world concerning the actual properties of individual organisms (about which we can and should be realists); classifications that don't respect these properties (e.g. the species 'weed') fail as species classifications; they are 'penalised' by the world, as Magnus puts it (2012, 50). But that isn't enough to support realism about species, as Stanford acknowledges. Species can be mind-dependent even if species classifications are constrained by the structure of the world. All that the appeal to constraints imposed by the objective nature of the world in these epistemic accounts of natural kinds shows is that these views are not full-blown conventionalist or nominalist views, something we have already noted.[50]

Mind-Independent Kinds

On this view there are natural kinds, and they are mind-independent. This includes metaphysical accounts, and some mixed accounts.

[50] Lemeire (2018) argues that even some apparently pragmatic or conventionalist categorisations defer to or are constrained by the world in a minimal sense. What natural kind classifications must defer to or be constrained by in order to qualify as realist accounts, he argues, is precisely *the natural kinds* in the world. And that requires that we give a prior metaphysical account of the nature of these kinds, to justify our epistemic criteria.

Mixed Accounts

While Boyd's version of HPC theory is an explicitly mind-dependent account, there is no logical connection between HPC theory as such and mind-dependence. It is coherent to defend the HPC account of natural kinds (as a mixed account) while holding that natural kinds so understood exist entirely mind-independently. This indeed appears to be Kornblith's view (1993). He defends a version of HPC theory that stresses metaphysics and epistemology equally, but it is clear from his discussion that he intends this to be a fully realist account of natural kinds: 'I have argued that real kinds in nature may be identified with homeostatic clusters of unobservable properties. Such an account allows us to locate real kinds … squarely in nature. It is nature which divides the world into kinds …' (*ibid.*, 56). Wilson et al.'s (2007) version of HPC theory, in their extended defence of the account, is also clearly intended to be a realist account. They interpret HPC theory as a natural extension of the new scientific realist paradigm initiated in the 1960s and 1970s, and they say nothing about natural kinds being mind-dependent (see also Williams 2011 for an explicitly realist version of HPC theory, *and* an explicit rejection of Boyd's anti-realism, which we don't typically get with realist versions of HPC theory: 'natural kinds exist prior to—and independently of—our knowledge of them' *ibid.*, 209).

Similarly, the more liberal BPC mixed view comes in both mind-dependent and mind-independent versions. Chakravartty's version in his (2007) is clearly a realist, mind-independent account, despite the fact that he wants to allow for the kind of pluralism and context- and interest-relative identification of natural kinds that people like Dupré and Boyd have emphasised. His point is precisely that the latter does not preclude realism about natural kinds (which for him is just the claim that properties cluster together—are 'sociable'—independently of us). In his more recent paper however (2023), he explicitly rejects natural kind realism, while continuing to endorse realism about the entities that make statements about kinds true—'Particulars and relations of similarity;

co-occurrences, clusters, correlations, and distributions of properties in spacetime; causal relations, mechanisms, and structures' (*ibid.*, 78)—a position similar to Boyd's.[51]

Metaphysical Accounts

Metaphysical accounts of natural kinds (e.g. Lowe 1989, 2006) are emphatically mind-independent, realist accounts. The paradigm example is essentialist views, both classical, and more recent (Bird 2018; Ellis 2001).

Several theorists (e.g. Magnus 2014; Bird and Tobin 2016) distinguish two questions that, they claim, have tended to be conflated in the literature on natural kinds: the question of what makes natural kinds natural (i.e. what distinguishes them from non-natural or conventional kinds) and the question of the metaphysical status of natural kinds, e.g. are natural kinds to be identified with universals, or sets, or are they sui generis entities? I recognise two questions here but I don't see them as two parallel, independent questions. Rather, the latter question about ontological status only arises once we have answered the former, naturalness question, in a certain way. Only if we give a metaphysical, rather than an epistemic, answer to the naturalness question does the question of the metaphysical status of natural kinds need answering. Only the strong realist about natural kinds has call to speculate about the metaphysical status of natural kinds.

An example of such speculation is Hawley and Bird (2011) who argue that natural kinds are *structured universals*. The project of attempting to ground natural kinds in metaphysical entities such as universals is a prime specimen of the realist stance (see also Ellis 2001). In this stance, as we have seen, the explanatory power that appeal to entities such as universals

[51] Slater (2013) also seems to defend a mind-dependent version of BPC theory. Chakravartty's rejection of natural kind realism in his (2023) seems to be based on the assumption that such a realism must hold that kinds are their own ontological category, over and above, and irreducible to other entities. It's not clear why the realist can't be a reductionist—i.e. can't hold that there are mind-independent kinds but what they are is nothing more than e.g. clusters of properties. Indeed this seems to be his view in his (2007). This might mean kinds are not part of 'fundamental ontology', not at the ground level of being, not primitive, unanalysable existents like, perhaps, objects, properties, processes, or causation. But does that make them unreal? Are only things that are metaphysically fundamental real?

is thought to have out-weighs the epistemic risk associated with ontologically committing ourselves to entities the positing of which is far removed from empirical evidence and thus empirical vulnerability.

Structuring our discussion has been the distinction—rough-and-ready no doubt, and, as we have seen, replete with borderline cases, but helpful nonetheless—between the empiricist stance, the SSR stance, the MoS stance, and the armchair metaphysics stance. So far in our discussion of species and biological kinds, the first three stances have dominated. The entities in question—species, higher taxa, natural kinds of various forms, and essences—have all plausibly been part of either the explicit or implicit subject matter of the sciences, with metaphysical inferences from science, in naturalistic mode, being to the fore. With the sort of theorising about the metaphysical status of natural kinds that Hawley and Bird engage in however, we begin to test the boundaries of the MoS stance, and potentially, depending on one's point of view (indeed, of one's favoured stance), risk embracing metaphysics of a non-naturalistic, armchair variety. Recall that as we move from the empiricist stance, to the SSR stance, to the MoS stance and beyond, the metaphysical inferences supporting our claims about what to include in a properly scientific ontology become more stretched, the links with the empirical science more tenuous, and thus the naturalistic credentials[52] of such inferences and posits more questionable. This is a kind of slippery slope that those who insist on metaphysics being naturalistic will likely want to get off at some point, before the link between the science and the metaphysics is deemed to have been stretched beyond breaking point.

Should the naturalist regard Hawley and Bird's proposal to analyse natural kinds in terms of universals as having crossed such a line into armchair metaphysics territory? Consider the question their view is an answer to. In his (2018), Bird, after claiming to have established strong realism about kinds (they are entities, and essentialism is true of them), goes on to ask, *what kind* of entities are they?

'Are natural kinds:

[52] This is the ontological, rather than the conceptual, form of naturalism.

(a) Identical or straightforwardly reducible to entities of a more general nature, for example sets or (non-kind) universals?
(b) Emergent entities, not reducible to other entities, but supervenient on them?
(c) Fundamental, sui generis entities?'

This *further question*, 'what kind of entities?', with potential answers invoking universals, asking about fundamental ontology, etc., does seem a reflection of the strongest form of the MoS stance, and perhaps one that veers into non-naturalistic territory. This is seen from the fact that this question is just not asked by those closer to the empiricist stance or weaker MoS stance. The reason, presumably, is it's hard to see how there could be a naturalistic answer to such questions: are we really to think the science itself could be of any relevance at all in answering them (Ereshefsky and Reydon 2023)?

One way to develop this objection would be to insist that universals, along with tropes, quiddities, haeccities, and the like, are *inherently* non-scientific, non-naturalistic, purely metaphysical entities. (My initial characterisation of the typical ontological commitments of the four stances suggested this view.) Sass writes:

> Say that an entity x is a *metaphysical* entity iff: either x is metaphysically primitive and irreducible, or the correct identity criteria for x reduce x to a type of entity quantified over *only* by a distinctively metaphysical theory... I take complex universals to be a paradigm example of "metaphysical entities." For my purposes, a "distinctively" metaphysical theory fulfills two criteria. (1) It posits at least some entities which are not regularly quantified over in the empirical research of the natural sciences; and (2) if the entities in (1) do any explanatory work at all vis-à-vis empirical hypotheses, such work could instead be done by a more modest ontology less dependent on entities not discussed in the scientific literature. Complex universals seem to feature only in such distinctively metaphysical theories. (2021, 11863)

I shall return to Sass's point about quantification over universals below.

One strategy here is to focus on observational consequences. The debate between nominalists and realists about universals has (as noted in

Chap. 2) always been a paradigm example of a dispute in which the rival positions appear to be strongly empirically equivalent: no possible observation could decide between them, seemingly. They appear to be equally consistent with all possible empirical evidence. Maclaurin and Dyke (2012, 291) define non-naturalistic metaphysics as 'any philosophical theory that makes some ontological claim … which, in principle, has no observable consequences'. The claim that universals exist, arguably, has no observable consequences, thus counts as a piece of non-naturalistic metaphysics.

Similarly, Ladyman and Ross (2007, 10) mention the debate about the existence of universals as a species of the armchair, a priori, scholastic metaphysics they reject. It is, they claim, an example of the type of debate that is carried out independently of and (often) in ignorance of science, and where the participants recognise that science has no, and could have no, relevance to the resolution of the debates. Central to their critique of this type of metaphysics is that it relies almost entirely on metaphysicians' intuitions.

If we examine Bird's (2018) and Hawley and Bird's (2011) claims and arguments, they do indeed appear to be vulnerable to this critique. Their defence of the natural kinds-as-complex-universals view is readily interpretable as a contribution to the traditional set of metaphysical debates surrounding the existence and nature of entities such as universals, quiddities,[53] properties, and so on. And while they do at times take their cue from science (e.g. facts from physics and chemistry about the nature and properties of methane molecules, electrons, and up quarks) at other times their reasoning appears largely or entirely a priori (Ereshefsky and Reydon 2023). Thus Bird advances modal claims about supervenience and possible worlds which are the stock-in-trade of intuition-based armchair metaphysics (e.g. 'Two worlds will not differ in their natural kinds without differing also either in the distribution of particulars and their non-kind properties or in the laws governing them' (ibid., 1416)) as well as making highly abstract logical points about sets, essences, properties and the like.

[53] I suggested that quiddities are paradigm examples of non-naturalistic metaphysical entities.

The views of the nature of natural kinds that Bird (*ibid.*) sees as necessary to engage with include those of Lewis (natural kinds as classes), Armstrong (natural kinds as universals), and Lowe (natural kinds as fundamental): all examples Ladyman and Ross give of non-naturalistic, armchair metaphysicians. In Hawley and Bird (*ibid.*) there are more references to works in traditional armchair metaphysics than there are to works in science or the philosophy of science. Thus Brigandt (2020) classifies Hawley and Bird's essentialist views on natural kinds (along with those of Ellis and Lowe) as part of analytic metaphysics, with its focus on limning the fundamental metaphysical structure of reality, in contrast with the philosophy of science approach, which is more naturalistic, in its insistence that we take seriously the range of (often non-essentialist) kinds on offer in the sciences.[54]

Yet there is another, more Quinean and holist, way in interpreting their project, one that they no doubt would prefer. If we think in abductive terms, it might be that universals-theory helps us to *explain* parts of science, in this case the part that deals in natural kinds. Positing them might not have empirical content or observable consequences as such or in itself, but it can perhaps form part of an overall explanatory ontological account of science that is, as a whole, answerable to the evidence. In this respect positing universals is perhaps no different in principle to positing natural kinds themselves, or essences, or the other implicit entities of science. We explain scientific practice by positing natural kinds, and possibly essences. We subsequently offer a *deeper* explanation by positing universals. Thus the simple route to writing off universals and their kin as non-naturalistic entities, entities that *as such* violate the norm of naturalised metaphysics, recommended by McLaurin and Dyke and Ladyman and Ross, is unpromising, given the availability of the Quinean-holist response.

[54] Ritchie (2010) notes that the methodology for metaphysics Bird favours, for instance in his (2007b), is very similar to the 'cost-benefit' approach characteristic of the sort of armchair metaphysics pursued by thinkers like David Lewis. Price (2009, 340) argues that Lewis' methodology, having to do with balancing criteria such as fruitfulness, simplicity, unity, and utility, cannot supply evidential reasons to believe metaphysical theories are true, as opposed to pragmatic reasons to adopt certain ways of speaking, in the sense of Carnap and external questions.

A different attempt to exclude universals appeals to another idea of Quine's—his criterion of ontological commitment (which we saw can be employed to support species realism). Sass (2021, 11867–9) argues that scientific theories quantify over natural kinds (and property clusters) but not over universals (see the quote from Sass above)—only 'metaphysical theories' quantify over universals—so Quine's science-criterion supports realism about the former but not the latter. But, as I noted in Chap. 2, science does not quantify over natural kinds *under that description* (unlike species, say). Sass may argue: 'science quantifies over species, chemical kinds, etc., and these *are* natural kinds, so to that extent science quantifies over natural kinds, just not under that description'. The problem is the friend of universals can say exactly the same thing about universals. If science quantities over species, if these *are* natural kinds, and if natural kinds *are* universals, then science quantifies over universals, just not under that description. Quine's criterion seemingly cannot distinguish between natural kinds and universals in this respect.

The problems with the strategies I've discussed suggest that the naturalistic critic of universals might do well not to press the objection that universals are by their nature purely metaphysical, non-scientific entities. We should stop looking for some objective criterion that will prove things like universals are beyond the pale in that they cross a clear, sharp, objective line, and rather focus on the *relative directness* of the inferences from the science, or, as we may put it, the relative degree of underdetermination of the ontology by the science. That science quantifies over natural kinds but not under that name is a *less* torturous inference from the science, and requires *fewer* distinctively philosophical concepts and a priori forms of reasoning than the claim that science quantifies over universals but not under that name. That species, chemical kinds etc. that science investigates are natural kinds is a metaphysical inference, and involves us in some philosophy; but the claim that *in addition* to this, natural kinds are universals, is a stronger, *more risky* metaphysical inference, and involves us in *more* distinctively philosophical and a priori reasoning, than the former claim. These are differences of degree, not kind. It is significantly *less* clear how the science could help us decide, even indirectly, whether natural kinds are universals, or universals of a particular type, than it is when it comes to e.g. whether natural kinds exist, and other facts about what they're like. The degree of underdetermination of the ontology by the science, and the consequent need for purely a priori reasoning, appears to be greater.

But that doesn't mean one cannot judge that at some point, these differences of degree as it were add up to a difference in kind: eventually on this continuum one may feel a line has indeed been crossed, the line where one judges that we have strayed too far from the science, our metaphysical inferences have become too adventurous, the epistemic risk too great, the empirical vulnerabilities too weak, for us to any more regard this as naturalised metaphysics, and for us to regard the deep ontological claims as constituting potential items of objective knowledge of the world. To use Chakravartty's metaphor, it is a smooth slippery-slope, but one we may choose to get off at some point.

The point—and this is Chakravartty's key insight—is that where precisely this line is, how 'far' from the science it lies (e.g. does it include or exclude universals? Or natural kinds for that matter?), is not something that can be determined objectively, for everybody, through the application of some clear, unambiguous criterion on which all reasonable people can be expected to agree (Slater 2017, 58–60). Hence, unsurprisingly, Ladyman and Ross's (2007) simple, 'objective' criteria that they proposed for judging whether some claim counts as a piece of naturalistic metaphysics have not been met with universal approval, including among naturalists. Rather, this sense of a boundary being transgressed is a necessarily partly subjective judgement call regarding which reasonable people (even those with naturalistic commitments) can legitimately and rationally disagree. It is a judgement that reflects one's values, i.e. it is stance-relative.

The focus in other words should be on stance-relative judgements of epistemic risk and explanatory power. Those of an empiricist bent will reject the idea that the putative explanatory power of universals outweighs the epistemic risk of positing something so distant from empirical evidence. And even those otherwise sympathetic to the MoS stance, who do value explanatory power alongside mitigating epistemic risk, might want to (as I suggested in Chap. 2) draw a distinction between natural kinds, and perhaps essences, on the one hand and universals (tropes, etc.) on the other: we can't really make sense of scientific practice without positing the former they may claim, our explanatory account of science would be severely impoverished if we (like the empiricists suggest) abjure belief in things like natural kinds and essences; and the epistemic risk here is less severe, given it is plausible that theories of essences and natural kinds enjoy a degree of empirical vulnerability—indeed Bird himself has argued we can have empirical knowledge of natural kind essences (2007a).[55] So the combination of explanatory power and epistemic risk is, perhaps, favourable here. But in the case of universals, it is perhaps implausible that our account of science would be greatly impoverished if we didn't posit universals (Sass *ibid.*, 11872); or at the very least, the feeling that our account *would* be so impoverished, that something important would be left out, without universals, can be attributed to the values inherent in the strongest version of the MoS stance, values that those who adhere to weaker versions reject. And the epistemic risk seems higher; it's less plausible that we can have empirical knowledge of universals. Indeed even if our account of science *would be* metaphysically impoverished unless we posited universals, it might be that this putative benefit is *outweighed* by the epistemic risk associated with positing them, even for those who adopt the MoS stance. All of these judgements are, of course, heavily stance-dependent. If strong realists value explanatory power to a greater degree than minimising epistemic risk, and universals are thought to offer us significant explanatory power in accounting for the role of natural kinds in science, then they will be happy to ontologically commit to them.

3.3 Natural Kinds and Epistemic Stances

To summarise the previous two sections: I have distinguished between metaphysical, mixed and epistemic accounts of natural kinds, and between conventionalist/nominalist/pragmatist (non)accounts, mind-dependent accounts, and mind-independent accounts. These two taxonomies are related as follows. Metaphysical accounts are mind-independent accounts. Epistemic accounts are mind-dependent accounts. Mixed

[55] See also Elder (2004).

accounts are in some versions mind-dependent accounts, and in other versions mind-independent accounts. And conventionalist/nominalist accounts are neither mind-independent not mind-dependent accounts. See Table 4.1.[56]

What of epistemic stances here? We have begun to discuss this in the previous section when we considered Hawley and Bird's view, now let's look at this in more detail. Realism about natural kinds requires (at least) the MoS stance, since the entities in question are part of the implicit subject matter of science (which, again, is not to say that one who has adopted this stance must be a realist about natural kinds). If realism about natural kinds is rejected on stance-based grounds (as being V-prohibited), this may be from the standpoint of the empiricist stance, or the SSR stance (which, recall, while permitting realism about the explicit entities of science, tends to adopt the characteristic empiricist attitude towards the implicit entities).

Which of the views we have considered fit with or are expressions of these stances? I suggest the following is plausible:

- 1. The strong realist, mind-independent, metaphysical account (e.g. Hawley and Bird *ibid.*, Lowe 1989, 2006) requires at least the strongest version of the MoS stance (but in some forms may require the non-naturalistic metaphysical stance).
- 2. Mind-independent mixed views (e.g. Kornblith 1993; Chakravartty 2007) require (at least) the MoS stance (but perhaps only a relatively weak version).

Table 4.1 Taxonomy of views on natural kinds

Mind-independent accounts	Metaphysical accounts Mixed accounts
Mind-dependent accounts	Mixed accounts Epistemic accounts
Neither mind-dependent not mind-independent accounts	Conventionalism/nominalism

[56] See Sass (*ibid.*), for a related but somewhat different taxonomy.

- 3. Mind-dependent mixed views (such as Boyd's (1991, 1999) and Franklin-Hall's (2015)), while not requiring the realist stance in terms of ontology, sometimes straddle the line—or perhaps more accurately, their adherents do—between the empiricist and realist stances in terms of values and attitudes.
- 4. Pure epistemic accounts (e.g. Ereshefsky and Reydon 2015) do not require the realist stance in terms of ontology, and are arguably motivated by, or expressions of, the empiricist stance.
- 5. Conventionalism/nominalism/pragmatism is typically, though not always, an expression of the empiricist stance.

1, 2, and 5 should be relatively uncontroversial. I will now say a word about 3. mind-dependent mixed accounts and 4. pure epistemic accounts, taking the latter first.

We have seen that epistemic accounts proffer mind-dependent, anti-realist accounts of natural kinds. They do not hold that natural kinds exist as part of the objective, mind-independent furniture of the world, that would have been there and possessed their properties, so to speak, whether humans had existed or not, and whether or not humans had had the epistemic aims and purposes they in fact have. Natural kinds are not part of our fundamental ontology of the world. Thus epistemic views do not, ultimately, involve ontological commitment to natural kinds. To that extent we don't here have ontological commitment to the relevant implicit entities of science, as per the realist stance. That is why epistemic views do not require the realist stance, but it doesn't yet tell us whether they are expressions of the empiricist stance.

The reason for interpreting them as potentially expressions of the empiricist stance has to do with their lack of interest in offering metaphysical explanations for the epistemic usefulness of natural kinds, which I mentioned above. Central to the empiricist stance is the rejection of the demand for metaphysical explanation when it comes to interpreting and understanding science. As I interpret pure epistemic accounts, the key idea behind them is the rejection of the suggestion that we should be trying to give a *metaphysical explanation* for the epistemic role natural kinds play in science, in terms of some account of the underlying metaphysical nature of natural kinds (see Ereshefsky 2018 for a clear expression of this view). Note that this rejection is a stronger stance than the claim, made by some mixed views such as the BPC view, that there is no single, unified metaphysical account to be had of natural kinds. On such views, there *are* metaphysical explanations for the epistemic utility of natural kinds, but these will differ from case to case, and form a heterogeneous set. On the pure epistemic view, as I read it, the very idea that we need or should be trying to obtain such metaphysical explanations in accounting for the epistemic value of natural kinds is a mistake. I trust that it is clear that such a view is redolent of the values and attitudes inherent in the empiricist stance. The view may take the form either of the claim that there is no true theory (whether unified or disunified) of the metaphysics of natural kinds (as per instrumentalist versions of empiricism), or that there is, or might be, such a true theory or set of theories, but since we do not have epistemic access to the facts that would make such a theory true, agnosticism is the appropriate epistemic attitude to take to all such theories (as per constructive empiricism[57]). I have tended to construe the view in the former way but the latter version is a possible view. Either way, the upshot is that philosophical theorising about natural kinds need not and should not take the form of the advancing of metaphysical-explanatory hypotheses, that posit metaphysical entities, and metaphysical goings-on, lying behind and accounting for the observable facts concerning natural kind-involving scientific practices.

What remains for philosophers of natural kinds to do once metaphysical explanation and speculation is abjured is perhaps best exemplified by the edited collection Kendig (2016), which explicitly presents itself as leaving behind purely philosophical speculation about the metaphysics of mind-independent natural kinds in order to, in naturalistic fashion, empirically explore the actual practices of 'kinding' to be found in the various sciences—how kinds and kind-classifications are actually *produced* and *used*—in accordance with the 'practice turn' in the philosophy of science (see Kendig's Introduction to the volume). One need not be an empiricist to embrace the practice turn—not all who endorse it reject

[57] Thus Valero (2023) says that his view is 'neutral with regard to specific metaphysical grounding claims', which makes it sound like the constructive empiricism version; similarly Ereshefsky (2018, 855) says his account is 'agnostic about the fundamental nature of natural kinds.'

metaphysics in general—but the suspicion of metaphysical speculation that is involved in at least some instances of it, and the focus on empirically investigating the actual practice of scientists, is consistent with and congenial to the empiricist stance. Thus Ereshefsky (2018, 855), in endorsing the practice turn, notes that it 'endeavors to understand the epistemic success of scientific practices rather than offer a metaphysics of science.'

We saw that taking conceptual naturalism seriously has led to successively weaker accounts of natural kinds (in terms of metaphysical commitment), with the logical end-point being purely epistemic accounts that prescind entirely from the project of metaphysical characterisation or explanation. When this involves not just anti-realism about natural kinds, but characteristically empiricist attitudes and values—i.e. when the claim that we should not or cannot provide a metaphysical account of natural kinds is based not just the view that *as a matter of fact* natural kinds are metaphysically heterogeneous, but at least in part on a dismissive attitude toward metaphysical explanation *as such*—the empiricist stance has entered the scene.

To be perfectly clear, I am not claiming that advocates of pure epistemic accounts would think of themselves as empiricists in general, or even that they *explicitly* adopt the empiricist stance with respect to natural kinds, or that there is no other possible interpretation of the view and its motivations. I am merely claiming that given the way the view is presented and defended, it seems to be consonant with, and is naturally interpreted as being an expression of, the values and attitudes characteristic of the empiricist stance as we have been understanding it.

When it comes to mind-dependent mixed views such as Boyd's version of the HPC account, we find, it seems to me, motivations and claims that combine elements of the empiricist stance and the realist stance, such that they do not belong straightforwardly in either. Like pure epistemic views, such mixed views do not hold that natural kinds exist as part of the objective, mind-independent furniture of the world, that would have been there and possessed their properties whether humans had existed or not, and whether or not humans had had the epistemic aims and purposes they in fact have. This is an anti-realist view of natural kinds, and to *that* extent, does not require the realist stance. On the other hand, on this view the world possesses a mind-independent causal structure that must be respected by natural kind classifications ('… accommodation between conceptual and classificatory practices and causal structures' is 'necessary for successful induction and explanation' Boyd 1999). This structure includes robust property clusters and underlying causal mechanisms that maintain them. Natural kinds arise or emerge as a result of the interaction of this causal structure with human epistemic aims and purposes. If there had been no humans, there would have been HPCs and causal mechanisms but not natural kinds. Thus we account for natural kinds partly by identifying the epistemic aims and purposes that are a necessary condition for their existence, and partly by identifying the objective features of the world 'out there' that provide the metaphysical ground for and constraint on natural kind classifications (even if the natural kinds themselves are not fully 'out there'). This is a *partially* metaphysical explanation for the epistemic utility of natural kinds: metaphysically stronger than pure epistemic views that abjure metaphysics altogether, but weaker than mind-independent strong realist views. And we may also note that, inasmuch as HPC theory is committed to entities other than natural kinds that are part of the implicit subject matter of science, such as properties, property clusters, homeostatic mechanisms, causation[58] etc., it to *that* extent requires (at least) the MoS stance.

Thus, although it is not as neat as one may wish, I see no alternative to classifying such views—or their adherents—as straddling the border, as it were, between the two stances, with a foot in each camp.[59] We can understand such a position as one that 'feels the pull' of values, attitudes and aims characteristically belonging to different stances. In particular,

[58] Magnus (2018) notes that HPC theory is committed to the reality of causal relations, and is to that extent metaphysically committed.

[59] Of course Boyd himself is well-known as, in general, a fervent defender of scientific realism and opponent of empiricism. But that doesn't mean that his account of natural kinds isn't in some ways (perhaps without him realising it) a reflection of certain values, aims and commitments more associated with the empiricist than the realist stance. Recall I am not in any way endorsing the absurd claim that one who adopts a theory of natural kinds that itself, in certain respects, reflects or expresses a particular stance, must thereby adopt or endorse only that stance (and its associated views) with respect to every (or any) other topic. As Strevens (2018) notes, one may be a realist with respect to some issues, and an empiricist with respect to others.

there is an openness to metaphysical explanation, but it is moderate and limited, and tempered by naturalistic and empiricist scruples.[60]

4 Conclusion

We have worked our way through various grades of ontic commitment with respect to species and biological kinds, and considered the epistemic stances underlying this commitment. Realists about explicit entities such as species, higher taxa and monophyletic clades implicitly accept the realist stance (typically either the MoS stance or the more cautious SSR stance). Some anti-realist views about these entities, such as nominalism, tend to be closely associated with empiricism, whereas others, such as the forms of pluralism defended by Ereshefsky and Stanford, are not (at least not obviously). When it comes to natural kinds, we have encountered a spectrum of views from nominalism/conventionalism, to pure epistemic views, to mind-dependent mixed views, to mind-independent mixed views, and finally mind-independent metaphysical views. As we move along this spectrum different stances come into play. The least ontologically-committed views, such nominalism/conventionalism and pure epistemic views, are often associated with and motivated by the empiricist stance; more strongly committed views such as the various mixed views, require the MoS stance if they posit mind-independent kinds (though possibly only in its weak form); while the most metaphysical forms of realism require either the strongest form of the MoS stance or the armchair metaphysics stance, depending on one's perspective.

This logical geography is summarised in Table 4.2 (talk of requirements and compatibilities are to be understood, as always, in terms of V-prohibitions and V-requirements):

Why does all this matter? To return again to the central theme of Part I, this analysis reveals the neglected role in these debates of the various non-belief elements of stances, such as goals, values and attitudes. The

[60] Magnus (2018) similarly argues that views like the HPC account do not fit neatly into either the empiricist or the realist stances. He seems to think this undermines the tenability of the distinction between these stances, but the fact that there are borderline cases does not mean a distinction is not real or important.

Table 4.2 The range of views and underlying epistemic stances with respect to species and natural kinds

Entities	View	Stance
Species	Realism	Requires realist stance (either SSR or MoS)
	Pluralist anti-realism	View is compatible with any stance but arguments for the view may presuppose the realist stance
	Nominalism	Often, but not always, motivated by empiricist stance. If species are concrete particulars nominalism compatible with species realism, and thus with the realist stance
Natural kinds	Mind-independent metaphysical accounts	Require MoS or armchair metaphysics stance
	Mixed accounts	Require MoS stance if mind-independent accounts. Mind-dependent accounts compatible with any stance but sometimes partly motivated by empiricist stance
	Pure epistemic accounts	Compatible with any stance but are arguably motivated by empiricist stance
	Nominalism/ conventionalism	Compatible with any stance but sometimes motivated by empiricist stance

purely epistemic account of natural kinds is, at least sometimes, motivated not (or at least, not solely) by compelling evidence or rational arguments against stronger (mixed, metaphysical and mind-independent) accounts of natural kinds, but in large part by a dismissive and disdainful attitude towards the realist project of explaining scientific practice concerning natural kinds by positing metaphysical-explanatory items underlying the phenomena. Realism about natural kinds is V-prohibited for such thinkers, with the epistemic risk of such a view (given its relatively low degree of empirical vulnerability) outweighing its putative explanatory power, meaning that, strictly speaking, whether there are arguments or evidence supporting particular metaphysical (or mixed) accounts of natural kinds just doesn't arise.

At the other end of the spectrum, strongly realist metaphysical accounts of natural kinds such as that defended by Hawley and Bird are only options for those who have adopted a stance on which heavy-duty metaphysical commitment is V-permissible, given the value it attributes to explanatory power. Once positing entities such as universals is considered fair play, once the associated epistemic risks are considered worth taking to get the potential benefits, working out the details of the strong realist view—what type of universals biological and other natural kinds instantiate and why—may require careful analysis and argumentation, but this only becomes relevant once the stance that permits such ontological commitment has already been accepted. Those who reject the stance (those for whom belief in such entities is V-prohibited) need not engage with those details.

In between these extremes we find more-or-less moderate realist (or partly-realist) views including species and higher-taxa realism that are allowed by the SSR stance with its characteristic V-permissions, and moderate forms of realism about natural kinds that go beyond the SSR stance and require a version of the MoS stance, but can be accepted by those who reject (as V-prohibited) stronger forms of the realist stance on which positing entities such as universals is permitted.[61] Again, assigning the various views to the various stances is not (I trust) an uninteresting exercise in pedantic meta-philosophical classification, but is potentially philosophically important in allowing us to grasp the overlooked but crucial role of values and attitudes—V-prohibitions and V-permissions—in the various debates; a role that is in an important sense prior to any consideration of argument, evidence and epistemic rationality, in working out what we should think when it comes to the metaphysics of species and biological kinds. And this, in turn, allows us to (a) either explain or predict[62] a certain degree of stalemate or impasse in the debates, and (b) reconcile ourselves to an unavoidable metaphilosophical pluralism or voluntarism with respect to the stance-foundations of the various positions on offer.

[61] Magnus (2018) puzzlingly suggests that the epistemic stance approach is only concerned with fundamental metaphysics, so is unable to capture these kinds of less metaphysically committed, more moderate views—what he calls 'middle range ontology'. Chakravartty (2018) rejects this claim; much of this chapter can be seen as a more extended response to Magnus.

[62] Depending on whether such a situation of stalemate or impasse has been independently recognised.

These themes shall be explored further in the next chapter when I look more closely at a topic we have touched on in passing in this chapter, that of essences and essentialism.

References

Atran, S. 1999. *Folkbiology*. MIT Press.

Bird, A. 2007a. A Posteriori Knowledge of Natural Kind Essences: A Defense. *Philosophical Topics*, Vol. 35, No. 1/2, Metaphysics: 293–312.

——— 2007b. *Nature's Metaphysics: Laws and Properties*. Oxford University Press.

——— 2018. The Metaphysics of Natural Kinds. *Synthese*, 195: 1397–1426.

Bird, A. & Tobin, E. 2016. Natural Kinds. *The Stanford Encyclopedia of Philosophy* (Winter 2016 Edition), Edward N. Zalta (ed.): https://plato.stanford.edu/archives/win2016/entries/natural-kinds/.

Boucher, S.C. 2022. Cladism, Monophyly and Natural Kinds. *Croatian Journal of Philosophy* XXII, No. 64: 39–68.

——— MS Conventionalism about species.

Boyd, R. 1991. Realism, Anti-foundationalism, and the Enthusiasm for Natural Kinds. *Philosophical Studies*, 61. 127–148.

——— 1999. Homeostasis, Species, and Higher Taxa. R.A. Wilson (ed.), *Species: New Interdisciplinary Essays*. Cambridge, MA: MIT Press. 141–185.

——— 2010. Homeostasis, Higher Taxa, and Monophyly. *Philosophy of Science*, Vol. 77, No. 5: 686–701.

Brigandt, I. 2003. Species pluralism does not imply species eliminativism. *Philosophy of Science*, 70 (5): 1305–1316.

——— 2009. Natural Kinds in Evolution and Systematics: Metaphysical and Epistemological Considerations. *Acta Biotheoretica* 57 (1–2): 77–97.

——— 2020. How to Philosophically Tackle Kinds without Talking About 'Natural Kinds'. *Canadian Journal of Philosophy* 52 (3): 356–379.

Chakravartty, A. 2007. *A metaphysics for scientific realism: Knowing the unobservable* (Cambridge University Press).

——— 2018. Feelings in Guts and Bones: Reply to Lewis, Magnus, and Strevens. *Metascience* 27(3): 379–87.

——— 2023. Last Chance Saloons for Natural Kind Realism. *American Philosophical Quarterly* 60 (1): 63–81.

Chang, H. 2016. The rising of chemical natural kinds through epistemic iteration. In C. Kendig (ed.). *Natural kinds and classification in scientific practice.* Routledge: 33–46.

——— 2022. *Realism for Realistic People: A New Pragmatist Philosophy of Science.* Cambridge University Press.

Crane, J.K. 2021. Two approaches to natural kinds. *Synthese* 199: 12177–12198.

Crane, J.K. and Sandler, R. 2011. Species concepts and natural goodness. In J.K Campbell, M. O'Rourke, & M.H. Slater (eds.), *Carving Nature at its Joints: Natural Kinds in Metaphysics and Science.* Cambridge, MA: MIT Press. 289–312.

Currie, A.M. 2016. The Mystery of the Triceratops's Mother: How to be a Realist About the Species Category. *Erkenntnis* 81 (4): 795–816.

Daly, C. 2010. *An introduction to philosophical methods.* Broadview Press.

Darwin, C. 1859. *On the Origin of Species by Means of Natural Selection.* London: John Murray.

Devitt, M. 2008. Resurrecting Biological Essentialism. *Philosophy of Science,* 75. 344–382.

——— 2021. Defending intrinsic biological essentialism. *Philosophy of Science,* 88: 1–16.

——— 2023. *Biological Essentialism.* Oxford: Oxford University Press.

Dupré, J. 1981. Natural kinds and Biological Taxa. *The Philosophical Review,* 1. 66–90.

——— 1999. On the impossibility of a monistic account of species, in R.A Wilson (ed.), *Species: New Interdisciplinary Essays,* (MIT).

——— 2002. Is 'Natural Kind' a Natural Kind Term? *The Monist,* vol. 85, No. 1: 29–49.

Elder, C.L. 2004. *Real Natures and Familiar Objects.* MIT Press.

Eldredge, N. and Cracraft, J. 1980. *Phylogenetic Patterns and the Evolutionary Process.* New York: Columbia University Press.

Ellis, B. 2001. *Scientific Essentialism.* Cambridge University Press.

Ereshefsky, M. 1992. Eliminative pluralism, *Philosophy of Science* 59, p. 671–90.

——— 2011. Mystery of mysteries: Darwin and the species problem. *Cladistics* 27: 67–79.

——— 2018. Natural Kinds, Mind Independence, and Defeasibility. *Philosophy of Science,* vol. 85, No. 5: 845–856.

Ereshefsky, M. and T.A.C. Reydon. 2015. Scientific Kinds. *Philosophical Studies,* 172: 969–986.

——— 2023. The grounded functionality account of natural kinds. In William C. Bausman, Janella K. Baxter & Oliver M. Lean (eds.), *From biological practice to scientific metaphysics*. Minneapolis: University of Minnesota Press.

Eronen, M.I. 2019. Robust realism for the life sciences. *Synthese* 196 (6): 2341–2354.

Fine, K. 1994. Essence and Modality. *Philosophical Perspectives* 8: 1–16.

——— 2001. The Question of Realism. *Philosophers' Imprint* 1: 1–30.

Franklin-Hall, L. 2015. Natural Kinds as Categorical Bottlenecks. *Philosophical Studies,* 172 (4): 925–948.

Ghiselin, M.T. 1974. A radical solution to the species problem. *Systematic Zoology* 23, 536–44.

Gould, S. J. 2002. *The Structure of Evolutionary Theory*. Cambridge, MA: Harvard University Press

Giere, R.N. 2010. *Scientific Perspectivism*. University of Chicago Press.

Gould, S. J. and N. Eldredge. 1972. Punctuated Equilibria: an Alternative to Phyletic Gradualism. T.J.M. Schopf (ed.). *Models in Paleobiology*. San Francisco: Freeman, Cooper & Co. 82–115.

Griffiths, P. 1994. Cladistic Classification and Functional Explanation. *Philosophy of Science*, 61: 206–227.

——— 1999. Squaring the Circle: Natural Kinds with Historical Essences. R.A. Wilson (ed.). *Species: New Interdisciplinary Essays*. Cambridge, MA: MIT Press. 209–228.

Haber, M.H. 2016. The individuality thesis (3 ways). *Biology and Philosophy* 31: 913–930.

Häggqvist, S. 2005. Kinds, Projectibility and Explanation. *Croatian Journal of Philosophy*, 13: 71–87.

Hawley, K. and Bird, A. 2011. What are natural kinds? *Philosophical Perspectives*. vol. 25 (1) 205–221.

Hazelwood, C.C. 2021. The species category as a scientific kind. *Synthese* 198 (Suppl. 12), 3027–3040.

Hull, D. 1976. Are species really individuals? *Systematic Zoology* 25, 174–91.

——— 1977. The Ontological Status of Species as Evolutionary Units. R. Butts and J. Hintikka (eds.), *Foundational Problems in the Special Sciences*. Dordrecht, Holland: D. Reidel Publishing Company. 91–102.

——— 1978. A Matter of Individuality. *Philosophy of Science*, 45. 335–360.

——— 1981. Kitts and Kitts and Caplan on Species. *Philosophy of Science*, 48. 141–152.

——— 1988. *Science as a Process: An Evolutionary Account of the Social and Conceptual Development of Science.* Chicago: University of Chicago Press.

Kendig, C. (ed.). 2016. *Natural Kinds and Classification in Scientific Practice.* Routledge

Khalidi, M.A. 2013a. Naturalizing Kinds. In Bana Bashour Hans Muller (ed.), *Contemporary Philosophical Naturalism and Its Implications*. Routledge.

——— 2013b. *Natural Categories and Human Kinds: Classification in the Natural and Social Sciences.* Cambridge University Press.

Kitcher, P. 1984. Species, *Philosophy of Science*, 51.

Kornblith, Hilary (1993), Inductive Inference and Is Natural Ground: An Essay in Naturalistic Epistemology (The MIT Press).

Ladyman, J. and D. Ross, with D. Spurrett and J. Collier. 2007. *Every Thing Must Go: Metaphysics Naturalized.* Oxford: Oxford University Press.

LaPorte, J. 2004. *Natural Kinds and Conceptual Change.* Cambridge: Cambridge University Press.

Lemeire, O. 2018. No purely epistemic theory can account for the naturalness of kinds. *Synthese* 198 (Suppl. 12): 2907–2925.

Lowe, E. J. 1989. *Kinds of Being.* Oxford: Blackwell.

——— 2006. *The Four-Category Ontology: A Metaphysical Foundation for Natural Science.* Oxford: Oxford University Press.

Maclaurin, J. and Dyke, H. 2012. What is Analytic Metaphysics For? *Australasian Journal of Philosophy*, vol. 19 (2): 291–306.

Magnus, P.D. 2011. Drakes, Seadevils and Similarity Fetishism. *Biology and Philosophy* 26: 857–70

——— 2012. *Scientific Enquiry and Natural Kinds: From Planets to Mallards.* Palgrave Macmillan.

——— 2014. NK≠ HPC. *The Philosophical Quarterly*, 64: 471–477.

——— 2018. Cautious realism and middle range ontology. *Metascience* 27: 365–70.

Massimi, M. 2022. *Perspectival Realism.* Oxford University Press.

Martínez, M. 2020. Synergic Kinds. *Synthese*, 197(5): 1931–1946.

Millikan, R.G. 1999. Historical Kinds and the "Special Sciences". *Philosophical Studies,* vol. 95, No. 1/2: 45–65.

Papineau, D. 2022. Swampman, teleosemantics and kind essences. *Synthese* 200, 509.

Price, H. 2009. Metaphysics After Carnap: The Ghost Who Walks? D. Chalmers, D. Manley and R. Wasserman (eds.). *Metametaphysics: New Essays on the Foundations of Ontology.* Oxford: Clarendon Press. 320–346.

Psillos, S. 2011. Choosing the realist framework. *Synthese* 180 (2): 301–316.
Quine, W.V.O. 1953. On what there is. In *From a Logical Point Of View*, (Harvard).
Ritchie, J. 2010. Naturalized Metaphysics. *International Journal of Philosophical Studies* 18 (5): 673–685.
Sass, R. 2021. An ontology of weak entity realism for HPC kinds. *Synthese* 198 (12): 11861–11880.
Schaffer, J. 2009. On what grounds what. D. Chalmers, D. Manley and R. Wasserman (eds.), *Metametaphysics: New Essays on the Foundations of Ontology*. Oxford: Clarendon Press. 347–383.
Slater, M. 2013. *Are species real? An essay on the metaphysics of species*. Palgrave Macmillan.
——— 2015. Natural Kindness. *British Journal for the Philosophy of Science* 66 (2): 375–411.
——— 2017. Naturalized Metaphysics and the Contention over the Ontological Status of Species. Slater, M.H. and Yudell, Z. (eds.). *Metaphysics and the Philosophy of Science: New Essays*. Oxford University Press. 55–80.
Soames, S. 2007. What Are Natural Kinds? *Philosophical Topics*, Vol. 35, No. 1/2: 329–342.
Spencer, Q. 2015. Genuine Kinds and Scientific Reality. In Kendig, C (ed.). *Natural Kinds and Classification in Scientific Practice*. London & New York: Routledge: 157–172.
Stamos, D.N. 2003. *The Species Problem: Biological Species, Ontology, and the Metaphysics of Species*, (Lexington).
Stanford, P.K. 1995. For pluralism and against realism about species. *Philosophy of Science* 62, p. 70–91.
Sterelny, K. and P. Griffiths. 1999. *Sex and Death: An Introduction to Philosophy of Biology*. Chicago: The University of Chicago Press.
Stoljar, D. 2010. *Physicalism*. Routledge.
Strevens, M. 2018. Explanation and reality: comment on Chakravartty. *Metascience* vol. 27: 371–378.
Valero, I. 2023. Bare Projectibilism and Natural Kinds: A Defense. *Croatian Journal of Philosophy*, 23 (2): 155–179.
Williams, N.E. 2011. Arthritis and nature's joints. In J.K Campbell, M. O'Rourke, & M.H. Slater (eds.), *Carving Nature at its Joints: Natural Kinds in Metaphysics and Science*. Cambridge, MA: MIT Press. 199–230.

Wilson, R.A. 1999. Realism, Essence, and Kind: Resuscitating Species Essentialism? R.A. Wilson (ed.). *Species: New Interdisciplinary Essays*. Cambridge, MA: MIT Press. 187–207.

——— 2005. *Genes and the Agents of Life: The Individual in the Fragile Sciences, Biology.* Cambridge University Press.

Wilson, R.A., M.J. Barker and I. Brigandt. 2007. When Traditional Essentialism Fails: Biological Natural Kinds. *Philosophical Topics*, 35 (1 & 2). 189–215.

5

Biological Kinds II: Essences and Essentialism

1 Introduction

In the previous chapter essences and essentialism were touched on briefly and in passing. In this chapter I shall examine this topic directly and in more detail. Along with natural kinds themselves, essences are the other prominent and much-discussed implicit entities in biology that are candidates for inclusion in a properly scientific ontology, deriving from scientific work on species and biological kinds. They are at best part of the implicit subject matter of science: it is plausible that science identifies, describes and theorises about particular essences (possessed by kinds and individuals), but investigation of the nature of essences *in general, as such* and *under that name* is not part of the program of any science; it is philosophers who concern themselves with such matters. To take an example: many philosophers think that being composed of H_2O is the essence of the kind *water*. It was scientific investigation that established that H_2O is the chemical composition of water, so to that extent, science has (empirically) discovered the essence of water. But *that* it is an essence;

This chapter has benefited enormously from the detailed and insightful comments from an anonymous reviewer for the *South African Journal of Philosophy*.

S. C. Boucher, *Empiricism, Naturalism and the Metaphysics of Biology*,
https://doi.org/10.1007/978-3-032-00555-7_5

what it is *in virtue of which* it is an essence *per se*; what it *has in common with* other essences *qua* essences, and so forth—these are matters for philosophy, not science, and are, it would seem, not straightforwardly empirical questions, but are at least partly conceptual and a priori.

Indeed, essences are a prime example of entities that can be theorized about from either a naturalist or non-naturalist standpoint. There is a long tradition of armchair speculation concerning essences, going back, of course, to Plato and Aristotle. Work on essences counts as potentially naturalistic, and thus as an expression of the MoS stance rather than the armchair metaphysics stance, to the extent that metaphysical inferences from science are to the fore, these inferences are relatively direct, they are based on a firm grasp of the relevant science(s), they indicate ways in which claims about the existence and nature of essences may be responsive to empirical evidence,[1] and a priori reasoning and speculation is minimal when it comes to defending the views, even given that it is likely that all claims about essences are underdetermined by scientific evidence and practice. The essentialist views that we shall consider in this chapter at least aspire to satisfy these naturalist demands. As ever, whether they are in fact sufficiently naturalistic is a matter of dispute, with some holding that positing essences of any kind is an unnecessary metaphysical add-on that can only disfigure a properly scientific ontology, and does nothing to enhance our understanding of the metaphysics of biology in particular.

As with natural kinds themselves, the empiricist stance does not license metaphysical inferences to the existence of essences (in biology or anywhere else), while the MoS stance does. Of course, if one has adopted the MoS stance, one is not thereby *compelled* to believe in essences, in biology or anywhere else; the stance rather *permits* one to include essences in one's ontology, if one believes the inferences from the relevant science to the existence of such entities are rationally compelling.

In this chapter I shall focus on species essences. I shall consider several prominent views on the ontology of species[2] (some of them related to or versions of views of natural kinds we encountered in the previous chapter),

---

[1] For naturalistic defences of the claim that we can gain a posteriori knowledge of natural kind essences see Bird (2007) and Elder (2004).

[2] Thus I pick up themes from the section in the previous chapter on SAI; that section did not focus directly on essences as such.

and address the question whether they are ontologically committed to species essences (i.e. whether on such views the relevant science supports the metaphysical inference to the claim that species essences exist, or, to put it another way, whether they think that we need to posit species essence to explain the relevant science) and thus whether they require (at least) the metaphysics of science (MoS) stance. This is distinct from the question, discussed in the previous chapter, whether certain views are committed to the existence of natural kinds. While discussion of natural kinds and of essences often goes together, they are conceptually distinct questions; one may, without making an obvious conceptual (as opposed to theoretical) error, be committed to natural kinds but not essences, in biology or in general (i.e. one may defend a non-essentialist account of at least some natural kinds), and one may be committed to essences but not natural kinds, in biology or in general (perhaps the entities possessing the essences are individuals rather than kinds). In fact, I shall argue in this chapter that the issue of essences and essentialism in biology cuts across the issue of natural kinds, and so can be discussed independently.

2 Two Approaches to Classifying Views as Realist or Anti-realist

In addressing these questions I shall take a somewhat different approach to the previous chapter on natural kinds. In that chapter the focus, when determining the ontological commitments of various views, was on whether they posited *mind-independent* natural kinds, whatever the precise account of natural kinds that was being offered. I did not myself defend a particular account of natural kinds, and use it determine whether certain views were committed to the existence of natural kinds *so understood.* In this chapter however, I will defend a particular account of essences, and use it to determine whether certain views on the ontology of species count as essentialist views, requiring at least the MoS stance. In Sect. 4 I shall respond to the worry that this leaves me open to the same charge of violating conceptual naturalism that has, as we've seen, been levelled at SAIers and others who have used a traditional philosophical conception of natural kinds as the basis for rejecting the existence of natural kinds in parts of biology.

There are in general two approaches one may take to classifying views (and theorists) as realists or anti-realists. The first is to more or less take the views and what their defenders say about them on face value with respect to their ontological commitments. That is, whatever account of Xs they accept, if they *claim* to be realists about Xs, then, (as long as their account of Xs satisfies some minimal criteria for being a legitimate account of Xs), we classify them as realists about Xs. This was, for the most part, my approach in the previous chapter. If a theorist claimed that their account of natural kinds was a mind-independent realist account, I classified the view as a realist view. The views were all legitimate attempts to capture the nature of natural kinds. Of course, if someone claimed that on their view all and only natural kinds were tomato sauce bottles, this would not count as a prima facie legitimate or reasonable view about natural kinds, so even if they claimed to be a realist about natural kinds as they understand them, we would not classify them as realists about natural kinds. But none of the views we considered were like this.[3]

The other possible approach is to begin by defending a quite specific account of Xs, and using that account to classify views and theorists as realists or anti-realists. On this view we *don't* take at face value the claims of proponents of particular views to be realists or anti-realists; we are prepared to over-rule these claims.[4] Someone might *think* they are a realist about Xs, but if their view implies that Xs *as we are understanding them* do not exist, we would classify their view as anti-realist about Xs. This approach might sometimes make sense for participants in the debates themselves. If A understands Xs as Ys, and B understands Xs as Zs, with B being a realist about Zs but an anti-realist about Ys, then A should perhaps classify B as an anti-realist about Xs, despite the fact that B claims to be a realist about Xs (as she understands them, i.e as Zs). For instance,

[3] To a certain extent I was willing to overrule what theorists thought: some proponents of the pure epistemic account of natural kinds regard their view as a realist view about natural kinds, but I argued the view is a mind-dependent and thus an anti-realist view.

[4] Much as Devitt (1984) argues that Rorty is a realist, despite appearances and what Rorty himself would claim, and Musgrave (1989) argues that Fine is a realist, despite appearances and what Fine himself would claim.

a libertarian might hold that a free action is by definition an uncaused action. A compatibilist holds free actions can be caused; she is a realist about such caused actions but an anti-realist about uncaused actions. The libertarian may then coherently claim that the compatibilist is an anti-realist about free actions, despite the fact that she (the compatibilist) claims to be a realist about them (as *she* understands them, i.e. as caused actions of a certain type). In the debate about natural kinds, Ellis (2001), for instance, is naturally interpreted as holding that anyone who does not think there are natural kinds in biology in his strong essentialist sense (e.g. HPCers) is not a realist about biological natural kinds, despite the fact they may think they are. This approach has implications for the role of epistemic stances, as we shall see later in the chapter.

It is hard to say in general which of these approaches is the correct one; each seems legitimate in different contexts, and for different debates. The fact that they give different verdicts on the question of which views count as realist views, and yet that they each seem legitimate, may suggest that the question of whether a view counts as a realist view about Xs has no objective, or at least no theory-neutral answer. Are compatibilists realists about free will? It depends what you think free will is, or can be. Are HPCers realists about natural kinds? It depends what you think natural kinds are, or can be. It would take us too far afield to pursue this difficult question here.[5] It suffices that we recognise that there are these two different approaches, and that it is hard to deny that each can be legitimately pursued, though not at the same time. In the natural kinds chapter, and in the following chapter on teleology, I am (for the most part) taking the first approach: classifying views as realist or anti-realist by (mostly) taking what their proponents say on face-value. In the current chapter on essences, I am taking the second approach: defending a particular account of essences, and using it to classify views as essentialist or non-essentialist, whatever their proponents think about this. Why the different approaches? It seems to me that it is possible and desirable to achieve consensus on the nature of essences. I make that case in this chapter. The nature of natural kinds, and teleology, on the other hand, are much more controversial and difficult topics, such that a more tolerant or permissive attitude with

[5] I pursue it in my (2021).

respect to classifying views as realist or anti-realist seems to be warranted. The view on natural kinds I'm attributing to Ellis, for instance—that anyone who doesn't think there are natural kinds in biology in *his* sense is not a realist about biological natural kinds—while coherent and defensible, strikes me as unappealingly rigid and dogmatic. We have had to adapt and develop our understanding of natural kinds (and teleology) in recent decades in response to a greater understanding of the role they play in science in particular. Yet essences are what they are, and what they have always been, I claim. But these judgments are, like so many others I discuss in this book, inconclusive, value-laden, and open to dispute.

Moreover, while it is true I am offering an account of essences on the basis of which to classify views as essentialist or non-essentialist, we will see that it is deliberately quite a minimal account. I do not wish to build into my *definition* of essences/essentialism claims with respect to which it seems to me it is permissible for essentialists to disagree, for example the claim that kind essences cause or help to explain observable properties characteristic of the kind, and the claim that essences must be intrinsic properties. One could reject these claims and still be an essentialist, I shall argue. Thus they do not belong in the definition of essences/essentialism. This liberal and permissive approach to defining essences/essentialism brings my approach in this chapter closer to the permissive approach I take in Chaps. 4 and 6 to natural kinds and teleology respectively.

3 What Is Essentialism?

Ereshefsky (2009) suggests that traditional essentialism about kinds has three components:

1. The essences are possessed by all and only members of the kind in question.[6]

---

[6] More precisely, if there is more than one essential property, we say that each property is possessed by all members of the kind, and all of the properties, collectively, are possessed by only members of the kind.

2. The essential properties are properties that *must* be possessed by all and only members of the kind.[7]
3. Possession of the essence causes/helps explain other properties associated with that kind.

This is meant to be a definition. As with essential properties themselves, these conditions are supposed to be individually necessary and jointly sufficient for a view to count as essentialist. That is, for a view to count as essentialist, the essences it posits must satisfy each condition, and positing properties or sets of properties satisfying of all three conditions suffices for a view to count as essentialist.

The second condition implies that the 'necessary and sufficient conditions' required by essentialism be understood in the strong modal, not the weak extensional sense. On the latter, 'Possession of A is a necessary condition for belonging to B' means nothing stronger than that all Bs have A. On the former, it implies that all Bs *must* have A. That is why, strictly speaking, talking just of necessary and sufficient conditions is inadequate here, as such talk is ambiguous between the above two senses. The causal condition and the explanatory condition are sometimes distinguished, and can be critiqued separately (see e.g. Nanay 2011). But they are closely connected—if the explanatory condition holds it is presumably in virtue of the fact that the causal condition holds—and I will continue to run them together.

How plausible is Ereshefsky's definition? (1) should be uncontroversial (although we shall see that it is denied by some authors). (2) gives us the *de re* necessity essences are thought to possess on standard contemporary (post-Kripkean) accounts. And it is needed to rule out including in the essence accidental properties that just happen, as a matter of contingent fact, to be possessed by all and only members of the kind. Suppose there is a geographical region which is occupied by all and only members of that species (the 'only' condition here is obviously unlikely, but accept it for sake of argument). Still, living in that geographic region presumably wouldn't count as part of the species essence, as it is probably an

[7] Strictly speaking, of course, (1) is redundant since it is implied by (2), but it will be helpful to retain the three conditions.

accidental, rather than a necessary, fact that it occupies that region. There is no necessity involved in their occupying that region. Relatedly, we want essential properties to be counterfactual-supporting, which (1) on its own does not accomplish.

(2) is not always included in the definition of essentialism. Ereshefsky elsewhere (2010b; see also Nanay 2011) does not include it. It might be thought redundant, given (3)—properties accidentally possessed by all and only members of a kind are unlikely to cause/explain observable properties associated with the kind. So given (3), (2) perhaps follows automatically. (Ereshefsky's insistence, which I discuss below, that (3) is needed to distinguish real from nominal essences suggests that this is how he is thinking about it.) But I will be arguing for the removal of (3) from the definition, so I require (2), to distinguish essences from accidental properties.

What about (3)? If essences exist do they cause/explain the traits that are associated with the kind? One worry might be that real life examples such as fool's gold, jadeite/nephrite, and animal mimicry, and imaginary examples such as the watery substance XYZ, show that the existence of an essence is not a necessary condition for the occurrence of some or all of the observable properties typically associated with the kind defined by that essence. But it would clearly be too strong a requirement on the causal/explanatory condition to expect that the relevant observable properties only be associated with the kind in question, and, in general, it may well be that X is able to causally explain Y without X being a necessary condition for Y. Possibly more of a worry is that as well as failing to be necessary for the occurrence of the relevant observable properties, the presence of the essence would not appear to be sufficient either, at least in the case of biological kinds.[8] On most versions of biological essentialism, for any observable trait T that we tend to associate with some species, there are real or imaginary cases of mutant members of the species who possess the essence that defines the species (whatever that happens to be) but fail to possess T. This may be more of a concern than the failure of

[8] Kornblith (1993) suggests that in the case of chemical kinds the existence of the essence is sufficient, in a fairly strong sense, for the existence of the observable properties we associate with the kind. I will not attempt to evaluate that claim here.

necessity, but, again, the requirement that the essence be universally accompanied by the relevant observable properties appears to be too stringent a constraint on the causal/explanatory claim.

There is another problem however. Clearly not *all* the observable properties typically associated with a kind are potentially explained by the kind's essence, only those which are somehow closely linked with the essence. Gold's malleability may be explained by its having atomic number 79, but it's being valuable to humans is explained by certain human customs and attitudes; water's being colourless may be explained by its having the chemical composition of H_2O, but the fact that it fills Earth's lakes and rivers would presumably be explained by certain facts about Earth's history, atmosphere and climate, not by the chemical composition of water. The question is how to delimit which properties are caused/explained by the essence in a principled, non-ad hoc way, i.e. in some way other than by saying that they are the properties that are closely linked to and explained by the essence. It is certainly not obvious how this could be done, which perhaps explains why no such specification is included in definitions like Ereshefsky's. Kornblith (1993), for instance, following Boyd (1991, 1999), suggests that the observable properties that are caused and explained by the essence of a kind are those that are an inevitable byproduct of the homeostatic mechanisms that underlie the kind (*ibid.*, 37; I say more about homeostatic mechanisms below). But since for Kornblith these homeostatic relationships are just what he means by 'essence', this doesn't solve the problem.

The inclusion of (3) in the definition of essentialism has been questioned by Okasha (2002, 204), who argues that, with respect to species, the explanation for an organism's morphological traits will not advert to any of the relational properties that, on the 'relational essence' view of species (that I discuss below), may be designated as the essence of the species to which it belongs, but rather to 'its genotype and its developmental environment'. Nanay (2011) has also argued that putative essences don't cause/explain the observable properties associated with a kind. But unlike Okasha he sees this as an argument against essentialism about kinds, because he accepts Ereshefsky's definition of essentialism. In effect he argues: (1) If kinds have essences they cause/explain the relevant observable properties of the kind's members; (2) but putative essences don't cause/explain the relevant observable properties; (3) so kinds don't have essences. In other words, nothing can (or nothing does) play the role essences are meant to play. Okasha accepts (2) for biological kinds, but blocks the argument by rejecting (1).

Ereshefsky says, in defense of the inclusion of the causal/explanatory condition:

> Essentialists, from Aristotle to Locke and from Kripke to Boyd, believe that essences figure centrally in explaining the traits typically found among the members of a kind. If we give up that explanatory component of essentialism, we give up a core feature of essentialism, a feature that distinguishes real essences from nominal essences. Nominal essences demarcate membership in a kind, but they do not explain the typical traits of a kind. Arguably, any approach to natural kinds that aims to capture the kinds of science should preserve this explanatory feature of kinds. (2010a, 683)

While Ereshefsky is right that the idea is a major part of the essentialist tradition, in what follows I will assume that the claim that essences (always or most of the time or some of the time) cause/explain the relevant observable properties is a *substantive* claim about essences, not part of the *definition* of 'essence'. Those such as Okasha who deny that (at least some) essences cause/explain the relevant observable properties are not, I claim, misusing the concept of 'essence'. They may be advancing a falsehood, but it is not an analytic falsehood. Even if one holds that essences in fact satisfy Ereshefsky's criterion (3)—that is, if one doesn't wish to 'give up [the] explanatory component of essentialism'—one should accept that one may deny that essences satisfy (3) and *still be classified as an essentialist*. So as far as a *definition* of essentialism goes we are left with Ereshefsky's first two criteria—positing essences that satisfy these two criteria is necessary and sufficient for a view to count as essentialist.

There is however a further condition that is sometimes added to the definition of essences and essentialism—the requirement that essential properties be intrinsic (Ellis 2001). I will call the construal of essentialism on which essential properties must be intrinsic the 'strict' construal, and the construal on which they need not be the 'non-strict' construal. As we have seen, if there is one lesson that has been learned from discussion about species and natural kinds in biology over the last few decades it is that the strict account of essences and essentialism does not fit the biological world. Biological kinds, most agree, do not have essences in the strict sense. But (closely mirroring the situation with natural kinds discussed in the previous chapter) there are two conclusions one may draw from this—one may conclude that, since the strict account is the correct account of essentialism, essentialism is false as an account of the biological world (this was the conclusion of the defenders of the species-as-individuals thesis such as Hull 1976 and Ghiselin 1974); or one may conclude that the strict account is the wrong account of essentialism, and that an appropriately modified version of essentialism may still be applicable to the biological world. The latter has increasingly become the orthodox view, and I will adopt it here. Essential properties can be either intrinsic or extrinsic.[9] More accurately, it is not necessary for a view to count as essentialist that all the essential properties it posits be intrinsic properties. So, at this point I am strictly-speaking not taking a stand on whether any essential properties are in fact extrinsic; my concern is with the definition of 'essentialism' and 'essences'—I argue that a view that posits extrinsic essences can still count as essentialist. In other words, some, most or all essences may *in fact* be intrinsic, but they are not so *by definition*. One who posits extrinsic essences may be mistaken, but if so they are mistaken on a question of substantive fact, they are not asserting an analytic or conceptual falsehood.

Although, as I noted, the non-strict construal of essentialism has become popular in the philosophy of biology, elsewhere in philosophy it is the strict construal that is still dominant, due in large part to the arguments and examples in Kripke's and Putnam's classic work on essentialism with respect to kinds and its focus on intrinsic microstructural properties. But this did not perhaps need to be the case. For instance, the widely accepted 'necessity of origins thesis', defended by Kripke, states, among other things, that a person's having the parents they have is an essential property of them. But this is a relational/historical property. So relational properties can, according to this tradition, be essential

[9] See Magnus (2012, 32–37) for a discussion and critique of the intrinsic-feature requirement on essences and natural kinds.

properties of *individuals*. Why could they not be essential properties of kinds as well? I.e. to be the individual you are you must possess certain relational properties; and to belong to the kind you belong to you must possess certain relational properties. Why should relational essences be allowed for individuals but not for kinds? Robertson and Atkins (2013), in what I take to be a fairly uncontroversial definition, define an essential property as follows: 'an essential property of an object is a property that it must have while an accidental property of an object is one that it happens to have but that it could lack.' This doesn't say the properties have to be intrinsic. We can rewrite it to define an essential property of a kind: 'an essential property of a kind is a property that members of the kind must have in order to be members of that kind'. There seems no reason for stipulating that such properties must *by definition* be intrinsic. As with the causal/explanatory requirement, even if we exclude the 'intrinsic' requirement from the *definition* of 'essence', it remains open for us to claim that as a matter of substantive fact, all essences, or perhaps more plausibly all essences *within a certain domain*, are intrinsic. It may be that all essences in physics and chemistry (but perhaps not in biology) are intrinsic, for instance. It is consistent with the purely definitional claim I have presented that all essences are, as a matter of substantive fact, intrinsic, though in fact I think this is probably not the case.

In next two sections I'll discuss two views on the ontology of species, and show that the way they've been construed by some theorists is precisely the opposite of the way they ought to be construed given the above definition of essentialism—the first view, the HPC view, counts as non-essentialist on the definition, yet has been construed as essentialist, and the latter, the SAI view, counts as essentialist on the definition, yet has (with some notable exceptions) been construed as paradigmatically anti-essentialist.

4 Essences and Views on the Ontology of Species

4.1 HPC Theory and Essentialism

Recall the homeostatic property cluster account (HPC) discussed in the previous chapter. We have considered different versions of HPC theory—some mind-dependent versions, some mind-independent—in relation to their account of natural kinds in general. In this chapter we shall consider their implications for species essences. It is mind-independent versions that shall be our focus; presumably if HPC natural kinds are mind-dependent, the essences that define them are also mind-dependent, and we do not have genuine ontological commitment to essences, even if HPC theory qualifies as essentialist. But if species qua HPC kinds are mind-independent, we have the possibility of commitment to mind-independent species essences.

The HPC view in all its versions explicitly denies that there are characteristics, either intrinsic or extrinsic, that are (or must be) possessed by every member of a natural kind, such as a species. The denial of intrinsic necessary conditions is today largely uncontroversial; it's the denial of *extrinsic* necessary conditions that is controversial in the contemporary context. The HPC view denies that, for example, belonging to a particular chunk of the genealogical nexus is a necessary condition for belonging to a particular species. So it clearly counts as non-essentialist on my definition. And indeed Ereshefsky (2009) had previously so characterized it.[10] Yet it is regarded as an essentialist view by some prominent defenders of the view (e.g. Boyd 1999; Griffiths 1999;[11] Kornblith 1993) and some opponents (Ereshefsky 2010a).

If the HPC view is construed as essentialist, clearly my definition of what it is for a view to be essentialist is being rejected—in particular, it's not thought to be a necessary condition of a view counting as essentialist

[10] Presumably because it violates his own definition of essentialism. Hence it is puzzling that in his (2010a) he seems happy to treat it as an essentialist view.

[11] Griffiths (1999) defends both the historical essence view and the HPC view, so he must regard the HPC view as consistent with essentialism.

that it posit necessary conditions for kind membership, that is, properties that are, indeed must be, possessed by every member of the kind.[12]

So what is the alternative definition of essentialism on which the HPC view is regarded as an essentialist view? Ereshefsky (2010a, 675) writes: 'Proponents of HPC theory see it as a form of essentialism because they believe that HPC kinds perform the inductive and explanatory roles of traditional essentialist kinds.' '...the aim of HPC theory is to capture groups of entities that share similarities that are projectable and sustain successful induction.' So is this condition—essences as sustaining our inductive and explanatory practices—being thought of as necessary and sufficient for a view to count as essentialist? In other words, one is an essentialist if and only if one posits entities that sustain certain of our inductive and explanatory practices? Although he doesn't say so explicitly, Kornblith[13] (*ibid.*) appears to regard HPC as an essentialist view because (1) the 'essences' it posits define real, as opposed to nominal kinds, in the sense of Locke; (2) the essences are unobservable, deep structural properties (either intrinsic or extrinsic[14]) that cause and help to explain superficial properties associated with the kind, and (3) the essences are useful for induction; for Kornblith it is the reality of natural kinds and their essences—the natural kind structure of the world—that *explains* the success and reliability of our inductive practices, and that is the focus of his book, which is after all primarily a work in epistemology. For Kornblith, then, these conditions are individually necessary and jointly sufficient for a view to count as essentialist. (1) is clearly a necessary condition of essences and essentialism, though is usually so obvious as to not need mentioning. I've argued (2) is not necessary. The necessity of (3) may be debatable. But in any case the conditions aren't jointly sufficient, on my view, as we are lacking Ereshefsky's (1) and (2).

Incidentally, Kornblith's view is an odd one, for he does not explicitly reject the traditional account of essentialism, despite his treating the HPC view as an essentialist view. His model of essentialism is the physics/

[12] Rieppel (2010, 663) notes that to classify the HPC view as essentialist 'requires a "refurbished" version of essentialism.'

[13] Other defenders of the HPC view have said similar things.

[14] Thus this qualifies as a version of the non-strict account of essentialism.

chemistry model—the model that works so well for physical and chemical kinds (e.g. the essence of water is H_2O). Yet that model is best understood as based on the strict construal of essentialism (essences as necessary and sufficient *intrinsic* properties), which the HPC account violates. Even if we connect the physics/chemistry model with the non-strict construal I have defended, HPC theory still violates the latter, in its rejection of necessary conditions for kind membership. Kornblith seems to suppose that the physics/chemistry model is consistent with the HPC account, so he must hold that the physics/chemistry model does not require necessary conditions for kind membership, which conflicts with the standard interpretation of it.

But to return to the main argument, on reflection it is quite surprising that some philosophers of biology are happy to call views such as the HPC account that dispense with necessary conditions 'essentialist'—that is, have accepted Kornblith's account of essentialism (or something very similar)—as this clashes with the way 'essentialism' is used everywhere else in philosophy, for example in the extensive debates about the general issue of essentialism in metaphysics and the philosophy of language. When Kripke and Putnam famously revived metaphysical essentialism, it was universally understood that what they were reviving was the view that kinds possess essences in the sense of properties that *must be possessed* for something to count as a member of that kind, just as individuals possess essences in the sense of properties they must possess to be that individual. (It was of course the 'must', with its implications of *de re* necessity discoverable a posteriori, that was revolutionary in the context of the analytic philosophy of the time.)

Now it will be said that Kripke and Putnam's views were ignorant of biology, and didn't apply to the biological world. It is true that they mistakenly believed (a) that the strict construal of essentialism was the correct construal, and (b) that essentialism so construed applied to the biological world. But (a) was incorrect because the strict construal mistakenly requires essences to be intrinsic, *not* because it requires essences to be necessary and sufficient conditions. Kripke and Putnam were, I maintain, wrong in thinking essential properties must by definition be intrinsic (and that they must by definition cause/explain observable properties); but they were right in thinking they must by definition be necessary and sufficient properties. The necessary and sufficient property condition is, I claim, definitional to essentialism in a way that neither the intrinsic property condition nor the causal/explanatory condition is.

We saw in the natural kind chapter that SAIers and others have been accused of violating the conceptual version of naturalism when they have appealed to the traditional philosophical conception of natural kinds in order to claim that there are no natural kinds in parts of biology. It may appear that I am doing something similar here in appealing to a certain philosophical account of essences to support my claim that HPC 'essences' are not true essences, and thus that *if* HPC 'essences' are the only essences to be had in biology,[15] there are in fact no essences in biology. As with natural kinds, would it not be more in keeping with naturalism to begin with the science, and craft an account of essences that enables us to vindicate the intuitive notion that there are essences in biology, even if they are not essences in the necessary-and-sufficient-condition sense recognised by the philosophical tradition? That account would presumably mirror the epistemic account of natural kinds (as per the Ereshefsky quote above): essences are understood as entities that play a certain epistemic role in science, where this conception is neutral on the metaphysical nature of such essences (e.g. whether they must be possessed by all and only members of the relevant kind).

I am happy to accept the naturalistic principle being invoked here (although it may be less of a flagrant violation of conceptual naturalism to defend an account of essences that has the implication that there are no essences in biology—or, to put it another way, that biological kinds (such as species) are not essentialist-kinds—than it is to defend an account of natural kinds that has the implication that many putative scientific kinds are not natural kinds). But we can honour it without going so far as to defend a purely epistemic account of essences. We should, I've suggested, relax our account of essences to the extent of allowing for extrinsic properties to count as essences (partly in response to developments in biological science, as conceptual naturalism requires), but not to the extent of allowing that essences may consist of non-necessary properties. The account I defend is thus intermediate (in terms of liberality) between

[15] If the historical essentialist view is true, this antecedent is false.

the strict traditional construal of essences, and the purely epistemic account. One argument for stopping short of the purely epistemic account is that it collapses the distinction between essences and natural kinds. On the epistemic account, essences are entities that sustain or underlie our inductive and explanatory practices. But this, of course, is precisely what the epistemic account of natural kinds says natural kinds are. It's not clear we need the concept of essence in addition to the concept of natural kind, if the two concepts are playing essentially the same role (and the two types of entity are playing the same role for science). In other words, what is gained, given the epistemic view, by saying that HPC kinds are not just natural kinds, but are essentialist kinds? It seems to be an unnecessary and redundant add-on (which is perhaps why even those who think HPC kinds are essentialist-kinds don't seem to think it matters very much, and is maybe just a verbal issue). To keep any content and purpose to the concept of essences, I claim, we need to retain the necessary property condition. Then perhaps we can have informative things to say about essences in science, e.g. that some scientific natural kinds are essentialist-kinds, and some are not.

4.2 Species as Individuals and Relational Essences

SAI, which we discussed in the previous chapter, is typically thought of as the quintessential non-essentialist position with respect to species. Defenders of SAI are not necessarily anti-essentialists in general. Indeed, we saw that it was the fact that they accepted the traditional essentialist account of natural kinds, which they were happy to concede may have been applicable in physics and chemistry, that they denied that species were natural kinds. So they held that while some kinds may be essentialist-kinds, species were not. They were historical individuals. But when we consider the nature of these species-individuals, we see that a form of *individual*-essentialism, i.e. an essentialism about parts and wholes, emerges.

According to SAI, the genealogical relationships an organism bears to other members (parts) of its species, and to a common species ancestor, are possessed by all and only members of that species, and must be possessed by them. To put it another way, being part of the particular genealogical nexus which defines the species is a property that is possessed, and must be possessed, by all and only members (parts) of that species. Thus a 'Martian tiger'—a member of a species that has independently converged on tiger morphology to such a degree as to be indistinguishable from Earth tigers—would not be a tiger, while a 'mutant tiger'—an animal born of tiger parents but lacking characteristic tiger morphology—would still be a tiger. In other words, according to SAI theory, similarity is neither necessary nor sufficient for conspecificity.

SAI is usually thought of as an anti-essentialist view because it denies species are natural kinds, and traditionally essences have been associated with natural kinds. Also, as I noted, defenders of SAI were assuming the strict account of essentialism which allowed only intrinsic properties. But we have seen reason to reject the strict model as an account of essentialism. Moreover, it seems clear that while kinds paradigmatically have essences (necessary and sufficient properties for kind-membership) individuals have them too—I am talking here not of the properties an individual must possess in order to be the individual it is, but of the properties the *parts* of the individual must possess in order to be part of that individual, in other words necessary and sufficient properties for being part of that individual. My hand is a part of me. In virtue of what is that the case? Presumably in virtue of certain spatial, causal and historical relations it shares with the rest of my body. These are the properties that are necessary and sufficient for something to count as a part of the individual that I am. Thus the fact that SAI claims species are individuals does not rule out SAI counting as essentialist. This point is not original to me but has been made by others such as Keller et al. (2003) and Rieppel (2006).

Since it is widely understood that the SAI view does posit necessary and sufficient conditions for organisms to belong to (form part of) a species, the fact that SAI has nonetheless often been designated as anti-essentialist implies that, contrary to my definition, the positing of necessary and sufficient conditions has not always been thought to be sufficient for a position to count as essentialist. When we combine this with my conclusion of the previous section, we find that the positing of necessary and sufficient conditions (properties that are, and must be, possessed by all and only members of a species) has often been thought to be *neither necessary nor sufficient* for a view to count as essentialist. I will return to this point below.

4.3 Relational Essence Theory

Ruse (1987), Griffiths (1999), Okasha (2002), LaPorte (2004) and others have argued that the view of species as natural kinds defined by an essence can be reconciled with the key insights of SAI theory, including the fact that species exhibit the features of change over time, historicity and cohesiveness (Koslicki 2008), if we recognise the category of extrinsic (relational) essence, that is, if we adopt the non-strict account of essentialism. As we have seen, SAI supporters argue that the character of species as unique, historical, spatiotemporally restricted, and evolving entities rules out any attempt to define them in terms of an essence. But I noted that the essences they have denied that species possess are essences in the strict sense—sets of necessary and sufficient *intrinsic* properties. They accept that what makes an organism a member of a species is its *relationship* to other organisms;[16] e.g. I am a member of *Homo sapiens* just in case I bear the right sort of genealogical relationship to other members of *Homo sapiens*. And Ruse, Griffiths etc. note that there is no reason why this shouldn't qualify as a definitional essence, albeit an extrinsic one. Species could then be thought of as natural kinds defined by such extrinsic essences. In naturalistic fashion, historical or relational essence theory draws and endorses metaphysical inferences from evolutionary theory to the existence of (extrinsic) species essences. The view only concerns species, it does not aim to provide a general account of essences or natural kinds. It is consistent with the acceptance of intrinsic essences in other areas, such as physics and chemistry. And it is consistent with any of the accounts of natural kinds we looked at in the previous chapter that allow that at least some natural kinds may possess purely extrinsic essences. Unlike in the debates about natural kinds in general, the issue of realism

[16] This relationship is typically construed as being genealogical (e.g. the relation to a common ancestor), as in the phylogenetic species concept, but as Okasha (2002) notes other candidate relations include the ability to interbreed with conspecifics, as in the biological species concept.

vs. anti-realism does not tend to arise in the discussions about historical essence theory. My guess is that most defenders would regard it as a realist view: would, that is, regard species as mind-independent, really existing natural kinds defined by mind-independent, really existing extrinsic essences. That is what I shall assume here, though I recognise the possibility that it could be defended in mind-dependent terms *a la* Boyd's or Franklin-Hall's views.

One possible view is that SAI theory and the relational essence view are equivalent—they are saying the same thing in different ways. That is, they agree on all the *facts* about species, and only differ over whether to *call* them individuals or kinds. This last decision would then be nonfactual and pragmatic.[17] But whether or not one thinks that they are strictly equivalent, it should be agreed that they have one important feature in common: they both posit extrinsic essences. Thus they both count as essentialist views. They should be *classified together* as essentialist views.

It is widely accepted (so long as one does not adopt the strict view of essentialism) that the relational essence view is an essentialist view. Ereshefsky (2010a) claims however that the relational essence view is not an essentialist view because the putative essences it posits do not cause/explain traits (or do so only weakly). Ereshefsky's argument is essentially a more specific version of Nanay's argument against essentialism in general (2011). Ereshefsky argues: (1) If the relevant relational properties constitute a species essence, then they cause/explain organisms' morphological traits; (2) they do not cause/explain those traits; (3) so the relational properties do not constitute a species essence. We saw above that Okasha accepts (2), but rejects (1). I also reject (1) (at least as a definitional claim) but am not here taking a stand on (2). Still, most would be happy to accept that relational essence theory is an essentialist view. What is more controversial is my claim that it and SAI theory should be classified together as essentialist views. But this follows straightforwardly from my definition of essentialism.

[17] Koslicki (2008, 793) suggests the issue of SAI could be a 'terminological' (non-substantive) issue. Such a view would have implications for realism and anti-realism about natural kinds, as discussed in the previous chapter. If the natural kind–individual distinction is just conventional, just a choice of words, this would seem to imply a fairly radical conventionalism (i.e. anti-realism) about natural kinds. For more on conventionalism—what I call 'neckercubing'—with regard to the metaphysics of species see my (2012).

It is also implicitly supported by Okasha in his (2002). Early in his paper Okasha sets aside the individual vs. natural kind issue, and notes that the argument of his paper doesn't depend on an answer to this question. He then goes on to argue that it clearly follows from evolutionary theory that species have relational essences—this is the case, he suggests, on any credible species concept (*ibid.*, 199–202). (Note again the *metaphysical inferences* he is recommending we endorse from evolutionary science to ontological commitment to species essences.) Thus everyone should agree, he claims, that species have relational essences, whether they are thought of as natural kinds or individuals (*ibid.*, 193).

4.4 Devitt's Essentialist View

The final view I will briefly discuss is Devitt's. Devitt (2008, 2021, 2023) defends the non-strict account of essentialism. (Although he expresses it differently, his account appears to be similar to the one I have defended.) Yet he argues that species, and other Linnaean taxa, have intrinsic essences. More precisely, he suggests that at least some essential properties of species are intrinsic. Insofar as he accepts that some essential properties of species are extrinsic, he departs from the traditional Aristotelian-Kripkean view. But his claim that some essential properties of species are intrinsic brings him close enough to the traditional view to justify his advertisement of his position as a resurrection of that view.[18] As with Historical Essence theory, Devitt's account of species essences is consistent with any general (realist) account of natural kinds that allows that at least some natural kinds possess essences in the sense in which Devitt claims species do.

Devitt's combination of views is unusual. Those who defend the non-strict account of essentialism typically deny that biological kinds have intrinsic essences (e.g. the relational essence people), whilst those who affirm that biological kinds have intrinsic essences typically defend the

[18] I will not go into his arguments for his view here, as I am concerned only with the nature of his view and its underlying commitments, not its plausibility. But see Barker (2010), Ereshefsky (2010a), Richards (2010), Okasha (forthcoming) and Lewens (2012) for critiques; see Devitt (2023) for replies to some of the critiques.

strict account of essentialism (e.g. Kripke and Putnam). The only other possible exception are those of the HPCers who reject the strict account of essentialism, and also hold that some essential properties *as they understand them* (which is not how Devitt (or I) understand them) are intrinsic.

Table 5.1 places four of the views we have looked at with respect to whether they accept the strict account of essentialism, and whether they hold that species have intrinsic essential properties. I have not included HPC theory as it cuts across these categories in ways that make it difficult to place in this taxonomy.

5 Taxonomy of Positions

What can we conclude from the previous section? I argued that on the most plausible definition of essentialism, the HPC view is a non-essentialist position and the SAI view is an essentialist position. Yet the HPC view has often been construed as essentialist, and the SAI view has often been construed as non-essentialist. In other words, the positing of necessary and sufficient conditions for species membership ought to be necessary and sufficient for a view to count as essentialist, but the way these views have tended to be classified indicates that the positing of necessary and sufficient conditions has been thought to be *neither necessary nor sufficient* for a view to count as essentialist. This reflects a mistaken view about essentialism, I claim, but more to the point for present purposes, it leads to a misleading taxonomy of views on the ontology of species.

Table 5.1 Classifying views on the ontology of species in terms of whether they accept the strict account of essentialism, and whether they hold that species have intrinsic essential properties

	At least some essential properties of species are intrinsic	Species do not have intrinsic essential properties
Strict account is the correct account of essentialism	Kripke/Putnam	SAI
Strict account is not the correct account of essentialism	Devitt	Relational essences

I want to propose a taxonomy for classifying views on the ontology of species. As I see it there are two independent dimensions along which views on the ontology of species divide:

(1) The issue of essentialism, as I have defined it.
(2) The issue of whether the properties that characterize/define a species, whether or not these are thought of as essences, are intrinsic properties, extrinsic properties, or both.

The importance of (1) should be clear. The importance of (2) is evident in the extended discussions over the last few decades about the inapplicability of the strict version of essentialism to biology. Although I have defended the broader, non-strict construal of essentialism, according to which essential properties may be either intrinsic or extrinsic, the distinction between intrinsic and extrinsic properties in this debate is important, and marks a major point of difference for different theories. In other words, although I have suggested that whether one holds that the properties that define a species are intrinsic or extrinsic is independent of, and orthogonal to, the issue of whether one is an essentialist, still the former issue is an important one in distinguishing different views. It is true the importance of this distinction has typically been bound up with the issue of essentialism—it has been thought to be important because the issue of essentialism has been thought to be important. But at times its importance has been made manifest quite apart from the issue of essentialism. For example, Hull suggests at one point that it is unimportant whether we call species 'individuals' or 'sets'—this is he suggests just a choice of language. What matters is that species are understood as *historical entities*, whether historical individuals or historical sets (1978, 336–7): 'The terms used to mark this distinction are not important, the distinction is.' (Note that he is talking here about *sets*; he regards *classes* as spatiotemporally unrestricted entities that figure in laws of nature, i.e. as non-historical entities.) Hull may of course be thinking of the non-historical conception as essentialist, but that seems a quite separate issue from the primary importance, in his mind, of the intrinsic-historical divide.

As another example, consider the dispute between Ereshefsky on the one hand, and Wilson et al. (2007) on the other, concerning whether the

HPC view can allow that species may be defined relationally. Whoever happens to be correct, the issue is clearly thought to be important, and seems here to be quite unrelated to the issue of essentialism—there is no suggestion, I take it, that whether the HPC view is an essentialist view hinges on this dispute.

These two dimensions are represented in Table 5.2, in which I have placed the five views on the ontology of species we have considered: the SAI view, the HPC view, the relational essence view, the Kripke/Putnam view, and Devitt's view.

6 Essentialism and Epistemic Stances

We are now finally in a position to return to epistemic stances. In the previous section I suggested a taxonomy for views on the ontology of species, emphasizing two dimensions. When it comes to classifying views in terms of epistemic stances, our interest is in the essentialism vs non-essentialism dimension: which views are ontologically committed to essences as implicit scientific entities, and thus which require (at least) the

Table 5.2 Classifying views with respect to the two dimensions of essentialism, and the issue of whether the properties that characterize/define a species are intrinsic properties, extrinsic properties, or both[a]

	Essentialist (necessary and sufficient conditions)	Non-essentialist (not necessary and sufficient conditions)
Species as entirely extrinsically defined kinds/individuals	SAI/Relational essences	
Species as entirely intrinsically defined kinds	Kripke/Putnam	
Species as kinds defined or characterised both extrinsically and intrinsically	Devitt	HPC

[a]The unoccupied cells on the upper right represent possible positions, but not ones that, as far as I am aware, have been defended in the literature. Although in practice HPC theory in its standard versions recognizes both internal and external properties as belonging to the property cluster that defines a kind, one could imagine versions of HPC theory that admit only internal, or only external, properties. These positions then would occupy those two cells

MoS stance. For the most part the MoS stance that is in play here is the relatively weak version. While there are clearly forms of essentialism, about biology and more generally, that require the stronger version of the MoS stance, and others that require the armchair metaphysics stance, the contemporary views on essences and the ontology of species we are considering (with the possible exception of the Kripke/Putnam view) are relatively metaphysically modest, are firmly based in science, and are relatively naturalistically respectable; in this respect they are akin to views such as the etiological account of biofunctions, and moderate-realist views of natural kinds.

In the previous chapter I claimed that the SAI view requires (at least) the MoS stance, due to its positing of species-individuals. We see now that SAI is also committed to (relational) species essences, further supporting this claim. SAIers are, it turns out, more metaphysically committed than they realise, inasmuch as they don't think of their view as essentialist, at least with regard to species (they may, as we saw, admit essences elsewhere). On the approach I am taking in this chapter (of not taking realist commitments on face value), one may be a realist about certain implicit entities of science, such as species essences, without realizing it, if one holds a view that has the implication that certain implicit entities of science exist, but one does not recognize that implication. Thus, SAIers clearly are not *motivated* in their (unrecognized) realism about species essences by a commitment to the MoS stance, inasmuch as they do not realise that they are committed to these implicit entities of science, and thus don't realise that the MoS stance is required by their view. This is not a problem for the epistemic stance approach, since the fact that, with respect to V-permissions and V-prohibitions, a certain stance is required by a certain view, does not mean that it is not possible or rational to accept that view without (at least explicitly) adopting the appropriate stance with respect to that issue. SAIers accept a view that, I have argued, involves, as a matter of fact, realism about species essences, and thus requires, in the sense of V-permissions, the MoS stance. But SAIers do not, on the whole, recognize that their view involves realism about species essences, thus are not aware that (at least) the MoS stance is, as we may say, a *V-requirement* of that view. Let us say that:

> Stance S, or one of a set of stances S+, is a V-requirement of view T if and only if T is ontologically committed to, or entails the existence of, entities belief in which is V-permissible according to S or S+.

Nothing about the epistemic stance approach implies that the V-requirements of a view must be clear and transparent to all who accept that view.[19] But does my claim that one can accept and argue for a view like SAI without realising it requires the MoS stance undermine my claim in the previous chapter that arguments for realism always presuppose the realist stance? No, since arguments for views such as species realism are *explicit arguments for realism about the relevant entities*; it is hard to see how such arguments would not presuppose the relevant stance. But arguments for SAI are not *explicit arguments for essence realism*; they are arguments for a view that I claim *implies* essence realism (with the implication not always being recognised). So there is no reason to think such arguments would have to presuppose the realist stance.

The relational essence natural kind view also requires (at least) the MoS stance, as it is, I've argued, also ontologically committed to essences; in this case the commitment is clearly recognized and embraced by defenders of the view. This is true also of the Kripke/Putnam view (as it is committed to essences, in this case intrinsic ones), and Devitt's view. That is, these theorists endorse metaphysical inferences from biological science to the existence of species essences, and they are all, I am assuming, realists about species essences.

Why does this matter? These views are only options for one (are only V-permissible) if one has adopted (at least) the MoS stance. Only if one values the explanatory power that ontological commitment to the implicit entities of science potentially brings, will one be moved to accept realism about essences in biology. Of course, within the realist camp, there are disagreements about the nature of essences: whether they are always intrinsic or can be extrinsic; whether they (sometimes, mostly or always) cause or explain observable traits; and so on. Indeed one point I have

[19] Another way the stance-requirements of a view may fail to be transparent is if the status of the relevant entities is not properly understood: S might explicitly accept realism about Xs, and Xs might be implicit entities of science, meaning that the MoS stance is required, but S might mistakenly think Xs are explicit entities of science, so think that only the SSR stance is required.

stressed is that such arguments are indeed arguments *within* the essentialist camp about the nature of essences: those who deny that essences are always intrinsic or that they always cause or explain observable traits can still be essentialists: their denials are not false by definition. Yet in a sense these arguments and disagreements are downstream from and presuppose stance choice. If one does not accept (at least) the MoS stance with its characteristic values and attitudes, one has no reason to take seriously (in the sense of V-prohibitions) any version of realism about essences, or engage in disputes concerning their nature.

Devitt in particular seems motivated in his views on species, natural kinds and essentialism by his general commitment to the realist stance. His forays into these matters began with work clarifying and defending forms of realism and related doctrines about species and biological natural kinds (e.g. 2011), work that grew out of his long-standing interest in clarifying and defending various forms of realism.[20] This then led to a series of papers (cited above) defending not just realism about species and biological kinds but biological essentialism also. While his earlier views on scientific realism (1984) may have only required the SSR stance, his more recent work defending realism about biological natural kinds and biological essences requires (at least) the MoS stance. Thus, in Devitt's work the close connection between views on species, essences and biological kinds on the one hand, and a commitment to certain epistemic stances that license certain ontological commitments and metaphysical inferences, on the other, is particularly clear.

I have argued the HPC view, even in its mind-independent form, is not as a matter of fact committed to essences. The 'essences' it posits are not true essences. Thus, in *this* regard, the view itself does not require the MoS stance. So we have the inverse of the situation with SAIers: HPCers are *less* ontologically committed than they think, inasmuch as they suppose that their view is essentialist. One may, it seems, fail to be a realist about certain implicit entities of science without realizing it, if one holds

[20] He argues however that many issues that have been thought to be realism issues in biology—thought, that is, to be issues about what things exist objectively and mind-independently—are in fact quite different issues concerning which kinds carve nature at the joints, and which kinds are causal-explanatory (2011). I don't think this way of thinking about it is correct, as I argued in my (2022), but cannot pursue the matter here.

a view that one thinks has the implication that certain implicit entities of science exist, but that implication does not in fact hold. This means that the HPC view with respect to essences may require a weaker stance than is often supposed. I noted above that nothing about the epistemic stance approach implies that the V-requirements of a view must be clear and transparent to all who accept that view. Just as a view may V-require a particular stance without this being recognized by defenders of the view, equally a view may *not* V-require a particular stance although its adherents believe it does.

Of course, a view may not, strictly speaking, require a particular stance, though its adherents accept or are sympathetic to that stance in general and with respect to other topics. In the natural kinds chapter I suggested HPC theorists straddle the line between the empiricist and MoS stances when it comes to natural kinds, endorsing values, attitudes and approaches embodied in each. A similar assessment is probably warranted when it comes HPCers' attitude to essences. Although I've argued the view is not committed to essences on the account of essences I have defended, many of its proponents think that it is, and are happy to embrace essences construed in a loose sense (appealing to a largely or purely epistemic account of essences[21]). HPCers are not, as we may say, *ideologically opposed* to essences as such, in biology or elsewhere—are not motivated in their views by a hostility towards implicit entities of science such as essences—the way that empiricists are.

I have focused in this section on the above views since they make explicit claims about the ontology of species that has enabled us, given the definition of essentialism I defended, to associate them with one or other epistemic stance. A view I have only mentioned in passing in this chapter that may be worth discussing briefly here is the epistemic account of natural kinds. What position does (or should) this view take on species essences (or essences in general)? In the previous chapter I suggested this view is an expression of the empiricist stance. Consistent with that, we find in those who defend epistemic accounts of natural kinds a fairly dismissive attitude towards essences. For a start, essentialist accounts of

[21] Although as with the epistemic account of natural kinds, it is questionable whether the purely epistemic account of essences counts as a realist view of the latter.

natural kinds are the paradigm example of strongly realist, metaphysical accounts of natural kinds that epistemic accounts define themselves against. But what we find, it seems to me, is not so much an anti-essentialist view according to which essences do not exist (in biology or in general), but more a lack of interest in, or a skepticism concerning the value of, positing essences as metaphysical entities that explain the epistemic role of natural kinds in science. It is the *demand for metaphysical explanation* that is being rejected. For instance, Ereshefsky and Reydon (2015), in their defense of an epistemic account of natural kinds, say almost nothing about essences, while in his (2010a, 684), Ereshefsky complains that (what he sees as) modern forms of essentialism 'multiply entities beyond necessity. Biologists explain the traits of organisms and provide the identity conditions of taxa without positing taxon essences.' While he might not put it this way, we can interpret this in terms of the characteristic empiricist objection to projects in the metaphysics of science: these posit supposedly explanatory entities unknown to (and unneeded by) science, such as essences.[22] Nothing is gained thereby when it comes to understanding scientific practice, and what we end up with is an unwieldy and unlovely inflationary metaphysics far removed from science.[23] Likewise Valero (2023), in defending what he sees as an epistemic account (the bare projectibilist view), claims that the *metaphysical neutrality* of the account—taking no stance on whether any natural kinds have essences and if so which ones[24]—is a virtue of it.

The attitude we can discern here—that once we have offered a purely epistemic account of natural kinds there is no *need* to posit essences as metaphysical-explanatory entities—is, of course, an attitude closely associated with the empiricist stance. The only way the epistemic

[22] One who adopts the SSR stance may also take this view of essences; we have seen that those who adopt this stance tend to take a characteristically empiricist attitude towards implicit entities of science.

[23] Similarly Stanford (2017), an anti-realist critic of metaphysics, complains that the metaphysician, in insisting on interpreting science in terms of metaphysical categories, adds nothing substantive (only verbal window-dressing in philosophical jargon) to the account of nature offered by the scientific theories themselves.

[24] Again, this metaphysical neutrality recalls the agnosticism of constructive empiricism: maybe some natural kinds have essences, maybe none do, but we do not need to answer such questions when giving a philosophical account of the relevant features of scientific practice.

account of natural kinds could incorporate essences, seemingly, would be to offer an epistemic account of essences as well, such as we have seen that some HPCers have defended: essences, like natural kinds, are those entities (whatever they are) that kinds possess and that contribute to successful induction, explanation and prediction in science. But, firstly, I have argued that this approach effectively collapses the distinction between natural kinds and essences, and secondly, as with epistemic accounts of natural kinds, purely epistemic accounts of essences may fall short of what is required for a fully realist account of essences as entities.

This completes our discussion of epistemic stances and biological kinds. In this chapter I have discussed species essences as an example of the implicit entities of science belief in the existence of which V-requires at least the MoS stance. I have looked at several prominent views on the metaphysics of species, and considered whether (given the account of essentialism I have defended) they involve ontological commitment to species essences, and thus whether they V-require the MoS stance. As was the case with the views on species and natural kinds discussed in the previous chapter, whilst species essentialism V-requires the MoS stance, the rejection of species essentialism need not involve (does not V-require) any particular stance, but it is plausible that certain views on species and essences, for instance those associated with the epistemic account of natural kinds, are in fact motivated implicitly by the empiricist stance. Thus we see, once again, the intimate connection between scientific ontology and value-laden epistemic stances.

References

Barker, M. J. 2010. Species intrinsicalism. *Philosophy of Science*, 77: 73–91.

Bird, A. 2007. A Posteriori Knowledge of Natural Kind Essences: A Defense. *Philosophical Topics*, Vol. 35, No. 1/2, Metaphysics: 293–312.

Boucher, S.C. 2012. *Empiricism, Metaphysical Stances and the Philosophy of Biology*. Ph.D. dissertation, University of Melbourne.

———. 2021. Biological Teleology, Reductionism, and Verbal Disputes. *Foundations of Science* 26 (4): 859–88.

———. 2022. Cladism, Monophyly and Natural Kinds. *Croatian Journal of Philosophy* XXII, No. 64: 39–68.

Boyd, R. 1991. Realism, Anti-foundationalism, and the Enthusiasm for Natural Kinds. *Philosophical Studies*, 61. 127–148.

———. 1999. Homeostasis, Species, and Higher Taxa. R.A. Wilson (ed.), *Species: New Interdisciplinary Essays*. Cambridge, MA: MIT Press. 141–185.

Devitt, M. 1984. *Realism and Truth*. Cambridge MA: Blackwell.

———. 2008. Resurrecting Biological Essentialism. *Philosophy of Science*, 75. 344–382.

———. 2011. Natural Kinds and Biological Realisms. In J.K Campbell, M. O'Rourke, & M.H. Slater (eds.), *Carving Nature at its Joints: Natural Kinds in Metaphysics and Science*. Cambridge, MA: MIT Press. 155–174.

———. 2021. Defending intrinsic biological essentialism. *Philosophy of Science*, 88: 1–16.

———. 2023. *Biological Essentialism*. Oxford: Oxford University Press.

Elder, C.L. 2004. *Real Natures and Familiar Objects*. MIT Press.

Ellis, B. 2001. *Scientific Essentialism*. Cambridge University Press.

Ereshefsky, M. 2009. Natural Kinds in Biology. *Routledge Encyclopedia of Philosophy Online*. Taylor & Francis Group. http://www.rep.routledge.com/

———. 2010a. What's Wrong with the new Biological Essentialism? *Philosophy of Science*, 77: 674–685.

———. 2010b. Species. *Stanford Encyclopedia of Philosophy*. 27 Jan. Stanford University Centre for the Study of Language and Information. http://plato.stanford.edu/entries/species/

Ereshefsky, M. and Reydon, T.A.C. 2015. Scientific Kinds. *Philosophical Studies*, 172: 969–986.

Ghiselin, M.T. 1974. A Radical Solution to the Species Problem. *Systematic Zoology*, 23. 536–544.

Griffiths, P.E. 1999. Squaring the Circle: Natural Kinds with Historical Essences. R.A. Wilson (ed.). *Species: New Interdisciplinary Essays*. Cambridge, MA: MIT Press. 209–228.

Hull, D. 1976. Are Species Really Individuals? *Systematic Zoology*, 25: 174–191.

———. 1978. A Matter of Individuality. *Philosophy of Science*, 45: 335–360.

Keller R.A., Boyd R.N., Wheeler Q.D. 2003. The illogical basis of phylogenetic nomenclature. *The Botanical Review* 69: 93–110.

Kornblith, H. 1993. *Inductive Inference and Its Natural Ground: An Essay in Naturalistic Epistemology*. Cambridge, MA: MIT Press.

Koslicki, K. 2008. Natural Kinds and Natural Kind Terms. *Philosophy Compass*, 3 (4). 789–802.

LaPorte, J. 2004. *Natural Kinds and Conceptual Change*. Cambridge: Cambridge University Press.

Lewens, T. 2012. Species, essence and explanation. *Studies in History and Philosophy of the Biological and Biomedical Sciences* 43: 751–7.

Magnus, P.D. 2012. *Scientific Enquiry and Natural Kinds: From Planets to Mallards*. Palgrave Macmillan.

Musgrave, A. 1989. NOA's Ark—Fine for Realism. *Philosophical Quarterly*, 39: 383–98.

Nanay, B. 2011. Three Ways of Resisting Essentialism about Natural Kinds. J.K Campbell, M. O'Rourke, M.H. Slater (eds.), *Carving Nature at its Joints: Natural Kinds in Metaphysics and Science*. Cambridge, MA: MIT Press.

Okasha, S. 2002. Darwinian Metaphysics: Species and the Question of Essentialism. *Synthese*, 131 (2). 191–213.

———. forthcoming. Does the anti-essentialist consensus about species rest on a mistake? Forthcoming in *Australasian Journal of Philosophy*.

Richards, R. A. 2010. *The Species Problem: A Philosophical Analysis*. Cambridge: Cambridge University Press.

Rieppel, O. 2006. The PhyloCode: a critical discussion of its theoretical foundation. *Cladistics*, 22: 186–197.

———. 2010. New Essentialism in Biology. *Philosophy of Science*, 77 (5). 662–673.

Robertson, T. and P. Atkins. 2013. Essential vs. Accidental Properties. *Stanford Encyclopedia of Philosophy*. Stanford University Centre for the Study of Language and Information.

Ruse, M. 1987. Biological Species: Natural Kinds, Individuals or What? *British Journal for the Philosophy of Science*, 38. 225–242.

Stanford, P.K. 2017. So long, and thanks for all the fish. In Slater, M.H. and Z. Yudell (eds.), *Metaphysics and the Philosophy of Science: New Essays*, Oxford University Press.

Valero, I. 2023. Bare Projectibilism and Natural Kinds: A Defense. *Croatian Journal of Philosophy*, 23 (2): 155–179.

Wilson, R.A., M.J. Barker and I. Brigandt. 2007. When Traditional Essentialism Fails: Biological Natural Kinds. *Philosophical Topics*, 35 (1 & 2). 189–215.

6

Biological Teleology

1 Introduction

Teleology (explanations that appeal to function, purpose, design, goals and so on) was expunged from the physical sciences during the Scientific Revolution of the seventeenth and eighteenth centuries. Aristotle's physics was thoroughly teleological, freely invoking goals, ends and purposes to account for the behaviour of matter. Not so post-Aristotelian physical science, where all physical and chemical phenomena were to be explained in terms of matter in motion, in accordance with non-teleological natural laws. In biology however, teleological thinking and language appeared to persist even beyond the Darwinian revolution. Once we have rejected an intelligent designer, it still makes sense to ask 'what is the function/purpose of the eye/the heart/the bird's wing etc.? What are these traits *for*?' Even if there is no designer, the organic world has the *appearance* of design: organisms are intricately adapted to, and well-suited for, their environments. Traits have functions that manifestly contribute to the organisms' 'ends' of survival and reproduction.

It was one of Darwin's great achievements to show how this can be explained naturalistically. We can explain what it means to say that traits

S. C. Boucher, *Empiricism, Naturalism and the Metaphysics of Biology*,
https://doi.org/10.1007/978-3-032-00555-7_6

have certain functions in a way that doesn't appeal to the intentions of an intelligent designer. Traits with functions are *evolutionary adaptations*. Natural selection has, through a blind, mechanistic process of variation, differential fitness and reproduction, produced *design-like* phenomena. Traits are shaped by natural selection for certain ends. As Lorenz puts it:

> If we ask 'What does a cat have sharp, curved claws for?' and answer simply 'To catch mice with', this does not imply a profession of any mythical teleology, but the plain statement that catching mice is the function whose survival value, by the process of natural selection, has bred cats with this particular form of claw. Unless selection is at work, the question 'What for?' cannot receive an answer with any real meaning. (Lorenz 1966, 9)

So much is uncontroversial today. Yet views on the *metaphysics* of function and design in the natural world vary widely. They include strongly realist views, weaker realist views (naturalistic reductionism) and anti-realist views. These are views in scientific ontology: they concern the metaphysical inferences we should draw from evolutionary science. Does the science license the inference that teleology and design exist in the natural world in a real, literal and irreducible sense? Or does it license rather the inference that teleology and design exist in the natural world in a real, literal but *reducible* sense? Or should we rather infer from modern evolutionary theory that teleology and design do not form part of the natural world at all? Again, we find that views on scientific ontology are largely a function, or outgrowth, of epistemic stances. As with the previous three examples (Chaps. 3, 4 and 5), while the question of the metaphysics of biological teleology has been much discussed, interpreting it in terms of the epistemic stance framework provides a fresh perspective that has the potential to get at the heart of the commitments underlying the rival positions that have been taken on this vexed topic.

2 The Philosophical Analysis of Functions[1]

Let us begin with orthodoxy. In 1970s, Larry Wright developed the 'etiological' account of functions (1973). He offered it as a general account of functions, not just biofunctions. His insight was that the function of an item is *the effect it has that explains why it exists.* The heart has several effects: it pumps blood, it makes a thumping sound, it takes up room in the chest. Why is pumping blood the function of the heart? Because it is the fact hearts pump blood that explains why hearts exist. The account clearly distinguishes between functions and fortuitous benefits (traits may have beneficial effects, but these are not functions of the trait unless the trait exists because it has these effects); and it successfully captures the normativity of function ascriptions, and thus the possibility of malfunction: a trait may fail to do what it *ought* to do, if it fails to perform the function whose successful performance in the past explains why it exists.

The account was further developed by Karen Neander (1991a, 1991b, 1995a, 1995b, 2017) and Ruth Millikan (1984, 1989a, 1989b, 1989c, 1993) with their notion of Proper Function. The role of selection was explicitly built into the analysis of Proper Function. According to this view, the function of a trait is the effect it has that explains why the trait was favoured by selection, and thus why it exists. That is, in the context of evolution, the function of trait T is X if and only if one of the effects of T is X, and this effect, by conferring a fitness advantage on the bearers of T in the past, helps explain the existence of T in present day populations.[2]

The contemporary etiological account comes in two broad versions, the ancient-history, and the recent-history account (Perlman 2004; Buller 1998). These divide over whether the current function of a trait is

[1] Sections 2 and 3 draw on my (2021).

[2] This is quite rough and would need to be articulated more carefully and precisely. For instance, we would need to specify whether it is trait tokens or trait types that have functions; and if the former, how it is that trait tokens may have functions even if they, for some reason, are unable to perform the function in question. Perhaps such a trait token has a function so long as sufficiently many other tokens of the trait type performed the function, for instance, and this performance contributed to fitness and selection. There is a further question of whether and how to distinguish between a trait *malfunctioning*, and a trait not performing its function where we would not necessarily want to say it is malfunctioning. These subtleties are important but need not detain us here.

determined by the distant, or the more recent, history of selection on the trait in the lineage in question (for discussion of the recent-history account see e.g. Godfrey-Smith 1994; Schwartz 2002; Kraemer 2014).

The etiological (or 'selected effects' (SE)) account, in one form or another, is the orthodox account of teleological biofunctions (Allen and Bekoff (1998) call it the 'standard line').[3] The other prominent analysis of functions in the literature is Cummins' 'causal role' account (1975), according to which the function of a component of a system is the contribution it makes to the capacities of the system as a whole. This is a non-historical account: the focus is on the role played by a part of a system in the context of the larger system to which it contributes, not on the explanation for the existence of that component in terms of a history of selection and reproduction. Importantly, on Cummins' account, the functions that are identified are relative to our explanatory interests and purposes.

There is an emerging pluralist consensus that these accounts are not rivals, but rather complementary: the etiological account captures function ascription in some areas of biology, the causal role account in others, and neither is reducible to the other (Amundson and Lauder 1994; Preston 1998; Bouchard 2013; Brandon 2013; Godfrey-Smith 1993; Millikan 2002; Perlman 2009).[4] But it is generally agreed that 'causal role' functions are not even potentially *teleological* functions (Griffiths 1992, 112; Matthen 1997, 24; Cummins himself accepts this (2002)). My focus here is on (putative) teleological functions.

[3] See Garson (2019) for a recent, thorough defence of a version of the selected effects view.

[4] Although see Neander (2017) and Longy (2009) for dissenting views. Some, such as Mossio et al. (2009), and Kitcher (1993), attempt to capture the insights of both approaches not through such a pluralist view, but through offering a general, 'unitarian' account of function, under which both the etiological and the causal role analyses can be subsumed.

3 Teleology, Realism, and Reductionism

3.1 Realism, Reductionism and Anti-realism in General

The standard account of functions and teleology in biology is a reductionist account. There is of course a large literature on the issue of what it is for one theory to be reducible to another, or for one property or fact to be reducible to another. But we may for the most part bypass these questions and proceed on the basis of an intuitive understanding of reduction and reductionism, as it is uncontroversial that the standard account of biofunctions is a reductionist account however that is spelled out, and nothing in what follows hinges on the details of the correct account of reduction(ism), or the correct philosophical analysis of the claim that teleology is reducible to non-teleology. On this intuitive account, statements that refer to the reduced entities or properties can be literally true, but what makes them true is facts about the reducing entities or properties.

In general, reductionist positions occupy an intermediate position between what I will call (for want of better terms) full-blooded realism and full-blooded anti-realism. The full-blooded realist holds that the disputed entities[5] exist in more or less the sense that they have traditionally been thought to, by common sense, folk theory or previous science. The entities cannot without losing their identity be reduced to some other class of entities. McGinn nicely, if somewhat colourfully, captures (a version of) this position:

> The...position is that (the relevant facts) are irreducible and indefinable and inexplicable, and we should cultivate an attitude of insouciance towards them. (They are) brute facts for which no explanation can be given...they are what they are and not another thing... reductionism stems from a misplaced monism, obsessive unification. We need to rid ourselves of the compulsion to oversimplify the world, to level it ontologically, and instead relax, indolently, into the *sui generis* variety of our given conceptual scheme.

[5] I shall for convenience talk in terms of 'entities', but this should, as I noted in the Introduction, be understood in a broad sense to include things like teleology.

After all, it was functioning perfectly well before we started to fret over it. (1993, 15)[6]

The full-blooded anti-realist agrees with the full-blooded realist that there can be no reduction of the entities or facts in question, but draws the opposite conclusion—since the entities cannot, on her view, exist in the sense they have traditionally been supposed to, the failure of reduction means that the entities do not exist in any sense. While the reductionist accepts that statements referring to Xs can be true, the full-blooded anti-realist is typically[7] an error-theorist about X talk, holding that discourse that refers to Xs is systematically false. The error theorist may then make one of two recommendations about the discourse. She may advise that it be eliminated (eliminativism); or she may advise that it be retained for its heuristic or pragmatic value (instrumentalism), in which case the sentences of the discourse will be understood as expressing 'useful fictions' (Boghossian 1990).[8]

'Eliminativism' is sometimes used to refer just to anti-realism: the view that the entities in question do not exist. But, in the context of non-reductionist anti-realism, we need a way to distinguish the error-instrumentalist from the error-eliminativist (both of whom are anti-realists), hence I will follow Boghossian (1990) in using 'eliminativism' to refer more narrowly to the error-eliminativist position. On this interpretation the recommendation is that the *discourse about Xs* (or perhaps the *concept X*—Burgess and Plunkett 2013) should be 'eliminated'

[6] As we shall see, there are two main varieties of full-blooded realism about teleology, the creationist and the Aristotelian. McGinn's description fits the latter, but not the former (since on the creationist view biological teleology is not *sui generis* and inexplicable, but is explained in terms of the intentions and goals of a divine intelligent designer).

[7] We will see that there are seemingly forms of strong anti-realism about teleology that are not error-theories. For now I focus on the error-theory version.

[8] Error theories need to provide an account of *why* there has been systematic error in the region of discourse in question. For instance, some error theorists about moral discourse have suggested that there are sound evolutionary reasons why people should have falsely believed that moral facts and properties exist. The error-instrumentalist has a ready-made answer to the 'why' question: if believing the false sentences has certain heuristic or pragmatic virtues, it is not surprising that such beliefs have flourished, despite their falsity. The error-eliminativist may have a harder time answering the 'why' question, for she does not share the instrumentalist's view about the benefits of the false beliefs. But other explanations for the systematic error are possible. (Of course, all that is needed for beliefs to become prevalent is a *perception* of a benefit, even if no such benefit exists.)

from our linguistic practices, not (or not only) that Xs should be eliminated from our ontology.

An alternative version of full-blooded anti-realism is nonfactualism. While error theories take the relevant discourse at face value, nonfactualism denies that the predicates of the discourse denote properties, and denies that the relevant sentences express judgments equipped with truth conditions, despite the fact that they appear to (Boghossian 1990). The sentences are therefore not to be thought of as systematically false, but rather as neither true nor false. As far as I know no anti-realists about teleology have interpreted teleological discourse in nonfactualist terms, so we can set this aside.

The three positions can best be distinguished in terms of their attitude to the following argument:[9]

P1. If X's cannot be reduced to another class of entities, Xs do not exist, and X-talk is systematically false.
P2. X's cannot be reduced to another class of entities.
C. Therefore, Xs do not exist, and X-talk is systematically false.

The full-blooded anti-realist (of the error-theory variety) holds that the argument is sound; the full-blooded realist accepts premise two but attempts to block the argument by denying premise one; and the reductionist accepts premise one (usually) but attempts to block the argument by denying premise two.

3.2 Realism, Reductionism and Anti-realism about Teleology

Full-blooded Realism about Teleology

Full-blooded realists about biological teleology hold that finalistic, teleological causation in nature is real and cannot be reduced to mechanistic, efficient, naturalistic causation. There are, broadly speaking, two ways of

[9] See Rudder Baker (1988) for a similar taxonomy of positions in the debate about intentionality.

spelling this out. The first is the traditional creationist idea that the natural world, including organisms and their traits, is the product of conscious, intelligent design. Thus the functions of the traits of organisms are grounded in conscious purpose and planning, that is, in the goals and intentions of conscious agents, in exactly the way that the functions of human artifacts are (Plantinga 1993). Plato's version of strong teleology is perhaps a variant (or metaphysical forerunner) of the creationist version; see Ariew (2002).

The second way of spelling out strong teleology is that associated with Aristotle.[10] This is often characterized as the view that, whether or not we invoke intelligent design, there is real, irreducible teleology in the natural world in the sense that there are objects, events or states at time t and events and states at time $t + n$ such that the objects, events or states at t are to be (irreducibly) causally explained by reference to the events and states at $t + n$. As Ayala puts it, on this conception 'future events are active agents in their own realisation' (1970, 14). Several people have argued in recent years however that this is a mistaken characterisation of Aristotle's position. Perlman (2004, 9), Walsh (2015, 196, 200) and Ariew (2002, 18), suggest that for Aristotle, future states *explain* earlier states without *causing* them. Final 'cause' is not a species of causation at all (Walsh *ibid.*). At least with respect to some authors (e.g. Walsh and Ariew) correcting this misinterpretation has been an important part of the project of rehabilitating a broadly Aristotelian conception of teleology, and making the case that it deserves to be taken seriously today by naturalistic philosophers of biology.[11] Backwards causation might be metaphysically beyond the pale—or simply incoherent—but (according to these authors) it is by no means obvious that backwards explanation is equally problematic. According to the neo-Aristotelians, this backwards explanation—explaining behavior and events in terms of their goals, explaining means in terms of their ends—is not, *pace* the reductionist orthodoxy, translatable without loss into statements referring only to history, fitness, and selection.

[10] We shall be concerned with Aristotelian teleology as applied to the biological world. Aristotle also, of course, defended a broader cosmological teleology, which is now only of historical interest (Ariew 2002, 20–21).

[11] Though we shall see that different things could be meant by 'naturalistic' in this context.

Finalistic explanations are not replaceable by causal, mechanistic explanations. The truth-makers for teleological statements are irreducibly teleological facts. There is objective goal-directedness built into nature, and any attempt to capture this goal-directedness in non-teleological, causal-mechanistic terms simply misses crucial features of the reality described.[12] It is only misplaced metaphysical scruples (Walsh *ibid.*) that leads people to embrace such an impoverished non-teleological language for nature. Misplaced since, on closer examination, the widely accepted view that irreducible teleology is metaphysically disreputable, and inconsistent with naturalism (being inseparable from vitalism, supernaturalism, and other 'spooky' entities and forces), is untenable. Walsh says of cybernetics research, 'This rich tradition shows us that goal-directedness is an unproblematic causal consequence of the architecture of an adaptive system' … 'goals are natural and observable. Teleology's reliance upon goals, therefore, should offer no impediment to naturalism' (*ibid.*, 195–196).

In response to Fodor's famously saying of intentionality, 'If aboutness [intentionality] is real, it must really be something else' (1987, 97), Searle replied 'You cannot reduce intentional content… to something else, because if you could [it] would be something else, and [it is] not something else… aboutness (i.e. intentionality) is real, and it is not something else.' (1992, 51). Similarly, with respect to teleology, the reductionist holds that if it is real, it must really be something else, while the neo-Aristotelian holds that it is real, and is not something else.

Walsh (*ibid.*, 199) defends 'an updated rendition of the Aristotelian account of teleological (final cause) explanation' in the context of advocating a recentering of the organism in evolutionary theory.[13] Modern Synthesis orthodoxy, with its focus on genes and its mechanistic metaphysics, had no place for teleology or goal-directedness. But once we have the organism as a functionally integrated and emergent whole firmly back in view, teleology is forced upon us, since it is so evident that organisms are purposive, goal-directed systems. The orthodox reductionist view 'casts teleological explanation as a cryptic form of causal explanation' (*ibid.*, 196). It gets rid of goals *as* goals, but in doing so 'risks losing

[12] See also Barham (2011).

[13] He updates Aristotle's view by cashing it in terms of counterfactuals and invariance.

the potential windfall to be had by a proper account of how goals can explain their means' (*ibid.*). Indeed he seems to assume that the reductionist account is not a realist account of teleology, in that he somewhat misleadingly presents the arguments for and against strong, irreducible teleology as arguments for and against teleology itself. This is perhaps unsurprising: from the standpoint of the neo-Aristotelian, reductionists about teleology are in fact anti-realists, despite what they say: 'reduced' teleology is, for them, eliminated teleology (Boucher 2021).[14]

Cameron (2004)[15] also defends a version of Aristotelian teleology, interpreted in terms of an emergentist metaphysics. Emergent properties are higher-level properties that arise out of and supervene on underlying properties and their interactions, but are not reducible to them, possessing independent causal powers of their own, thus allowing for a kind of 'downward causation'. I shall discuss Cameron's views and arguments at length in the following section.

The two versions of strong teleology are logically independent. The Aristotelian conception does not require there to be an intelligent designer; indeed on some interpretations of Aristotle, he did not invoke conscious design at all in this connection; the strong teleology in question was simply built into the structure of reality in a *sui generis* way.[16] Conversely, on the creationist version, there seems no more reason to suppose that teleology and design in the natural world involves backwards explanation than there is in the familiar case of designed human artifacts. As has often been noted (e.g. Bunge 1999, 46), while in a loose sense the existence and character of a pencil, or a computer, at time t, is to be understood in terms of the states or effects of the artefact at time $t + n$, in a stricter sense, the correct explanation for their existence and

[14] Mossio et al. (2009), in their 'organisational' account of biological functions, combine a Walsh-type holistic and non-reductionist account of the complex systems that *have* functions with a reductionist account of the *functions themselves*. They are clear that they are seeking to 'naturalise' teleology by reducing it to the causal capacities of organisationally closed, differentiated, self-maintaining systems: 'Organisational closure grounds teleology and normativity in a naturalised way' (*ibid.*, 824). I don't interpret them as holding (as Walsh does) that complex, self-maintaining, highly organised systems are the locus of irreducible teleology. Thus, for my purposes in this chapter, their account would be regarded as a realist-reductionist one.

[15] See also Walsh (2013).

[16] On this view teleology comes apart from design.

character, need only advert to the mental states of the human agents responsible for their design and manufacture, which exist prior to the existence and character of the artefact. If there is nothing metaphysically mysterious about designed human artifacts, it is no more metaphysically weird to suppose that organic beings are the product of conscious design.

There seems little point in wrangling over which of these conceptions best captures 'the' idea of strong biological teleology. Better to just accept that there are these two, logically distinct, notions, and be careful not to conflate them. Lewens (2004), in seeking to reduce all of teleology to the 'artefact model', is arguably guilty of this (see also Matthen 1997). Teleology is not exhausted by the artefact analogy. Aristotle's teleology seemingly had little if anything to do with the artefact analogy, but was still teleological in virtue of its positing of objective goals, ends and purposes (Ariew 2002, 18). So the idea that we've naturalised teleology when we've 'examine(d) the similarities between the processes that underlie the production of organisms and artifacts' (Lewens 2004, Kindle Locations 193–194) strikes me as wrong. The artefact analogy, historically important as it has been, is only one part or aspect of teleological thinking.

Full-blooded Anti-realism about Teleology

Full-blooded anti-realism about biological teleology comes in two forms. According to the first, teleology cannot be reduced to mechanistic/efficient causation, and we should thus be error-theorists about function talk, either of the eliminativist, or the instrumentalist, variety. On the face of it, Searle is a full-blooded anti-realist: '…in addition to its various causal relations, the heart does not have any functions.' (1992, 238; see Dennett 1995, who also interprets Searle as an error-theorist; see Perlman 2004, 11[17]). His argument for this view is that function ascriptions are value-relative: we only attribute to the heart the function of pumping blood (such that hearts that do this well count as good hearts) because *we value* the end to which this contributes, i.e. health and survival. Whether

[17] Perlman (2004, 11) suggests Searle has a non-reductive naturalist account of functions, which is misleading; given Searle doesn't think functions exist in nature at all, he should not be said to be giving a *naturalistic account* of functions.

Searle is an error-instrumentalist or error-eliminativist is not entirely clear, although what he says about the value that function-talk has for us given our various interests and purposes might suggest the instrumentalist version. Another version of full-blooded anti-realism is the view, often expressed, that function and design talk when applied to the natural world is purely *metaphorical*—that is, literally false, but useful and illuminating in a metaphorical sense. This is best understood as a version of the error-instrumentalist view.

Lewens (2004) seems to defend a version of this view. Teleology and function talk is useful in biology, he argues, not because of the role of selection, as many think, but because of the similarity between organisms and artifacts. It follows, presumably, that traits do not have functions in an objective sense, independently of us: it's just illuminating *for us* to think and talk in terms of functions and purposes, because *we see* this similarity between traits and *things we make*. (He admits this is a 'deflationary', or 'pragmatic' view of functions, which does suggest full-blooded anti-realism, of the error-instrumentalist variety.) Suppose humans had never existed. Then there would be no artifacts. So there would be no similarity between organisms and artifacts, and no one to notice this similarity. So the 'artifact model' would not apply; so traits would not have functions or purposes, and there would be no teleology in nature, on Lewens' view. It is striking that Lewens focuses on when 'artifact thinking would be of use', where 'teleological language would be appealing' etc. Realists about functions would want to know which functions *actually exist*, not just when it is appealing or useful to use functional language.

It is tempting to attribute the error-eliminativist view to strong critics of the focus on functions within the adaptationist program such as Reiss (2009), although of course their critiques of teleological and adaptationist thinking are rarely put in the form of an explicit error theory such as we find presented and defended in other areas of philosophy such as metaethics. I include here only the strongest critics of teleological and adaptationist thinking, according to whom all function talk is defective, and a hangover from the days of Natural Theology. More moderate critics of adaptationism, such as Gould and Lewontin (1979) would count as reductionists on my view. There is no reason to think that Gould and Lewontin hold the radical view that statements that attribute functions to traits are systematically false, or deny that teleology can be reduced to mechanistic, efficient causation in accordance with the standard etiological (or some other reductionist) account. In my view it is a mistake to interpret the critique of adaptationism associated with such thinkers as involving the rejection of teleological notions altogether. To put it another way, nothing about the reductionist view of biofunctions in itself entails adaptationism. As Lewens has argued convincingly (2004), the idea that there is a central and indispensable role for teleological language, function-talk, and solutions-to-problems talk in biology survives the critique of adaptationism launched by Gould, Lewontin and their cothinkers, even if we accept much of the substance of that critique, as Lewens does.

The second form of full-blooded anti realism holds that in a sense there are biofunctions[18]—and thus that sentences referring to biofunctions can be true—but they are not a mind-independent part of nature, rather they are constructed by our social and linguistic conventions and practices. They are mind-dependent. They exist after a fashion, but are imposed on nature by us, not 'found', pre-existing, in nature. Gayon makes the point that Wright, the originator of the etiological account, which is usually interpreted as an unambiguously realist account (see following section), was not actually a function-realist, as he held that functions were not an objective part of nature, but were 'relative to our explanatory purposes' (Gayon 2013, 75). So understood, the view is clearly analogous to the mind-dependent accounts of natural kinds we discussed in the previous chapter.

Reductionism about Teleology

Finally, reductionists assert that teleology can be reduced, via Darwin's theory of evolution by natural selection, to mechanistic, efficient, non-finalistic causation, typically by way of one or other version of the etiological (SE) theory of biofunctions. On this view statements that refer to the function of a trait can be literally true, but what makes such

[18] Thus this is a somewhat less 'full-blooded' form of anti-realism than the previous version. We may prefer to call it 'moderate anti-realism'.

statements true are entirely non-teleological facts about fitness, selection, and so forth. The etiological account is not the only reductionist account of biofunctions. The propensity account of Bigelow and Pargetter (1987), for instance, according to which the function of a trait is the effect it has that contributes to the fitness—the propensity to reproduce—of the organism that bears it, irrespective of selective history, is also a reductionist account. It has not been anywhere near as popular as the etiological account, hence I am setting it aside. But most of what I say about the etiological account's relation to reductionism, realism and teleology, applies *mutatis mutandis* to the propensity account.

It is widely thought that reductionist accounts such as the etiological view are realist accounts of teleological functions.[19] 'In all its versions, the etiological account assumes a realist stance towards functions: Those are not only elements of our descriptions of nature; they exist in the biological and ecological domains' (Huneman 2013, 105; see also Cameron 2004).[20] The idea is that on the etiological (or other evolutionary-reductionist) accounts, whether a trait has a function, and the function it has, is an objective matter, and is independent of our interests or explanatory purposes (Bigelow and Pargetter 1987). Natural items would have had functions whether there were humans (or other agents) around to take an interest in them or not (Cameron 74). This (it is said) contrasts with the more anti-realist Cummins account, on which the function of an item is relative to our explanatory interests, aims and practices.[21]

On this view reductionism about teleology is like other classic naturalistic reductions in the history of science, such as that of water to H_2O, or of heat to molecular motion, or lightning to electrical phenomena. Just as these successful reductions in no way showed that water, heat, or lightning failed to exist, the reduction of teleology to efficient

[19] We have seen that Wright was an exception to this.

[20] Huneman (2013) argues that the etiological account requires a weaker kind of realism about functions that is typically supposed. But he doesn't dispute the standard claim that the etiological account is fundamentally a realist account of teleological functions.

[21] As I noted in the previous chapter, I am in this chapter mostly taking at face value the realist commitments of the views I'm discussing, in particular reductionist views. In my (2021) I discuss the fact (alluded to above) that from the standpoint of the strong realist, reductionists about teleology could be interpreted as being anti-realists. In denying the existence of irreducible teleology, they are, she may claim, rejecting teleology altogether.

causation in no way shows that teleology fails to exist. Or so the reductionist-realist argues.

In his (2017) Dennett gives expression to this view that the naturalistic reduction of design[22] and teleology is not an eliminativist view, but a firmly realist one. He notes that there is a fear among Darwinians (even adaptationists such as Dawkins) that admitting there is genuine design and teleology in the natural world would give comfort to creationists. Thus Dawkins speaks of 'designoid' rather than 'designed' features of organisms. In fact, Dennett argues, it is this very reluctance to admit what is blatantly obvious to any observer—that 'the biosphere is utterly saturated with design, with purpose, with reasons'—that plays into the creationists' hands, in implying that evolutionists have reason to be embarrassed about, and must seek to explain away, the adaptive intricacies of nature, as their theory cannot account for them. Much better, Dennett suggests, to openly accept that 'there can be design, real design, as real as it gets...' in a Darwinian world without a designer.[23]

4 Realism about Biological Teleology and Epistemic Stances

I have distinguished between full-blooded anti-realism, reductionism, and full-blooded anti-realism, with respect to biological teleology. In this section I shall connect these views with epistemic stances. I regard biological functions as part of the implicit subject matter of science, in that although biologists refer to functions, it is not the business of science to investigate the nature of functions or teleology *as such*. Discussions of the various accounts of functions we have canvassed fall squarely within the philosophy literature. Even if one could argue that the etiological account

[22] See Allen and Bekoff (1998) on the relation between function and design. They argue convincingly that in biology function is necessary but not sufficient for design. See Buller (2002) for a critique of Allen and Bekoff. See also Lewens (2004) and Kitcher (1993). All these authors accept Dennett's view that there is real design in the natural world, but disagree about its relation to function ascription.

[23] Dennett has, it seems to me, held this view for a long time, which makes Perlman's categorisation of him as an eliminativist about teleology (2004, 12) hard to understand.

of functions is in some sense implicit in Darwin's theory of natural selection, its explicit articulation and defense was undertaken by philosophers, and biologists for the most part have not needed to bother themselves about the details of these extensive philosophical debates in order to carry on with their functions-involving scientific work.[24] (One is unlikely to find many articles defending or critiquing views such as the etiological account in scientific journals.) Thus it is (at least) the metaphysics of Science (MoS) stance that is in play here: can we or must we infer to the existence of teleological biofunctions understood as part of the implicit subject matter of biology?

The fact that scientists refer routinely to functions and purposes means there is a difference between these and, say, natural kinds and essences, which are much more philosophical terms of art ('teleology' is perhaps more a philosopher's term of art). This creates an added burden for the full-blooded anti-realist about biofunctions (at least of the error-theory variety). If it turned out there were no natural kinds, such that all statements of the form 'X is a natural kind' were false, scientific discourse need not be much affected, seeing as it is philosophers, not, on the whole, scientists, who make such claims. But if it turned out that there were no biological functions, such that all statements of the form 'X is the function of trait T' were false, one would need to endorse an error-theory for large swathes of scientific discourse, a significant cost.

4.1 Reductionism

The paradigm contemporary reductionist account, the etiological (SE) account, seems on the face of it require a relatively weak form of the MoS stance. The entities being inferred to—mind-independent teleological functions understood as selected effects—are implicit entities of science, and seem to satisfy the naturalistic constraint on metaphysical inferences. The 'distance' from the science is modest, the entities or properties in question are not purely metaphysical-explanatory posits, and the arguments for their existence and nature are based on a detailed

[24] As always, there are exceptions, such as the work of Mayr (1961) and other biologists on teleonomy vs. teleology.

understanding and analysis of the relevant biological theories. A priori reasoning is minimal, and the position seems to enjoy a degree of empirical (or quasi-empirical) vulnerability (see, for instance, Lewens' argument (2004, loc. 1192) that the etiological account is undermined by cases in nature (real or hypothetical) in which function and selection come apart, or Matthewson's argument (2020) that the account is undermined by the fact that natural selection comes in degrees, whereas etiological functions are all-or-nothing, such that the etiological account 'is in tension with the biological facts').[25]

But in philosophy, things are rarely so simple, and in below we shall consider an argument for the claim that the naturalistic credentials of the reductionist orthodoxy have been drastically oversold.

4.2 Full-blooded Anti-realism

Full-blooded anti-realism about biological functions, in either its mind-dependence form, or its error-theory form, clearly does not require the realist stance, and is *consistent with* the empiricist stance (and, for that matter, the SSR stance, if functions and teleology are implicit entities of science).[26] But, just as we saw with respect to natural kinds, not having certain mind-independent implicit entities in one's ontology does not on its own mean that one has not adopted the MoS stance, since one may adopt the latter yet accept anti-realism about *some* implicit entities of science, perhaps including teleological functions. With respect to the empiricist stance, we have seen that being an empiricist involves not just anti-realism, but above all a certain rejection of metaphysical explanation. To be an empiricist with respect to the biofunctions debate would thus be to endorse *both* of the following: (1) anti-realism about teleological functions; (2) the rejection of the idea that we can or should explain the relevant data by positing teleology as a metaphysical reality lying

[25] The view is analogous to the view discussed in the previous chapter that there are natural kinds but they are 'nothing over and above' sets of properties and their relations (Chakravartty 2007). This is opposed to the strong realist view that natural kinds are a sui generis category, irreducible to other types of entity (Lowe 2006).

[26] I set aside Cummins' mind-dependent account, as by his own admission, it is not even potentially a teleological account.

behind the phenomena. Endorsing (2) clearly involves rejecting full-blooded realism about teleology, with its various metaphysical posits of dubious naturalistic respectability (a creator god, sui generis teleology, emergent properties etc.). It is less clear, as I noted above, that the reductionist accounts can be convicted of trafficking in problematic metaphysical-explanatory posits, so endorsing (2) may not on its own distinguish empiricists from realist-reductionists. What does distinguish them is their attitude to (1): empiricists are anti-realists, while reductionists are, at least on the face of it, realists, about biofunctions. What is clear however is that one argument for realism that plays a key role for both reductionists (see Neander 1991b) and full-blooded realists (see Cameron 2004) is that teleology is *explanatory* for *science*. This is not the claim that the *philosopher's* realism about teleology explains the phenomena, the way the philosopher's more general scientific realism explains the success of science according to the No Miracles Argument, but rather that, at the first-order level, biologists are able to explain why organismic traits have the form they do by citing their evolutionary function. Without this appeal to function biological science would be explanatorily impoverished. This explanatory criterion of reality is, of course, internal to the realist stance, and is not characteristically endorsed by empiricists.

We may also note that the error-instrumentalist account, on which there are no biofunctions but it is useful to speak *as if* there are, has much in common with classical empiricist and instrumentalist accounts of scientific discourse, on which the observable phenomena *as if* there are certain theoretical entities, but that is as far as we ought to epistemically commit ourselves. The realist characteristically insists that the best explanation for the fact that things are as if entity E exists is that entity E does indeed exist. The empiricist demurs. Similarly here, the realist, including the reductionist-realist, about biofunctions, may argue that the best explanation for the acknowledged usefulness, indeed indispensability, of function-talk in biology, for the fact that things are *as if* traits have functions, is that traits really do have functions (and to that extent even the reductionist can be understood as offering metaphysical explanations of a sort). But this abductive move will not impress the empiricist *qua* empiricist.

4.3 Full-blooded Realism

There are versions of full-blooded realism that are clearly expressions of the armchair metaphysics stance, such as Intelligent Design Theory. A creator-God is not a scientific entity, even implicitly, so this version of full-blooded realism goes beyond the MoS stance, into non-naturalistic territory. This is uncontroversial. Things are less clear when it comes to our other forms of full-blooded realism however.

Aristotelian Teleology and Two Forms of Naturalism

Whether Aristotelian and sui generis forms of teleology are consistent with naturalism has been a topic of discussion. Ariew (2002) suggests for instance, that Aristotle's teleology, at least in its explanatory version, is more naturalistic than has been supposed. He distinguishes it sharply from the unambiguously non-naturalistic Platonic form of teleology (and its creationist successors) with which it has been unfairly associated, and notes that despite Aristotle positioning his account of teleology in opposition to the mechanistic 'materialist' view, it is, like Aristotle's philosophy more generally, naturalistic in fundamentals. This is the case, Ariew suggests, partly because it eschews any appeal to objective values or 'goodness' (2002, 8). Walsh (2015, 200–1) makes a similar point. This is quite unorthodox however. Standardly it is thought that teleology (especially strong teleology) does indeed require the postulation of ends that are good or valuable in themselves (Bedau 1998). This is the idea that teleology is inherently normative: to say that F is the function of X is to say that performing F is what X is *meant* to do, what it is *supposed* to do. And this in turn requires, seemingly, that F contributes to some end or goal that is good or desirable (such as survival or reproduction). Then one can either bite the bullet and accept there are such objectively valuable states in nature (see Dennett 1984; Ruse 2003; Bedau 1998), thus vindicating teleology, or one can reject the existence of such states and teleology along with them (Searle 1992).[27]

[27] I discuss this in Boucher (2021).

But the issue I want to focus on here is a different one. In discussions such as Ariew's, 'naturalism' and 'naturalistic' often have a somewhat different meaning from the meaning we've given them. When the naturalistic credentials of these views are discussed, what is at issue is typically whether the view is consistent with a *naturalist metaphysical picture of the world*: roughly the view that the natural world of space and time is all there is and there are no supernatural entities or forces (call this MN for Metaphysical Naturalism[28]). We have been concerned however with the related but somewhat different question of whether positing certain entities is consistent with a *naturalistic constraint on metaphysical inferences* (call this NC for Naturalistic Constraint): are the entities 'close enough' to, in sufficiently close inferential or explanatory contact with, our best current science to be palatable to those who wish to do metaphysics in a naturalistic spirit ('naturalistic' as in 'based on our best science', not as in 'not supernatural')? These two questions will often line up. Entities that violate MN (such as supernatural entities) will almost certainly violate the NC.[29] Entities that satisfy the NC will almost certainly satisfy MN. But there may be entities that satisfy MN but violate the NC. Positing entities such as universals, as we have seen, arguably conflicts with the NC, in that universals are 'too distant' from, theorizing about them too unconstrained by, contemporary science. Yet there are theories of universals, such as Armstrong's (1978)—or, for that matter, Aristotle's—on which they are *immanent* in the world and thus may be consistent with MN. As another example, Ladyman and Ross (2007) complain that much contemporary metaphysics presupposes an outdated classical Newtonian physics. Such an approach to metaphysics would satisfy MN, since nothing in Newtonian physics is supernatural, presumably. But since it is not based on our best current science, it would violate the NC.

Since we have now introduced three non-equivalent forms of naturalism in this book, it would be well to pause briefly to define them, so that we do not confuse them:

[28] See Boucher (2020) for a critique of the intelligibility of MN.

[29] This distinction didn't come up in the natural kinds chapter, since unlike the teleology debate, there is no question in the natural kinds debate that all the relevant entities are consistent with MN.

- *Conceptual Naturalism*: when one is offering a philosophical account or definition of some scientific or scientifically-relevant concept, one should be guided by, and not depart too radically from, the way scientists in the relevant field think about and deploy the concept, or related concepts, and surrounding theories.
- *Naturalistic Constraint on Metaphysical Inferences* (NC): when engaging in ontology one should, as far as possible, refrain from positing entities too far removed from, and too unconstrained by, empirical, scientific evidence.
- *Metaphysical Naturalism* (MN): the natural world of space and time is all that exists.

It is the latter two that are relevant to our present discussion. The key point for now is that even if Ariew is right that the Aristotelian form of teleology he is discussing is consistent with MN—because, for Aristotle, teleology (like universals) is immanent in the world, not transcendent as it is for Plato—this doesn't settle whether it is consistent with the NC. Ariew (2002) notes that strictly speaking Aristotle's teleology is inconsistent with Darwin's theory of natural selection; for instance, Darwin claimed there was no purpose or design involved in the *source* of functional adaptations (since the variations among living things that were the raw material on which natural selection worked arose randomly with respect to the adaptive needs of the organism); something Aristotle would have denied (*ibid.*, 29). Clearly no metaphysics that conflicts with such a fundamental component of contemporary science as Darwin's theory would satisfy the NC. Elsewhere however, Ariew insists that Darwin's theory, and all functional explanation in modern evolutionary biology, is fundamentally Aristotelian in spirit. Darwin's *arguments*—if not all of his conclusions—are Aristotelian (*ibid.*, 8). For instance, Darwin accepted Aristotle's teleological postulate that organismic traits 'can be explained in terms of [their] *usefulness*' in the struggle for survival and reproduction (*ibid.*, 28). And he also accepted Aristotle's notion that individual organisms are sites of teleology inasmuch as they strive for self-preservation as an end or goal (*ibid.*).

Thus Ariew's view is that while certain features of Aristotle's teleology, as he articulated it, don't fit comfortably with modern materialist-mechanist science or Darwin's theory, in a broader sense it is not inaccurate to say that Darwin's theory was an instance or expression of Aristotle's immanent, naturalistic teleology. After all, no neo-Aristotelian account of teleology is likely to endorse everything Aristotle, writing in a pre-Darwinian, pre-modern-science age, asserted about functions and teleology in the natural world. The question is whether the spirit—the broad outline, fundamental core assumptions, and basic pattern of argument—of Aristotle's account survives the rejection of some of the details. If so, it can perhaps be rendered compatible with the NC.

Walsh (2015), for instance, defends the consistency of a broadly Aristotelian, non-reductionist realism about teleology with naturalism in both senses:

> It is incumbent on any claim on behalf of teleology to show that it doesn't make any illicit metaphysical commitments, and that, the completeness of mechanism notwithstanding, teleology has some indispensable explanatory role to play. The case against teleology is compelling.[30] Yet, addressing the charges one-by-one will help build an understanding of teleology and purpose, and of how the natural sciences can avail themselves of genuine, legitimate teleological explanations. (*ibid.*, 189)

His view is that irreducible teleology, with its real, objective goals, its explanations of means in terms of their ends, and its irreplaceability with efficient-causal explanations, need not involve any spooky, non-natural entities or forces, as we saw above; but more than this, as an ontological claim it captures and helps to account for (in a way that reductionist or eliminativist views cannot) central features of contemporary biological theorising, and thus satisfies the NC. There is also, he would insist, an appropriate degree of empirical vulnerability, and ontic responsiveness to science. It is because evolutionary biology itself is in need of a radical overhaul, with mechanistic modern-synthesis biology needing to give way to a more holistic organism-centred view, that Aristotelian teleology is (or should be) back in business. In this sense the metaphysics follows

[30] Note that, as I indicated above, he is using 'teleology' here to mean some sort of strong teleology such as Aristotle's. He seemingly doesn't have reductionist-realism in mind. The arguments against teleology he discusses are arguments against strong, unreduced teleology.

the science: reductionism about teleology may have served for the mechanistic gene-centred paradigm; but that paradigm is degenerating (to use Lakatos' term).[31] The alternative, progressive, holistic paradigm Walsh favours demands a more robust, more metaphysically inflationary, less austere conception of teleology. Or so he claims.

Cameron's Emergentism

The most interesting attempt to establish the consistency of irreducible teleology with the NC is found in Cameron (2004), and it will be worth considering at some length. Cameron claims that his emergentist view counts as naturalistic in both the metaphysical and the epistemological-methodological sense. Let us grant that it is consistent with MN. On the face of it, it is more questionable whether Cameron's emergentism satisfies the NC than is the case with the reductionist view. Many would regard reduced teleology—e.g. functions as selected effects—as closer to the science, more a matter of 'taking science at face value', less metaphysically inflationary or speculative, thus more in the spirit of the NC—than the kind of strong teleology, cashed in terms of a recondite emergentist metaphysics unknown to science or scientists, defended by Cameron. Cameron however disputes this. He argues that his view is in a way *more* responsive to, more ontically deferential to, the science, thus more consistently naturalistic in the methodological sense, than the reductionist view. Biology clearly needs teleology: the explanatory power of especially evolutionary biology would be greatly compromised were it not able to avail itself of teleological explanations (*ibid.*, 73). Thus, naturalism requires that we be realists about teleology, *however metaphysically weird or mysterious or irreducible it may turn out to be*. In particular, it requires that we be realists about teleology even if the attempt to reduce it to efficient causation fails, as he suggests it does (*ibid.*, 84). Reductionists are at best *conditional* naturalists. Their realism about teleology is conditional on it being given a naturalistic reduction to more acceptable or familiar

[31] I am using 'paradigm' fairly loosely here. Whether these and related theories/approaches count as paradigms (initiating scientific revolutions) in the full Kuhnian sense I leave aside. For discussion of this issue see Godfrey-Smith (2007) and Jablonka and Lamb (2007).

phenomena. If the reduction fails, we should turn eliminativist. But our naturalism should not be conditional—shouldn't be hostage to successful philosophical analysis and ontological domestication—in this way. If an entity or force or process is indispensable for a branch of science in good standing, naturalism requires that we include it in our ontology, even if this offends against certain entrenched metaphysical scruples. If science cannot do without reference to irreducible entities, then the naturalistic philosopher has no choice but to embrace them.

An analogy Cameron doesn't use but that might support his case comes from physics. Modern physics posits all sorts of metaphysically weird entities and forces, from the irreducible indeterminacy of quantum mechanics to curved spacetime, from wave-particle duality to quantum entanglement. Naturalism seems to require that, other things being equal, we commit to these or include these in our ontology, even if they can't be interpreted in terms of or reduced to more familiar or metaphysically unproblematic entities, forces or processes. Irreducible teleology is in the same boat, arguably.

One possible response the reductionist (who accepts the etiological account) could make is that far from being an artificial or purely philosophically motivated imposition on or reinterpretation of the science, the etiological account in fact captures the way biologists themselves tend to understand biofunctions. The account thus makes explicit what is already implicit in evolutionary biologists' conception of teleology. This is why Neander (1991a) can claim that the etiological account is a kind of conceptual analysis of the way biologists use the concept of function. And it is arguably biologists, as much as reductionist philosophers, who have wanted to avoid the problematic metaphysics of unreduced teleology, who have wanted to cash out teleology in naturalistic (efficient-causation) terms, to render it ontologically respectable. Thus the analogy with physics breaks down. Physicists have, on the whole, embraced the physical entities and forces mentioned above, with all their apparent counterintuitive metaphysical implications. If our theories imply that some particles are in a superposition of states, or that some quantum systems are entangled, then we have to learn to live with that: that's just the way the world is, mind-boggling though it may be. This is clearly not the attitude biologists take towards the metaphysics of teleology. To the extent that they reflect on such matters at all, they are very likely to agree with the reductionist philosophers that to be scientifically respectable, teleology must be reducible to non-teleology. So the idea that emergent, irreducible teleology is, so to speak, *right there in the science* the way quantum entanglement with its holistic metaphysics is right there in the science, is implausible. *Teleology* may be right there, embedded in the science, indispensably. But the metaphysics of teleology is an open question, and it is much less plausible, it seems to me, that this question is settled by the science than it is that the metaphysical status of the entities and forces in physics is settled by the science (though of course there are philosophical debates about this as well).

Cameron may well agree with this. That the correct philosophical account of biological teleology is not settled by the science is something he would likely accept. But, he would claim, it cuts both ways. Perhaps emergent teleology as such is not implicit in the science. But the claim I alluded to above, that the etiological account merely makes explicit and rigorous a conception of function that is already implicit in the biology, is, he'd argue, no more plausible. In this I think he is basically correct. The etiological account was a hard-won philosophical insight, that took until Wright's landmark paper in 1971 to be properly articulated. The Millikan-Neander version of the account is rightly lauded as a major breakthrough in contemporary philosophy of biology precisely because it required careful, sustained philosophical analysis and argumentation, it wasn't so to speak lying there in the biology literature just waiting to be picked up and articulated by philosophers. The fact (if it is a fact) that biologists on the whole share the reductionist inclinations and attitudes of the naturalistic philosophers with respect to teleology does not imply that the etiological (or any other reductionist) account can be straightforwardly read off the science.

So granting that neither the emergentist view nor the reductionist view is implicit in or can be read off the science,[32] Cameron's claim about naturalism can be reformulated in the following way. Whether one is a consistent naturalist depends not just on one's actual view, but also on

[32] This is consistent with the reductionist view being metaphysically *closer to* the science than the emergentist view in the sense in which we have been understanding that notion.

one's so to speak *fall-back* view, revealing one's deeper commitments. Most philosophers with considered views on biological teleology are reductionist realists. That view in itself, insofar as it is involves realism, may be consistent with the NC. But suppose the reductionist project failed. What should one's fall-back view be? Cameron suggests that naturalism dictates it must be non-reductionist realism. (Of course for Cameron this is his actual view, not a fall-back view, as he thinks the reduction does fail.) Yet the fall-back view of the reductionists is very likely to be full-blooded anti-realism. If realism about teleology and an austere reductionist metaphysics with no place for emergent or irreducible entities or forces were incompatible, it is the former that must be abandoned. Yet this, Cameron objects, would involve rejecting, on philosophical grounds, an entity/force/causal factor in good scientific standing, which is a clear violation of the NC.

This is reminiscent of Searle's dispute with Fodor over intentional realism (mentioned above). Fodor famously stated that he was a physicalist-reductionist and an intentional realist, but that if this position turned out to be untenable, his fall-back position would be eliminativism about intentionality (Cain 2002). We should jettison intentional realism before we jettison physicalist-reductionism. Searle's response to this was that we can be much more confident that intentional realism is true than that physicalist-reductionism is, so in the event that the two are incompatible (as he thought they were) we ought to embrace non-reductionist realism about intentionality. This, with a naturalist twist, is essentially Cameron's position on teleology. We can be more confident that realism about teleology is true (largely due to its role for evolutionary biology) than we can be that a strict naturalist-physicalism with no place for emergent or sui generis features is true.

The claim I am suggesting Cameron would make about the fall-back position of the reductionist is more speculative than in the case of Fodor. *Would* her fall-back position be full-blooded anti-realism? Part of the problem is that the view that the etiological account represents a successful naturalistic reduction of teleology is so widespread that the question of fall-back positions tends not to arise. They're not needed! My sense is that most would find either fall-back view—full-blooded anti-realism or non-reductionist realism—about equally unattractive. Indeed it is precisely the unpalatability of each of these views that has motivated so much recent work attempting to articulate and make plausible forms of reductionist realism. The feeling has been that teleology must be real, given its ineliminable role in evolutionary biology; but that given how unappealing *strong* realism about teleology is, some form of naturalistic reduction without elimination has *got to* be true. Compare compatibilism about free will. Compatibilists are motivated by the sense that while determinism has to be true, we must also have free will. Therefore there must be *some* way of reconciling the two. Of course we can ask the compatibilist what her fall-back position would be, should she become convinced they were not compatible: hard determinism or libertarianism. And she might have a preference for one over the other, but her main response is likely to be that she hopes it doesn't come to that, as each is highly unappealing!

Now Cameron will object at this point that it is only because strong realism has been identified almost exclusively with supernaturalism (e.g. creationism) that it has seemed so unappealing (*ibid.*, 86). My sense however (and again this is somewhat speculative) is that the type of sui generis, emergentist view defended by Cameron would be considered by the orthodox almost as unappealing as supernaturalism (even if it is granted that it is consistent with metaphysical naturalism).[33] After all, roughly this sort of view has always been available in the form of Aristotle's strong realism about teleology, and despite a small number of philosophers in recent years attempting to rehabilitate a version of this view (as discussed above), for the most part it has not been considered a live option in the contemporary debate.

In any event, as I noted, whether the fall-back position of the reductionist-realists would be or should be full-blooded anti-realism or non-reductionist realism, and whether, if it was the former, this would represent an unacceptably conditional form of naturalism, will only arise if reductionist realism turns out to be untenable. As things stand,

[33] Just as most philosophers of mind find Searle's non-reductionist strong realism about intentionality unappealing, even though it clearly is not a supernaturalist view; it is rather quite similar to Cameron's emergentism about teleology.

reductionist realism is the overwhelmingly dominant view, and, we are supposing, it is consistent with the NC.

I conclude that Cameron's argument that emergentism satisfies the NC better than reductionist realism since the NC entails that realism about teleology is so compulsory that it must be maintained whatever metaphysical status teleology has, fails.

We began with the intuition that Cameron's view is further down Chakravartty's slippery slope, is less empirically vulnerable or responsive to the science, and thus more questionably an expression of the MoS stance, than the reductionist (e.g. the etiological) view. We saw that Cameron's arguments that this gets things backwards, were unpersuasive. Further support for the intuitive view can be found in Cameron's own discussion (*ibid.*, 78). Cameron makes the important point that the difference between his view and the orthodox view involves more than the type of entities involved, or whether the views are reductionist or emergentist. There are also important epistemological and methodological differences in approach. Key arguments Cameron offers against the etiological account rely on intuition. These include the intuition that swamp-man's heart has a function, and what he calls our 'categorial intuitions' (*ibid.*, 75–76) that a temporal cum efficient-causal relation is just not the sort of thing with which teleology can be identified, recalling the claim made by some non-physicalists about consciousness, such as Goff (2017), that physicalism is not just wrong, but a kind of category-mistake, since third-person, objective, physical objects and properties are just not the sort of thing with which first-person, subjective, phenomenal properties and states can be identified.[34]

Yet traditional philosophical methodology in which a priori reasoning, conceptual analysis and the appeal to intuitions have a central role is treated with scepticism by the orthodox. They tend, explicitly or implicitly, to endorse a more thorough-going naturalistic methodology for metaphysics, one in which folk intuitions about, say, whether swamp-man's heart would have a function are not treated as evidential, or at least, as not carrying much evidential weight. Recall that where we locate a

[34] Ruse argues (1973, 195–6) similarly that one can perhaps *replace* a teleological description with a non-teleological one, but one is not thereby *translating* a teleological description.

view or approach along the slippery slope often depends not just on the type of entities in question, but the type of methodology endorsed for metaphysics. A crucial component of the naturalistic approach to metaphysics is the eschewing of appeal to folk intuitions (Ladyman and Ross 2007). If our most theoretically plausible metaphysics of science is in certain respects counterintuitive to the folk, then so much the worse for those intuitions. Cameron on the other hand subscribes to the more traditional approach, for which our intuitions—including those of the folk, not just the scientific 'experts'—about swamp-man or category errors give us convincing a priori reasons to abandon the etiological view.

Cameron, we can say, has a foot in both camps. He states (*ibid.*, 84) that the methodology he wishes to employ in arguing for his view involves both deference to the ontic authority of science (which forces realism about teleology upon us), and deference to folk intuitions (which undermines the etiological account). Thus we should, I suggest, locate Cameron's view as straddling the line separating the MoS and armchair metaphysics stances, much like the forms of strong realism about natural kinds we discussed in the previous chapter such as Hawley and Bird's, due both to the character of the entities/forces/phenomena endorsed (emergent teleology) and to the methodology for metaphysics advocated. Like Hawley and Bird, Cameron appeals to very general, largely a priori philosophical considerations when articulating and defending his view, in this case to do with causation.

In general, strong realist views on teleology straddle the line between naturalist and non-naturalist metaphysics, such that whether the entities they posit—emergent properties, sui generis teleology—are consistent with naturalized metaphysics, is a matter about which there can be, and has been, legitimate, rational disagreement. As with the case of Hawley and Bird, I would argue that we are unlikely to establish whether such views satisfy the NC through the application of a simple all-or-nothing criterion or set of criteria on which all reasonable people can be expected to agree. It will again be a matter of degree: inferences to emergent properties and sui generis teleology are less direct, more epistemically risky, than inferences to functions as understood on, say, the reductionist etiological account. It is less clear how the science could help us decide whether emergent properties and sui generis teleology exist than is the case for functions understood as selected effects. The degree of underdetermination of the ontology by the science appears to be greater. As with universals, whether the naturalist regards emergent properties and sui generis teleology as too far removed from scientific evidence, too 'purely metaphysical', to be worth the epistemic risk, is a value-laden judgement call.

5 Conclusion

We have examined three broad positions on the metaphysics of teleology: full-blooded anti-realism, full-blooded realism, and reductionism. And we have examined the ways in which each is or could be an expression of, or require, a particular epistemic stance. Full-blooded anti-realism may be a product of the empiricist stance (or the SSR stance); at least, this is the case insofar as it is motivated explicitly or implicitly by a rejection of the putative virtue of explanatory power that stronger stances associate with committing to the reality of teleological functions. We have seen that the instrumentalist version of full-blooded anti-realism, according to which the phenomena are *as if* traits have functions, is one way of avoiding committing to realism about functions, with the claim that 'the best explanation for the fact that the phenomena are as if traits have functions is that they really do have functions' being dismissed as issuing from realist values and commitments the empiricist rejects.

The reductionist view, in committing to the existence of functions in the natural world partly due to their explanatory role, is an expression of the realist stance. Two features of the most prominent version of reductionism, the etiological view, indicate that it requires only a relatively weak version of the MoS stance. Firstly, its reductionism about functions entails that while it is (in most cases) a realist view about functions, the functions it commits to are, ontologically, nothing over and above selected effects that can be characterised in non-teleological terms. So it invokes no particularly inflationary metaphysical items. Its ontology is relatively 'close' to the science (at least when compared with full-blooded realism). It is ontologically committed to certain implicit entities of science, but its analysis of their nature shows them to be not at all metaphysically abstruse. Secondly, as Cameron notes, its defenders typically accept naturalism about philosophical methodology with its suspicion of intuition, thought experiments, and a priori reasoning.

The full-blooded realist view also commits to the existence of functions in the natural world partly due to their explanatory role, and thus is also an expression of the realist stance. But, as Walsh notes, its defenders lack the 'metaphysical squeamishness' of the reductionists. Thus they are happy to posit irreducible, sui generis teleology. Their metaphysics is more inflationary, more 'distant' from the science; they are more ontologically committed. Thus this view requires at least a strong form of the MoS stance, and, depending on one's perspective, may risk veering into armchair metaphysics territory. Whether irreducible, sui generis teleology is able to satisfy the NC is, as we have seen, controversial. In Cameron's version a further reason for supposing that the view is an expression of the armchair metaphysics stance is his explicit endorsement of armchair methods—intuition, thought experiments, and a priori reasoning—as being central to his defence of the view (though we have seen that he regards his view as being more naturalistically respectable than the orthodox reductionist view).

We can thus set up a kind of decision-tree. If you reject the realist stance, in particular its valuing of explanatory power, then, other things being equal, you ought to adopt full-blooded anti-realism about teleological functions (this is also true if you accept the SSR version of the realist stance). Belief in the reality of teleological functions is V-prohibited for you. If you accept the MoS stance and its valuing of explanatory power, you may (but don't have to) be a realist about teleological functions: belief in them is V-permissible for you. If you do accept realism about them, and you are metaphysically squeamish (i.e. you only endorse a weak version of the MoS stance), then, other things being equal, you should accept one or other reductionist account: strong realism is V-prohibited for you. If you accept realism about them, and you are not metaphysically squeamish (you accept a strong version of the MoS stance), then you may adopt (one or other version of) full-blooded realism (it is V-permissible for you).

This does not mean, of course (to reiterate another central theme of Part I), that one's precise views about the nature of teleological functions will or should be dictated by one's stance. Argument and evidence still play a vital role. If one is a metaphysically squeamish realist, then one is likely to be attracted to reductionism, but *which* of the many versions of reductionism one accepts will depend on how one assesses the relevant arguments and evidence. If one is a non-metaphysically-squeamish realist, then one may accept full-blooded realism, but which version of this one accepts will also depend on how one assesses the relevant arguments and evidence. The point is simply that—as we have seen throughout Part I—these questions of argument and evidence are *downstream from stance choice*. Unless you value explanatory power (more precisely, the putative explanatory power associated with committing to implicit scientific entities), you have no reason to be moved by—or even consider—any arguments or evidence favouring one or other realist view about functions. None of these positions is an option for you (in the sense of V-prohibitions) unless you have already accepted at least the MoS stance. Within the realist stance, unless you have embraced a strong version of the stance that is happy to invoke relatively inflationary metaphysical items, you have no reason to be moved by, or even consider, any arguments or evidence favouring one or other form of full-blooded realism. Walsh makes this clear when he implies that the account of irreducible teleology he offers will only be of interest to those who lack metaphysical squeamishness.

The same points apply with respect to the methodological dimension of stances. If one accepts a stance that permits only naturalistic methods, then one may only use such methods to argue for one's view: one will of course not be moved by non-naturalist, armchair or a priori arguments for rival views. But which views are best supported by application of the methods one's stance permits or encourages may be a matter of considerable controversy, as the extensive debates about functions *within* the naturalist-reductionist camp attest to.

So while argument and evidence play an important role, they are in a sense secondary to one's choice of stance. If there are no rationally convincing, objective, stance-independent reasons to value explanatory power, then there is no rational, stance—independent compulsion for any particular form of realism about teleology to be an option for you. The evidential arguments for the positions will not move you unless you have already adopted the stance that values explanatory power. When

Cameron says eliminativism can't account for or explain the biological facts (*ibid.*) this simply begs the question: the empiricist has no interest in *explaining* the biological facts or giving a metaphysical underpinning for the empirical success of evolutionary biology. And if there are no rational, objective, stance-independent reasons to not be metaphysically squeamish, there is no rational compulsion for any particular form of strong teleology to be an option for you.

These results are, I suggest, hard to deny, yet their implications are rather startling. It would appear that (and we may let this serve as the ultimate upshot of Part I) there are bound to be severe limits to the effectiveness of argument and evidence in these debates (given my claim at the end of Chap. 2 about values and the limits of rational argument); more precisely, that rational argument and appeals to evidence can only take place *within a* stance, that is, only come into play once a stance has been adopted, and thus that there is little prospect of those occupying rival stances rationally convincing one another to change their view.[35] Now this stalemate or impasse-like situation is of course very familiar in the traditional argument between scientific realists and anti-realist empiricists. As we have seen, it was precisely this impasse, this inability for realists and empiricists to convince one another, this persistent rational disagreement, that the epistemic-stance analysis was intended to explain, and, in a sense, reconcile us to. Such a situation is not as familiar in the debates in the metaphysics of biology that I have discussed (though one senses it here and there). In these areas, the suggestion about stance-based limits to rational argument is more like a prediction than an explanation. If my analysis is correct, we should *expect* to find the same limits to rational resolution, the same stance-based impasses, the same persistent rational disagreement, in these debates, as it is widely-understood that we do find in the standard realism-empiricism debate.

[35] This will inevitably recall Kuhn on the incommensurability of scientific paradigms, but I have argued elsewhere (2014) that while there are some similarities between stances and paradigms, the disanalogies are more significant.

References

Allen, C. & Bekoff, M. 1998. Biological Function, Adaptation, and Natural Design. In Allen, C., Bekoff, M., and Lauder, G. V., (eds.) *Nature's Purposes.* Cambridge, Mass.: MIT Press. 571–587.

Amundson, R., and Lauder, G. V. (1994). Function without Purpose: The Uses of Causal Role Function in Evolutionary Biology. *Biology and Philosophy* 9: 443–469.

Ariew, A. 2002. Platonic and Aristotelian Roots of Teleological Arguments. In Ariew, A., Cummins, R., and Perlman, M., eds. *Functions: New Essays in the Philosophy of Psychology and Biology.* Oxford: Oxford University Press. 7–32.

Armstrong, D. M. 1978. *Universals and Scientific Realism. Vols. 1 & 2.* Cambridge University Press.

Ayala, F. (1970). Teleological Explanations in Evolutionary Biology. *Philosophy of Science* 37: 1–15.

Bedau, M. 1998. Wheres the Good in Teleology? In. Allen, C., Bekoff, M., and Lauder, G. V., (eds.) *Natures Purposes.* Cambridge, Mass: MIT Press. 261–291.

Barham, R. 2011. *Teleological Realism in Biology.* Ph.D. dissertation, Graduate Program in the History and Philosophy of Science, Notre Dame, Indiana.

Bigelow, J., and Pargetter, R. (1987). Functions. *Journal of Philosophy* 84: 181–196.

Boghossian, P.A. 1990. The Status of Content. *Philosophical Review*, 99 (2). 157–184.

Bouchard, F. 2013. How Ecosystem Evolution Strengthens the Case for Functional Pluralism. In Huneman, P. (ed.), *Functions: Selection and Mechanisms.* New York: Springer.

Boucher, S.C. 2014. What is a philosophical stance? Paradigms, policies and perspectives. *Synthese*, 191(10): 2315–2332.

———. 2020. Methodological naturalism in the sciences. *International Journal for Philosophy of Religion*, 88 (1): 57–80.

———. 2021. Biological Teleology, Reductionism, and Verbal Disputes. *Foundations of Science* 26 (4): 859–88.

Brandon, R.N. 2013. A General Case for Functional Pluralism. In Huneman, P. (ed.), *Functions: Selection and Mechanisms.* New York: Springer.

Buller, D. 1998. Etiological Theories of Function: A Geographical Survey. *Biology and Philosophy* 13: 505–527.

———. 2002. Function and Design Revisited. In Ariew, A., Cummins, R., and Perlman, M., (eds.). *Functions: New Essays in the Philosophy of Psychology and Biology.* Oxford: Oxford University Press. 222–243.

Bunge, M. 1999. *The Sociology-Philosophy Connection.* New Brunswick, N.J.: Transaction.

Burgess, A. and Plunkett, D. 2013. Conceptual Ethics I and II. *Philosophy Compass* 8 (12): 1091–1110.

Cain, M.J. 2002. *Fodor: Language, Mind and Philosophy.* Cambridge UK: Polity.

Cameron, R. 2004. How to be a Realist about *Sui Generis* Teleology Yet Feel at Home in the 21st Century. *The Monist* vol. 87 (1). 72–95.

Chakravartty, A. 2007. *A Metaphysics for Scientific Realism: Knowing the Unobservable.* Cambridge: Cambridge University Press.

Cummins, R. (1975). Functional Analysis. *Journal of Philosophy* 72: 741–764.

———. 2002. Neo-Teleology. In Ariew, A., Cummins, R., and Perlman, M., (eds.). *Functions: New Essays in the Philosophy of Psychology and Biology.* Oxford: Oxford University Press. 157–172.

Dennett, D.C. 1984. *Elbow Room: The Varieties of Free Will Worth Wanting.* Cambridge, Mass.: MIT Press.

———. 1995. *Darwin's Dangerous Idea.* New York: Norton.

———. 2017. *From Bacteria to Bach and Back: The Evolution of Minds.* W. W. Norton & Company.

Fodor, J.A. 1987. *Psychosemantics.* Cambridge, Mass: MIT Press.

Garson, J. 2019. *What Biological Functions Are and Why They Matter.* Cambridge University Press.

Gayon, J. 2013. Does Oxygen Have a Function, or Where Should the Regress of Functional Ascriptions Stop in Biology? In Huneman, P. (ed.), *Functions: Selection and Mechanisms.* New York: Springer.

Godfrey-Smith, P. (1993). Functions: Consensus Without Unity. *Pacific Philosophical Quarterly* 74: 196–208.

———. 1994. A Modern History Theory of Functions. *Nous* 28: 344–362.

———. 2007. Is it a Revolution? Symposium paper on Jablonka and Lamb's *Evolution in Four Dimensions. Biology and Philosophy* 22: 429–437.

Goff, P. 2017. *Consciousness and Fundamental Reality.* Oxford University Press.

Gould, S. J., and Lewontin, R. (1979). The Spandrels of San Marco and the Panglossian Paradigm: A Critique of the Adaptationist Programme. *Proceedings of the Royal Society*, vol. B205: 581–598.

Griffiths, P. 1992. Adaptive Explanation and the Concept of a Vestige. In Griffiths, P. (ed.), *Trees of Life.* Kluwer, 111–131.

Huneman, P. 2013. Weak Realism in the Etiological Theory of Functions. In Huneman, P. (ed.), *Functions: Selection and Mechanisms*. New York: Springer.
Jablonka, E. & Lamb, M. J. 2007. The expanded evolutionary synthesis—A response to Godfrey-Smith, Haig, and West-Eberhard. *Biology and Philosophy* 22: 453–472.
Kitcher, P. 1993. Function and Design. *Midwest Studies in Philosophy* 18: 379–397.
Kraemer, D.T. 2014. Revisiting Recent Etiological Theories of Functions. *Biology and Philosophy*, 29: 747–759.
Ladyman, J. and D. Ross, with D. Spurrett and J. Collier. 2007. *Every Thing Must Go: Metaphysics Naturalized.* Oxford: Oxford University Press.
Lewens, T. 2004. *Organisms and Artifacts: Design in Nature and Elsewhere.* London: MIT Press.
Longy, F. 2009. How Biological, Cultural, and Intended Functions Combine. U. Krohs & P. Kroes (eds.), *Functions in Biological and Artificial Worlds: Comparative Philosophical Perspectives*. MIT Press.
Lorenz, K. 1966. *On Aggression*. Methuen Publishing.
Lowe, E. J. 2006. *The Four-Category Ontology: A Metaphysical Foundation for Natural Science*. Oxford: Oxford University Press.
Matthen, M. 1997. Teleology and the Product Analogy. *Australasian Journal of Philosophy*, 75 (1): 21–37.
Matthewson, J. 2020. Does proper function come in degrees? *Biology and Philosophy* 35, 39.
Mayr, E. (1961). Cause and Effect in Biology. *Science* 134: 1501–1506.
McGinn, C. 1993. *Problems in Philosophy: The Limits of Inquiry*. Oxford: Blackwell.
Millikan, R. G. (1984). *Language, Thought, and Other Biological Categories.* Cambridge, Mass: MIT Press.
———. (1989a). An Ambiguity in the Notion of Function. *Biology and Philosophy* 4: 172–176.
———. (1989b). Biosemantics. *Journal of Philosophy* 86: 281–297.
———. (1989c). In Defense of Proper Functions. *Philosophy of Science* 56: 288–302.
———. (1993). *White Queen Psychology and Other Essays for Alice*. Cambridge, Mass.: MIT Press.
———. 2002. Biofunctions: Two Paradigms. In Ariew, A., Cummins, R., and Perlman, M., eds. *Functions: New Essays in the Philosophy of Psychology and Biology.* Oxford: Oxford University Press. 113–143.

Mossio M., Saborido C., & Moreno A. 2009. An organizational account of biological functions. *British Journal for the Philosophy of Science* 60(4): 813–841.
Neander, K. (1991a). Functions as Selected Effects: The Conceptual Analysts Defense. *Philosophy of Science* 58: 168–184.
———. (1991b). The Teleological Notion of Function. *Australasian Journal of Philosophy* 69: 454–468.
———. (1995a). Explaining Complex Adaptations: A Reply to Sobers 'Reply to Neander'. *British Journal for the Philosophy of Science* 46: 585–587.
———. (1995b). Misrepresenting and Malfunctioning. *Philosophical Studies* 79: 109–141.
———. (2017) Functional Analysis and the Species Design. *Synthese* 194: 1147–1168.
Perlman, M. 2004. The Modern Philosophical Resurrection of Teleology. *The Monist* vol. 87 (1). 3–51.
———. 2009. Changing the Mission of Theories of Teleology: Dos and Don'ts for Thinking about Function. U. Krohs & P. Kroes (eds.), *Functions in Biological and Artificial Worlds: Comparative Philosophical Perspectives.* MIT Press.
Plantinga, A. 1993. *Warrant and Proper Function.* New York: Oxford University Press.
Preston, B. (1998). Why Is a Wing Like a Spoon? A Pluralist Theory of Functions. *Journal of Philosophy* 95: 215–254.
Reiss, J. (2009). *Not by Design: Retiring Darwin's Watchmaker.* University of California Press.
Rudder Baker, L. 1988. *Saving Belief: A Critique of Physicalism.* Princeton, NJ: Princeton University Press.
Ruse, M. 1973. *The Philosophy of Biology*. London: Hutchinson.
———. 2003. *Darwin and Design: Does Evolution Have a Purpose?* London: Harvard University Press.
Schwartz, P.H. 2002. The Continuing Usefulness Account of Proper Function. In Ariew, A., Cummins, R., and Perlman, M., eds. *Functions: New Essays in the Philosophy of Psychology and Biology*. Oxford: Oxford University Press. 244–260.
Searle, J.R. 1992. *The Rediscovery of the Mind.* Cambridge MA: MIT Press.

Walsh, D.M. 2013. Mechanism, Emergence, and Miscibility: The Autonomy of Evo-Devo. In Huneman, P. (ed.), *Functions: Selection and Mechanisms*. New York: Springer.

———. 2015. *Organisms, Agency and Evolution*. Cambridge: Cambridge University Press.

Wright, L. (1973). Functions. *Philosophical Review* 82: 139–168.

Part II

Metaphysical Stances

7

Introduction to Part II

1 Metaphysical Stances

Part I was concerned with *epistemic stances*: epistemic attitudes to scientific ontology. In Part II we turn to *metaphysical stances* as they feature in biology: metaphysical positions construed as stances that can act as presuppositions of and influences on biological science. We shall consider three debates in the metaphysics of biology: that between functionalists and structuralists, that between punctuationists and gradualists, and that between reductionists and holists. I shall argue that in each case, the relevant positions ought to be construed as metaphysical stances rather than factual beliefs.

Recall the key features of stances in van Fraassen's sense:

- They are not reducible to beliefs; they are *adopted* rather than believed, like an approach or policy.
- They are *non-propositional*; so they are not to be thought of as true or false.
- They are largely *values-driven*. One adopts a stance which coheres well with one's values, both epistemic and non-epistemic.

S. C. Boucher, *Empiricism, Naturalism and the Metaphysics of Biology*,
https://doi.org/10.1007/978-3-032-00555-7_7

- They are pragmatically justified in part in terms of their fruits; one adopts a stance at least partly on the basis of the *consequences* of doing so.[1]

When it comes to metaphysical stances, along with the above, I think support can be found in van Fraassen's writing for the following views on the relation between metaphysical stances and beliefs/theories:

- The relationship between a metaphysical stance and its associated beliefs is *contingent*, in the sense that, at least in principle, any metaphysical stance is logically compatible with any belief or set of beliefs (although some combinations of beliefs and stances will certainly be more natural than others).[2]
- Metaphysical stances and empirical scientific hypotheses neither entail one another, nor inductively support one another.[3]
- Metaphysical stances are (as I discussed in the Introduction) capable of (positively or negatively) influencing empirical science, although only in the context of discovery, not in the context of justification.
- Pluralism will often, perhaps typically, be the appropriate attitude to take to such metaphysical stances: since the question of truth and falsity drops out, adopting an attitude of pluralism—combined perhaps with an ideally non-dogmatic *preference* for a particular stance—becomes rationally permissible.

---

[1] See Boucher (2018a) for a detailed discussion of the question of how a stance can be rationally assessed.

[2] As I noted in Chap. 2, van Fraassen (2004, 173, 176) suggests that there is a certain 'pragmatic incoherence' involved in adopting a stance while denying certain characteristic beliefs naturally or typically associated with it. The stance does not logically conflict with the denial of the beliefs, but one who adopts the stance and denies the beliefs is like one who intends to become a hippopotamus while believing that that is not possible: the intention and the belief do not logically conflict, but one who has that intention with that belief is in a state of pragmatic incoherence akin to Moore's example of one who states 'P but I do not believe that P' (*ibid*; see also Rowbottom (2005, 206, 210)). On the relation between stances and the beliefs associated with them see Ladyman (2004, 139), Mohler (2007), and Boucher (2018b).

[3] This point (along with the previous one) follows trivially from the fact that stances are not beliefs, and are nonpropositional, so cannot stand in relations of entailment or justification. But it also coheres with van Fraassen's more general views on the relationship between science and metaphysics; see e.g. van Fraassen (1996, 149, 175).

I shall argue that in each of the debates I discuss, construing the positions as stances that satisfy the above conditions can help make sense of certain features of the debates, and helps us to appreciate the positive role such stances may play in science. As I noted in the Introduction, one need not be an empiricist to agree with this assessment, but the stance interpretation is particularly congenial for the type of empiricist who wishes to honour the role metaphysical assumptions play in science while avoiding the need to accept a central role for substantive metaphysics in science, at least when construed in the traditional inflationary factual-beliefs sense.

In making the case for the stance view I shall pursue two (related) argumentative strategies, one historically-oriented and one epistemically-oriented. The former is an application of van Fraassen's argument (mentioned in the Introduction), based on Hempel's Dilemma, for construing materialism as a stance. The argument takes as its starting point the historical continuity of the relevant metaphysical views and the fact that they appear to be associated with quite different, and incompatible, scientific theories and factual beliefs at different historical periods. Chapter 8 applies this argument to functionalism and structuralism. The second, more epistemic, argument, infers from the fact that (a) candidate factual theses with which we might identify the metaphysical position in question appear to be false, and (b) the positions nonetheless appear to play a positive role in science, to the conclusion that they should be understood as stances (rather than any of the candidate factual theses). The inference is far from deductive, but I hope to show that it is reasonable, even compelling, given various assumptions. Chapters 9 and 10 apply this argument to punctuationism, gradualism, holism, and reductionism in biology.

2 The Context of Discovery and the Context of Justification[4]

van Fraassen's views on the relation of metaphysics to science echo the familiar distinction in the philosophy of science between the context of discovery and the context of justification.[5] As this distinction will feature prominently in Part II, and it is not uncontentious, I need to say a word about it here. The context of discovery is the context in which scientists come up with theories and hypotheses, while the context of justification is the context in which those theories and hypotheses are tested, and come to be accepted or rejected. One of the most famous statements of it is Popper's: '[T]he act of conceiving or inventing a theory seems to me neither to call for logical analysis nor to be susceptible to it… the question of how it happens that a new idea occurs… may be of great interest to empirical psychology; but it is irrelevant to the logical analysis of scientific knowledge' (Popper, quoted in Ladyman (2002, 75)).

van Fraassen appears to hold that metaphysical commitments may influence scientists in the context of discovery, may assist, that is, in the formulation of specific scientific hypotheses. But they play no role in the context of justification—when scientists are testing a theory, and trying to decide whether or not it's acceptable, all that matters is whether it's empirically adequate; the question of its origins is irrelevant. '[A]n empirical hypothesis introduced into science for reasons, and perhaps with putative credentials, coming from outside that science, will so to speak *forget its origins* … so that its fortunes within science will soon have little or nothing to do with those origins' (1996, 157). I noted above that Popper defended the distinction. He also seems to have employed it in a similar way to van Fraassen (e.g. Popper 1963, 37–8. See Ladyman 2002, 72). Popper accepts the importance of metaphysical theories in inspiring and helping to generate scientific theories, which is in line with van Fraassen's view that metaphysical commitments in science may play a potentially valuable role 'inspiring' scientists in the context of discovery.

[4] This section draws on my (2019).

[5] For classic statements and defenses of the distinction, see Hempel (1966, 3–18), Popper (1963, 42–59), and Reichenbach (1938). See Schickore and Steinle (2006) for an extended discussion.

In early twentieth century philosophy of science the distinction between the two contexts was, for its defenders such as Popper and Reichenbach, a means by which both to recognise, and neutralise, the non-rational and subjective elements that it was understood play a role in scientific theorising, but that seemed to escape analysis in terms of logic and inference. Scientists may be inspired by dreams or religious beliefs, may be suddenly struck by inspiration, may pay excessive heed to unsupported hunches and guesses, may be subject to biases, and so forth. It was conceded that the nonrational elements play a role in the process of science, but it was claimed that this role (in the context of discovery) was logically distinct from the rational part (in the context of justification). Thus the nonrational part could in effect be bracketed, and, for the purpose of analysing the logic of scientific inference, ignored. There was no reason to think it in any way infected the rational part. As Longino (1990, 172) puts it:

> This distinction enables positivists to acknowledge the play of subjective factors in the initial development of hypotheses and theories while guaranteeing that their acceptance remains untainted, determined not by subjective preferences but by observed reality. The subjective elements that taint its origins are purged from scientific inquiry by the methods characteristic of the context of justification: controlled experiments, rigorous deductions, etc.

This approach was thrown into question by Kuhn's analysis in the early 1960s.[6] Far from being safely sealed off from the properly rational process of scientific inference, Kuhn suggested that non-rational and subjective elements intervened at every level, ruling out the possibility of giving a general account of scientific methodology that showed it to be a logical, purely rational process (Kuhn 1996). Thus the distinction between the two contexts fell into disrepute.[7]

[6] Others who rejected the distinction included Hanson (1971) and Feyerabend (1975).

[7] The distinction has also been challenged from, as it were, the opposite direction. Whereas Kuhn highlighted the non-rational or values-based elements contaminating the context of justification, others have argued that the process of discovery may be susceptible to rational scrutiny and analysis no less than the context of justification: something like a 'logic of discovery' may be feasible after all (Schickore 2022).

A version of the distinction can perhaps be defended however. Firstly, Kuhn's arguments may still presuppose a distinction between rational and nonrational components of scientific method, which may be all that defenders of the distinction require.[8] Even if nonrational and subjective elements do *in fact* intervene at every level of scientific methodology, we can still point out that they shouldn't, and still attempt to bracket them *normatively* when characterising the scientific method normatively.[9] Secondly however, we can seemingly disassociate the distinction from the logical empiricist and falsificationist philosophies of science with which it is usually associated. We can accept the general consensus in the philosophy of science that the project of characterising *the* scientific method is hopeless, while still insisting that it makes sense to distinguish between the two contexts. I'd suggest that the insight informing the distinction between the two contexts—that the origin of a hypothesis is irrelevant to the question of its truth-value, or whether it's justified; essentially that we should be wary of committing the genetic fallacy—is a valid insight, and continues to be so even once we have abandoned the approaches and views—such as the idea that the context of justification is a values-free-zone—with which it is typically associated (Ladyman 2002, 75–6).

As Hoyningen-Huene (1987) points out, the distinction has traditionally been associated with (and has served the purpose of legitimating) several other distinctions e.g. between discovery and justification understood as temporally distinct processes; between the proper domain of the philosophy of science (context of justification) and the domain of other disciplines such as sociology and psychology (context of discovery); and between the domain of logical analysis (justification) and the domain of empirical description (discovery). But the distinction is, Hoyningen-Huene argues—and I think he is correct—only contingently related to these other distinctions, thus we need not accept or reject them all as a package-deal. Hoyningen-Huene suggests that at bottom the distinction

[8] See Siegel (1980, 309–313) for this sort of response.

[9] Kuhn responds to this sort of defence of the distinction in his (1977), in which he argues that the contamination of the context of justification by subjective, individual, and value-based elements from the context of discovery, is not a contingent matter from which we can abstract to give an admittedly idealised account of the scientific method, but follows necessarily from the very nature of the context of justification itself.

reduces to the relatively uncontroversial distinction between the normative and the factual; the former pertains to the context of justification, the latter to the context of discovery. We should distinguish factual questions concerning the origin of theories or hypotheses from normative questions concerning whether they are justified. So long as this distinction is not conflated or saddled with other, more problematic distinctions, he suggests the distinction between the contexts is legitimate.

We may compare the history and fortunes of the distinction to another celebrated distinction, that between analytic and synthetic statements. In Quine's famous critique of the latter distinction (1951, 1953), he connected it with other distinctions, such as that between the a priori and the a posteriori, the necessary and the contingent, the epistemic and the pragmatic, and between science and philosophy. The general view on which these are all hard and fast distinctions, with strict correspondences between them, reflected a distinctive empiricist philosophical outlook that he was concerned to dislodge. But the conclusion that the analytic-synthetic distinction had no value or role whatsoever for philosophy or science has more recently come to be seen by some philosophers as premature (e.g. Sober 2000).[10] Much the same could be said, I'd argue, for a suitably modest and qualified version of the distinction between the two contexts. Undoubtedly it was, as originally formulated and defended, a component part of a set of distinctions and concepts that reflected a particular logical empiricist outlook on science that cannot perhaps be accepted in quite that form today. But we need not throw out the baby with the bathwater. For my purposes, all I require is that we recognise a rough and ready, vague and context-dependent, and perhaps somewhat idealised, distinction between factors that, as a matter of fact, influence the identification and formulation of scientific theories and hypotheses, and factors that are judged to be (normatively) relevant to their rational acceptance by the scientific community. If my particular employment of the distinction in the chapters that follow, with respect to the role played by metaphysics in science, is thought to be illuminating, that can further support the general validity and value of the distinction.

[10] Although see Williamson (2007, chs. 3 & 4) for a detailed discussion and critique of different versions of and arguments for analyticity.

References

Boucher, S.C. 2018a. Stances and epistemology: values, pragmatics and rationality. *Metaphilosophy* 49 (4): 521–547.

———. 2018b. What is the relation between a philosophical stance and its associated beliefs? *dialectica* 72 (4): 509–524.

———. 2019. An Empiricist Conception of the Relation between Metaphysics and Science. *Philosophia,* 47 (5): 1355–1378.

Feyerabend, P. K. 1975. *Against Method: Outline of an Anarchistic Theory of Knowledge* (London: New Left Books).

Hanson, N. R. 1971. The idea of a logic of discovery. In N. R. Hanson, *What I Do Not Believe and Other Essays* (Dordrecht: Reidel). 288–300.

Hempel, C.G. 1966. *Philosophy of Natural Science.* Englewood Cliffs, NJ: Prentice-Hall.

Hoyningen-Huene, P. 1987. Context of Discovery and Context of Justification. *Studies in the History and Philosophy of Science,* 18 (4): 501–515.

Kuhn, T. 1977. Objectivity, Value Judgment, and Theory Choice. Reprinted in M. Curd and J.A. Cover (eds.). 1998. *Philosophy of Science: The Central Issues.* New York, NY: W.W. Norton & Company. 102–118.

———. 1996. *The Structure of Scientific Revolutions.* Chicago: The University of Chicago Press.

Ladyman, J. 2002. *Understanding Philosophy of Science.* London: Routledge.

———. 2004. Empiricism Versus Metaphysics. *Philosophical Studies,* 121 (2). 133–145.

Longino, H. 1990. Values and Objectivity. Reprinted in M. Curd and J.A. Cover (eds.). 1998. *Philosophy of Science: The Central Issues.* New York, NY: W.W. Norton & Company. 170–191.

Mohler, C. 2007. The Dilemma of Empiricist Belief. B. Monton (ed.). *Images of Empiricism: Essays on Science and Stances, with a Reply from Bas C. van Fraassen.* Oxford: Oxford University Press. 209–228.

Popper, K. 1963. *Conjectures and Refutations.* London: Routledge and Kegan Paul.

Quine, W.V.O. 1951. On Carnap's Views on Ontology. *Philosophical Studies,* 2 (5). 65–72.

———. 1953. Two Dogmas of Empiricism. In Quine, W.V.O. *From a Logical Point of View.* Harvard University Press.

Reichenbach, H. 1938. *Experience and Prediction.* Chicago: University of Chicago Press.

Rowbottom, D.P. 2005. The Empirical Stance vs. The Critical Attitude. *South African Journal of Philosophy*, 24(3). 200–223.
Schickore, J. 2022. Scientific Discovery. *The Stanford Encyclopedia of Philosophy* (Winter Edition), Edward N. Zalta & Uri Nodelman (eds.), URL = https://plato.stanford.edu/archives/win2022/entries/scientific-discovery/
Schickore, J. and F. Steinle. 2006. *Revisiting Discovery and Justification. Historical and Philosophical Perspectives on the Context Distinction*. Dordrecht: Springer.
Siegel, H. 1980. Justification, discovery and the naturalizing of epistemology. *Philosophy of Science* 47, 297–321.
Sober, E. 2000. Quine's Two Dogmas. *Proceedings of the Aristotelian Society* 74: 237–280.
Van Fraassen, B.C. 1996. Science, Materialism and False Consciousness. J.L. Kvanvig (ed.). *Warrant in Contemporary Epistemology: Essays in Honor of Plantinga's Theory of Knowledge*. Lanham, Md.: Rowman and Littlefield. 149–181.
———. 2004. Replies to Discussion on *The Empirical Stance*. *Philosophical Studies*, 121 (2). 171–192.
Williamson, T. 2007. *The Philosophy of Philosophy*. Blackwell.

8

Functionalism and Structuralism

1 Introduction

In this chapter I consider the broad perspectives in biology known as functionalism and structuralism,[1] as well as a modern version of functionalism, adaptationism. Adapting van Fraassen's argument for treating materialism as a metaphysical stance, I will in the first part of the chapter offer an argument for construing functionalism and structuralism as metaphysical stances also. The argument draws especially on Gould's (2002) insights concerning functionalism and structuralism, in particular their apparent historical continuity from the pre-Darwinian period until today. In the second part of the chapter I consider Godfrey-Smith's distinction between empirical and explanatory adaptationism, and suggest that while the former is an empirical scientific hypothesis, the latter is closely related to the functionalist stance.

This chapter draws on my (2015).

[1] Also sometimes called 'formalism'.

S. C. Boucher, *Empiricism, Naturalism and the Metaphysics of Biology*,
https://doi.org/10.1007/978-3-032-00555-7_8

One doesn't have to be an empiricist to accept the arguments I will offer. However construing views such as functionalism and structuralism as metaphysical stances is congenial to empiricism. It allows the empiricist to do justice to the role played by metaphysical assumptions and commitments in science, while upholding the traditional empiricist thesis that science does not, or should not, incorporate metaphysics in the traditional sense.

2 Functionalism and Structuralism Past and Present

In 1979 Gould and Lewontin published their extremely influential *Spandrels* paper (1979; see special issue of *Biology and Philosophy*, Forber 2009). This paper, which critiqued what Gould and Lewontin saw as a narrow and damaging 'adaptationist programme' in evolutionary biology, kicked off a large and often vitriolic debate in biology and philosophy between adaptationists and constraint advocates over the power of and limits on selection (see, as a small sample of the classic sources, Dennett 1995; Dupré 1987; Hull and Ruse 1998, Part 1; Orzack and Sober 2001; Gould 1997, 2002). However as a number of people have noted, this controversy is really just the latest manifestation of a philosophical debate that goes back centuries (see e.g. Russell 1916; Gould 2002; Grene and Depew 2004; Appel 1987). The debate is over the question of whether in the organic world function is prior to structure, or structure prior to function. Before Darwin, functionalists, such as Cuvier (1800–1805) and Paley (1802) held that the 'conditions of existence' principle was the most important principle for biology. The structure of organic forms was subordinate to the functional use to which it was put; structure was to be explained in terms of function—function was primary. Structuralists on the other hand, such as Geoffroy Saint-Hilaire (1818), von Goethe (1790) and Agassiz (1857), held that structure was primary; the functional use of organic structures was a secondary overlay on the fundamental structural basis the nature of which was to be explained *independently* of the functional use to which it was put. These thinkers were most struck by the 'unity of type' found in the organic world—the fact that all organic designs appear to be variations on a limited number of common underlying themes, types, patterns or forms.

This debate was transformed by the Darwinian revolution, but it persisted in an evolutionary context. In this new context, the functionalist position was occupied by those who held that adaptation and selection were primary; traits were to be explained in terms of their evolutionary function and adaptive significance (e.g. Dawkins 1976, 1982, 1996; Dennett 1995). The structuralist position was occupied by those who held rather that the distribution of organic forms in morphospace was to be explained in large part by *internal* factors, such as developmental and structural constraints (e.g. Gould 1997, 2002). They held that organic structures were often to be explained by nonselective factors—that is, structure was (often) to be explained independently of function.[2]

Gould in particular has stressed the historical continuity of the functionalist-structuralist dichotomy, before and after the Darwinian revolution. It is true, he notes, that Darwin's theory of evolution by natural selection may have, to some degree, reconciled the 'conditions of existence' and 'unity of type' principles, but, he argues, this 'scarcely rendered the old dichotomy irrelevant' (2002, 278).

> Several themes pass right through this great revision [Darwinism], only altering their terms and explanations. Formalism [structuralism] vs. functionalism may be the most prominent and persistent of issues too grand even for evolution to undo (or fully resolve). Paley and Agassiz once fought this battle …; Dawkins and Goodwin … pursue the same conflict. (*Ibid.*, 279; see also 261)

> This dichotomy set the major debate of pre-Darwinian biology: does God reveal himself in nature primarily by the harmony of taxonomic structure, or by the intricacies of particular adaptations? This dichotomy continues to define a major issue in modern evolutionary debates: does functional adaptation or structural constraint maintain priority in setting evolutionary pathways and directions? (*Ibid.*, 252)

[2] For some recent discussions of the controversy see Novick (2023), Winther (2015), Brown (2023), Sultan (2015), Boucher (2025), and Amundson (2021).

As I said, I am not concerned in this chapter to take a side in the debate between functionalists and structuralists. My concern is rather with the prior question, what sort of positions are functionalism and structuralism? I will argue they are metaphysical stances in van Fraassen's sense. To establish this I will adapt van Fraassen's argument (1996, 2002) for treating materialism (physicalism) as a stance rather than a factual belief, which I spelled out in the Introduction.

3 Functionalism and Structuralism as Metaphysical Stances

Recall from the Introduction that van Fraassen suggests that only by construing materialism as a metaphysical stance can we escape from Hempel's dilemma (Hempel 1980). Van Fraassen asks the question, how should we define materialism? Perhaps as the view that 'matter is all there is' (van Fraassen 2002, 49). But until we say what matter is this has the feeling of vacuity. Perhaps we should define materialism as the view that 'everything is composed solely of atoms or elementary particles' (*Ibid.*, 52). van Fraassen points out that this will not work: 'When Newton introduced forces in addition to bodies, did he deny [materialism]? Forces are not composed of particles. When Huyghens's waves-in-the-ether theory defeated Newton's particle theory of light, was that a setback for materialism? Surely not, although the ether was a continuous medium, not particulate' (*ibid.*). Materialism has sometimes been defined instead as the view that everything that exists is located in spacetime. But, van Fraassen notes, materialism was not refuted when the Copenhagen school of quantum mechanics said that elementary particles lack well-defined spatial locations. It begins to look as if materialism is a view that is capable of surviving the overthrow of any particular scientific theory or worldview, and thus cannot hope to obtain its empirical content from the empirical content of current science. Materialists thus face 'a dilemma': when materialism is 'tied to current scientific theories, [it] must die with those theories; but if not, it seems to lack content all together' (*ibid.*, 53). The problem, van Fraassen argues, is with the interpretation of materialism as a factual claim. Materialism might be something worthwhile, but if so, it is not as a contentful factual belief, but rather as a stance.

His argument can be set out as follows:

1. If materialism has content as a factual belief, the understanding of 'physical' must either be tied to current physics or future physics.
2. If it is tied to current physics, materialism is very likely to turn out false.
3. If it is tied to future physics, materialism is empty.
4. Thus, if materialism is a factual belief, it is, very likely, either false or empty.
5. Therefore, materialism shouldn't be construed as a factual belief.
6. If it is not a factual belief, it is a stance.
7. So, materialism is a stance.

Those who defend the meaningfulness and validity of materialism as a doctrine/belief would typically deny either premise (2) or premise (3). Vicente (2011) for instance, denies premise (2), while Dowell (2006) denies premise (3). Melnyk (1997), however, accepts both these premises, but still regards materialism as a factual claim that is worthy of belief, thus he rejects the inference to (5). Others deny (1), as they claim that materialism can be defined independently of physics, by reference to an intuitive 'paradigm case' understanding of the physical (Donahue 2024). It is also possible to accept the argument to (5) but reject (6): maybe materialism is neither a belief nor a van Fraassian stance. For my part, I believe none of these objections succeeds and the argument goes through (Boucher 2020), but I do not need to defend that claim here. Whether or not the argument about materialism is convincing, I will argue that a similar argument with respect to functionalism and structuralism in biology *is* convincing.

3.1 Divergent Beliefs, Shared Stance

We have seen that functionalism and structuralism come in pre-evolutionary and evolutionary versions. They have each survived the

Darwinian revolution in biology in just the way materialism survived successive revolutions in physics. As with empiricism and materialism (according to van Fraassen), it appears that what functionalists and structuralists over the centuries have had in common is not a set of beliefs about the world but a distinctive attitude or stance. Pre-evolutionary functionalists and structuralists had very different beliefs about the natural world, and very different theories, from evolutionary functionalists and structuralists. Pre-evolutionary functionalists such as Cuvier (1800–1805) and Paley (1802) would characterise their functionalism in terms of God's designing each part of an organism so that the organism was exquisitely well adapted to its environment (the pre-evolutionary 'conditions of existence' principle) while Darwinian functionalists such as Dawkins (1976, 1982, 1996) and Dennett (1995) would characterise their functionalism in terms of the work of natural selection in building adaptations. It is clear that the role of teleology in biology (the understanding of purpose, design and function) was radically altered in the Darwinian revolution. Whether we think of Darwin as naturalizing teleology (as some argue, e.g. Lennox 1993, 1994) or as dispensing with it altogether (as others argue, e.g. Ghiselin 1994) isn't important here. What is clear is that after Darwin, biological functions were understood in terms of history and selection. The function of a trait is the effect it has for which it is an adaptation, and for Darwinians adaptations are by definition built by natural selection. As I discussed in Chap. 6, this is sometimes called the etiological account of function—as it's often expressed, a trait's function is the effect it has which explains why there was selection for that trait (Wright 1973; Millikan 1989; Neander 1991)—and it obviously did not predate the Darwinian revolution.

The same goes for pre-evolutionary and evolutionary structuralists. The pre-Darwinian understanding of 'unity of type' typically involved ahistorical, transcendentalist ideas about abstract, other-worldly, sometimes Platonic 'archetypes' underlying and constraining organic diversity (Ruse 1979, 94–132; Gould 2002, 251–342). These archetypes, or ideal forms, tended to be thought of as ideas or themes in God's mind (e.g. Agassiz 1857), and taxonomic relationships among organisms were thought to reflect God's orderly plan for the natural world.

The Darwinian explanation of 'unity of type' and homology on the other hand appeals to history and inheritance. All arthropods share the same basic body plan not because they are variations on an abstract, pre-existing archetype, but because they have inherited that body plan from a common ancestor, and it has been retained by all subsequent arthropod lineages. Later structuralists such as Gould, Lewontin and others would also invoke the notion of constraints on natural selection to account for the phenomenon of unity of type (about which I will say more below).[3] Again, these ideas—history, phylogenetic inheritance from common ancestors, constraints on selection—did not predate the Darwinian revolution.

So the Darwinian revolution brought about a revolutionary shift in the understanding of both function and structure in the natural world. And yet the old and the new thinkers clearly have something in common. Paley and Dawkins manifestly have something like a commitment to functionalism in common; Dawkins is happy to identify with aspects of the Paleyan tradition, despite Paley's anti-evolutionism (Dawkins 1987). As for structuralism, Gould (2002, 1089–1122) claims that modern-day evolutionary developmental biology, with its strongly structuralist implications, has 'vindicated' the general approach of Geoffroy, the pre-eminent pre-Darwinian structuralist. And as Gould notes in the quote above, the modern disputes (in an evolutionary context) between functionalists (or, at least, those strongly influenced by functionalism) such as Dawkins (1976, 1982, 1996), Dennett (1995), Maynard Smith (1987), Cronin (1991), Williams (1992) and Ruse (2003) on the one hand, and structuralists (or those influenced by structuralism) such as Gould (2002), Lewontin (Gould and Lewontin 1979), Goodwin (1994), and Kauffman (1993) on the other, recall, in their broad philosophical outlines, the pre-Darwinian disputes between functionalists such as Paley (1802) and Cuvier (1800–1805) on the one hand, and stucturalists such as Geoffroy Saint-Hilaire (1818), von Goethe (1790), Agassiz (1857) and, in some of

[3] Indeed accepting the bare notion of inheritance from a common ancestor as accounting for observed unity of type is not sufficient to make one a structuralist, since the idea is fully compatible with Darwinian functionalism, just as the bare notion that selection explains adaptation is (at least on most accounts) not sufficient to define functionalism or, later, adaptationism (see below).

his moods, Owen (e.g. his work on the 'vertebrate archetype'—Owen 1848; see Gould 2002, 312–330, and Ruse 1979, 116–125), on the other.

We can understand this, I suggest, in terms of beliefs and stances. The pre-evolutionary functionalist's *belief* that local adaptations are the most significant and revealing feature of the organic world for those who would look for the mark of God's hand in nature is plainly a very different belief from the modern functionalist's belief that evolutionary pathways are set by selection and are not significantly restricted by historical, structural and developmental constraints. But there is seemingly a common philosophical *stance* here—both approach the natural world in a certain way, from a certain perspective, with a tendency to focus on, emphasise, and find meaning and significance in certain features, and de-emphasise others.

Gould writes:

> We often blunder in our historical understanding by assuming that evolution must be an ultimate watershed, marking a complete break between a bad before and an enlightened after. In fact, many continuities pass right through Darwin's rupture of history, with evolution only providing a different explanation for unaltered principles and phenomena. The good ship Dichotomy – Unity of Type vs. Conditions of Existence – entered the Darwinian current by converting its terms from a debate about God's primary mode of self-expression in nature to an argument about constraint and adaptation in evolution … (2002, 252–253)

The mistake Gould refers to—thinking there can be no substantial or interesting continuity across the Darwinian boundary—can perhaps be explained as resulting from a lack of appreciation of the stance/belief distinction. With the distinction in hand, we are able to make sense of both the continuities and the discontinuities across this divide.

3.2 An Objection and a Reply

It might be argued of course that what the old and new thinkers have in common *is* a belief or set of beliefs about the natural world: functionalists believe that, for example, function is prior to structure, and structure is to be explained in terms of function; structuralists believe that structure is to be explained independently of function. But the friend of stances will ask whether these beliefs, at that level of abstraction and generality, have sufficient content to do the job required of them of expressing the central commitment shared by functionalists and structuralists of different eras. What is meant by 'function', 'structure'—not to mention 'prior'—here? Can we give a specification of their meanings that is sufficiently determinate to render the beliefs in question contentful and truth-evaluable, yet sufficiently general that they abstract away from the theories of the natural world—the theories of function, structure etc.—accepted by the thinkers in question, and with respect to which the old and new thinkers differ?

We can, in other words, pose a dilemma for those who wish to characterise the positions as beliefs. Either the concepts used to express the beliefs will remain un-filled-out, in which case the positions will arguably be so vague and under-specified that they will lack content as beliefs, and thus ought to be conceived of rather as stances; or the concepts will be given content in terms of the specific theories of function and structure accepted at the time, in which case the positions will be submerged into those different theories and the unity between the old and new thinkers will be lost.

The general problem for the stance theorist is that once we explicate the content of a stance—once we articulate the attitudes and so forth it is characterised by—there seems, at least sometimes, to be a way in to construing it as a belief. So if we assert that functionalism as a stance is an inclination or disposition or commitment to approach the biological world by focusing on function and according it explanatory primacy, the stance sceptic might complain that this is not fundamentally different from asserting that functionalism is the belief that function is prior to structure, or something along those lines. This issue will arise for all positions construed as stances. For instance, if we say that materialism as a stance at time t is a tendency/inclination to defer ontologically to physics as constituted at time t, is this really different from saying that materialism is the belief that, say, all of reality can be captured in the language of present-day physics?[4]

[4] Thank you to Daniel Stoljar for urging this objection to me.

The stance theorist replies that while it will always be possible to artificially construct a statement of a belief with which to characterise the position in question, the belief that is constructed is unlikely to adequately capture what proponents of the position have in mind, and is unlikely to do full justice to the value that its proponents believe it possesses. The key thought here is that positions-as-beliefs must have sufficient content to be evaluable as true or false; yet when they are given this content they seem to lose their identity, and become something other than what their proponents thought they were. Consider materialism. It seems that defining materialism-as-a-belief in terms of future physics renders the position contentless; but if we try to give it sufficient content to be evaluable as non-trivially true or false by building into it assumptions about present-day physics, we turn it into a position that will very likely turn out to be false, but more to the point, we turn it into something quite different from what its advocates intended it to be. Whatever materialism is, it is surely not a position whose fate is tied to that of the specific theories that comprise contemporary physics. (And this approach would entail that materialism in the past was tied to the fate of the physical theories of the past, in which case it would have died with those theories, which it plainly didn't.) Similarly, characterising functionalism as the belief that function is prior to structure does not seem to give it enough content to be really meaningful; yet building in assumptions about (say) the nature of function and adaptation undermines functionalists' firm intuitions that they stand in a meaningful functionalist tradition that has survived revolutionary changes in theory. For example, any mention of natural selection in the characterisation of Dawkins' functionalism will undercut his belief—widely shared, and surely justified—that he and pre-Darwinian thinkers such as Paley are united in their functionalist commitments. (Yoking materialism to contemporary physics also rules out the existence of a materialist tradition that pre-dates contemporary physics.) Positions-as-stances on the other hand do not possess propositional content, and are not evaluable as true or false, so these problems simply do not arise.

On this way of looking at the matter, the apparently radical and revisionary stance conception is actually seen to be *conservative* in its aims and motivation. Conceiving of metaphysical positions as beliefs that must be given specific propositional form often forces us to characterise them in ways that diverge considerably from the intentions and commitments of their proponents. Interpreting them as stances on the other hand allows us to preserve their essential nature and function. Hence van Fraassen's claim (2002, 60) that only the stance conception 'gives materialism its due'. Thus, although the stance conception undoubtedly clashes with theorists' *explicit, conscious* views about the character of these positions, at a deeper level, it aims to represent and do justice to the implicit, unconscious understanding of them revealed by the history of debates in which they figure. In other words, theorists might *claim* the positions are factual beliefs, but their actual practice of proposing, debating and defending them suggests that they are *treating them* as van Fraassian stances. Or so the stance theorist argues.

3.3 Summary of the Argument

Summarising the discussion of Sect. 3, the general argument for interpreting functionalism and structuralism as stances is as follows. Functionalism and structuralism have survived the Darwinian revolution. How can we give a characterisation of these positions that renders this fact intelligible? If the positions were tied to theories about the nature of function and structure, they wouldn't have so survived. But if they were not so tied, they must be lacking in empirical content, and so be empty, if interpreted as beliefs. Yet their role in biological theorising gives us reason to deny that they are empty. Thus we should interpret them as metaphysical stances.[5]

Or, schematically:

[5] Rose Novick (2023) takes issue with my interpretation of functionalism and structuralism as stances. She argues they should be seen rather as 'explanatory strategies', proposing this as a compromise between my account (which is, she claims, overly abstract) and the view defended by Winther (2015) that they are Kuhnian paradigms (which is, she claims, not abstract enough). She argues that while the stance interpretation captures the continuity of these views over time and across theory change, it doesn't 'elucidate the kinds of empirical evidence relevant to settling structure/function disputes' (*ibid.*, 11). See Boucher (2025) for a reply.

1. Functionalism and structuralism have survived the Darwinian revolution.
2. If functionalism and structuralism have survived the Darwinian revolution, they are not essentially tied to theories about the nature of function and structure.
3. If they are not essentially tied to theories about the nature of function and structure, and they are beliefs, then they are empty and uninteresting.
4. They are not empty and uninteresting.
5. So they are not beliefs.
6. If they are not beliefs, they are stances.
7. So they are stances.

The argument is deductively valid. (1) and (4) should be uncontroversial. In this section I have defended (2) and (3). If (6) is accepted also, the argument goes through.

One apparent difference between this argument and the classic formulation of the materialism argument I presented above is that this argument refers to the past, whereas the materialism argument in its typical form refers to the future. But I trust it is clear that they are making the same basic point: the metaphysical stances in question cannot be essentially yoked to particular theories (whether in the past, the present or the future) lest they share their fate (and lest their historical continuity be lost). We have seen that van Fraassen's version of the materialism dilemma (2002) focuses more on materialism's past that its future. This means that the argument about functionalism and structuralism doesn't depend on the likelihood of future revolutionary paradigm shifts in biology (which might be thought less likely than in physics), since the claim is that whatever happens in the future, the stance view is still needed to make sense of the past and present.

4 Empirical and Explanatory Adaptationism

Setting aside for the moment the above argument, and the question of stances vs. beliefs, a natural way of understanding the dispute between functionalists and structuralists is as a dispute about *emphasis*. Both functionalists and structuralists (at least of the Darwinian variety) recognise that both internal and external factors—structure and function—play an important role in evolution. But functionalists emphasise external over internal factors, and structuralists emphasise internal over external. The issue is not whether internal or external factors enjoy exclusive sway or exercise total control over the evolutionary process, but about the *relative importance* of each, in relation to the other. Thus Gould notes that the functionalist orthodoxy does not deny that internal constraint plays a role, but it systematically marginalises and downplays it, acknowledging it only as a secondary, unimportant and subsidiary phenomenon. He thinks constraint deserves to be highlighted as centrally important, not confined to the margins.

But these sorts of disagreements about emphasis are open to two possible interpretations. The first takes them to be empirical, factual disagreements: the two sides are disagreeing about the types of phenomena, facts and processes that are objectively most common in nature, that are causally responsible for more things, that happen at a higher *relative frequency* than other things.[6] On the second interpretation, the disagreements are nonfactual and nonempirical; they concern the most interesting phenomena, facts and processes where 'interesting' is interpreted in terms of human preferences, purposes and values. On the latter interpretation, the two sides may agree on the facts that obtain, and only disagree about which are most (subjectively) interesting; on the former they don't agree on the facts.

Godfrey-Smith's (2001) distinction between empirical and explanatory adaptationism can be understood against this background. He defines them as follows:

---

[6] See Gould (2002, 255–6) for a discussion of the form that such arguments about relative frequency tend to take.

> **Empirical Adaptationism**: Natural selection is a powerful and ubiquitous force, and there are few constraints, except general and obvious ones, on the biological variation that fuels it. To a large degree, it is possible to predict and explain the outcome of evolutionary processes by attending only to the role played by selection. No other evolutionary factor has this degree of causal importance.
>
> **Explanatory Adaptationism**: The apparent design of organisms, and the relations of adaptedness between organisms and their environment, are the *big questions*, the amazing facts, in biology. Explaining these phenomena is the core intellectual mission of evolutionary theory. Natural selection is the key to solving these problems; selection is the *big answer*. Because it answers the biggest questions, selection has unique explanatory importance among evolutionary factors.[7] (*Ibid.*, 336)

He distinguishes a third kind of adaptationism, 'methodological adaptationism'. This is the view that '[t]he best way for scientists to approach biological systems is to look for features of adaptation and good design. Adaptation is a good 'organising concept' for evolutionary research' (2001, 337). It is purely a 'policy recommendation'; it 'recommends a heuristic, and no more' (ibid., 337–8; see also Resnik 1997). He argues that this view is logically independent of both empirical and explanatory adaptationism, however empirical adaptationism may support methodological adaptationism, and the success of the latter may lend support to the former (ibid., 341–2). I shall set aside this form of adaptationism for the remainder of the chapter, as it is less relevant to our concerns.

Godfrey-Smith (*ibid.*) suggests that while empirical adaptationism is a contingent and straightforwardly empirical claim about the natural world, explanatory adaptationism is a more unusual combination of science and philosophy. The distinction maps onto the distinction I drew above. Empirical adaptationists and their opponents disagree about how causally important natural selection has been in evolution, and how objectively common adaptation is in the natural world. Thus their disagreement is factual and, presumably, empirically decidable. Explanatory

[7] Lewens (2009a) distinguishes seven kinds of adaptationism. His discussion builds on Godfrey-Smith's, and can in the main be thought of as adding nuance to Godfrey-Smith's distinctions, rather than replacing them.

adaptationists and their opponents on the other hand (to put it somewhat tendentiously) disagree about how interesting and intriguing adaptation and apparent design are. Explanatory adaptationists are struck by the beauty, intricacy and efficiency of functional adaptation, thus it stands out for them as the primary explanatory target of evolutionary biology.[8] Their opponents do not share this sense of wonder or judgment of importance; they are more (or equally) struck by other features of the natural world. Godfrey-Smith suggests that we may be justified in suspicions that this is not at bottom an objective, scientific question, but one pertaining to values and psychological preferences:

> Although many might agree that we find apparent design in nature interesting, this is apparently just a fact about our psychology. As onlookers, we are puzzled by some things and untroubled by others, but why should we take this to reflect differences between objectively puzzling and objectively unpuzzling states of affairs in nature itself? … If this objection is correct … it turns out that explanatory adaptationism … is not a scientific position at all, but just a set of preferences that some people happen to hold. (Godfrey-Smith 1999, 188)[9]

Thus Dawkins notes that 'large quantities of evolutionary change may be non-adaptive, in which case these alternative theories may well be important in parts of evolution, but only in the boring parts …' (Quoted in Godfrey-Smith, *ibid.*, 340.) Gould (2002, 1020) has also complained that Dawkins' and Dennett's hostility towards theories such as punctuated equilibrium and species selection derives purely from their lack of personal interest in anything other than complex adaptation:

> They regard punctuated equilibrium as trivial because our theory doesn't speak to the restricted subset of evolutionary questions that, for them,

[8] It should be noted that, according to Godfrey-Smith, Darwinian explanatory adaptationists are not just fascinated by apparent design. They regard it as a special problem for a naturalistic, secular worldview—a special, unique challenge for such a worldview, that must be answered for that worldview to be viable (Godfrey-Smith 2001, 349–51).

[9] If explanatory adaptationism is interpreted as a stance however, we can agree with Godfrey-Smith's objector that it is not a scientific position, without dismissing it as 'just a set of preferences that some people happen to hold'. See below.

> defines an exclusive domain of interest for the entire subject ... Does nothing but adaptive organismal design excite Dawkins' fancy ...?

Consider as a further illustration the following. Gould remarks (1983, 156–7) that with respect to the famous Japanese crab that has apparently evolved a resemblance to a samurai warrior on its back, he finds the 'crab's ability to do such an uncrablike thing in the first place' more interesting and intriguing than the hypothesised selection process that led to the trait being evolved (i.e. Japanese fisherman consistently tossing back crabs with the face-like pattern for centuries; Gould is sceptical of this selection scenario, but his point is that even if it is correct he finds it less striking than the variation among crabs that made it possible). Thus he thinks that those who have focused on the selection regime at work have been 'marvelling at the wrong item (or at least failing to give equal time to another remarkable aspect of the case)' (*ibid.*). The internal developmental flexibility that allows traits such as face-like patterns to be offered up to selection 'grabs me far more' than anything selection might do with such variation (*ibid.*).

Is Gould here arguing against empirical adaptationism or explanatory adaptationism? In other words, is he engaged in a factual, empirical debate about causal importance, or a nonfactual debate about subjective values and preferences? His use of expressions like 'marvelling' and 'grabs me' and 'remarkable' are strongly suggestive of the latter interpretation. We may say in the case of the samurai crab that both the variation (the internal, structural constraints) and the selection (the external force building the adaptations) are potentially interesting, and equally necessary for evolution, and that the question of which of these stands out for one as the more striking and fascinating phenomenon is not an objective scientific question, but rather a question of personal interests and preferences.

Godfrey-Smith (*ibid.*, 338–9) notes that empirical adaptationism is consistent with the denial of explanatory adaptationism and explanatory adaptationism is consistent with the denial of empirical adaptationism. (He suggests Dawkins and Dennett may well be explanatory adaptationists without being empirical adaptationists.) So they do not entail one another; whether they may inductively *support* one another is a question I will return to.

Explanatory adaptationism has two components: the first is the claim that design and adaptedness are the big questions, or amazing facts, about the natural world, and the second is the claim that natural selection is the big answer. These two components are distinct. This is especially clear when we notice that the first claim, but not the second, pre-dates Darwinism and evolutionary theory. Pre-Darwinian thinkers such as Paley are in agreement with Darwinian explanatory adaptationists that adaptation and apparent design are the big questions.[10] Dawkins is quite explicit that he accepts Paley's judgment here (1987), as was Darwin; as Gould notes (2002, 253), 'Darwin never abandoned Paley's conviction that adaptation must be designated as the primary phenomenon of natural history'. Where the disagreement lies is over what the answer is: the latter claim it is natural selection, the former claim it is God. (Godfrey-Smith (2001, 337) acknowledges that it is 'possible in principle' to hold the view that apparent design and adaptation are the big questions, without holding that selection is the big answer, but he doesn't seem to notice, at least in this paper, that this is precisely the position of pre-Darwinian functionalists.)

Our discussion here dovetails with the discussion in the previous section of functionalism and structuralism. Indeed I want to suggest that the first component of explanatory adaptationism—the claim about adaptation and design being the big questions—may be identified with, or, perhaps more accurately, thought of as a component of, functionalism as we

[10] See Godfrey-Smith (1999): here he makes the point that explanatory adaptationists have accepted the questions posed by natural theology about adaptation, but have given a different answer. He notes that one may object that we should reject not just the answers given by natural theology, but the questions they asked as well (see also Reiss 2009). This objection can be interpreted as a denial of the possibility of genuine continuity across the Darwinian boundary: to the extent that evolutionists are worrying about design, they are not really being consistent evolutionists, but are still clinging to the terms of argument of natural theology. So there can be no genuine commonality of adaptationist/functionalist stance between natural theologians and Darwinians. The two worldviews differ not just in their beliefs (the answers they give) but in their stances (the questions they ask). I find this argument unconvincing. It is not clear to me why the question, 'why should organisms be well adapted to their conditions of life and present the appearance of having been the product of conscious design?' should be thought of as an illegitimate relic of a bygone creationist age; it (and the pressing need for a rational scientific answer to it) is surely still one of the first things that strikes anybody today when contemplating the natural world.

have understood it. The distinction I drew between, on the one hand, functionalism as a stance shared by thinkers like Paley and Dawkins, and on the other, the particular theories of (beliefs about) function and adaptation that are *not* so shared, maps reasonably well on to the distinction between the first and second components of explanatory adaptationism. This also suggests an interpretation of the structuralist stance as, perhaps among other things, the position that unity of type constitutes the big question, or amazing fact, about the natural world, which it is the core mission of biology to explain. This stance is shared by pre-Darwinian and Darwinian structuralists. Where they disagree is over what the explanation of unity of type is; the former think it is God's plan for Creation, the latter think it is the existence of different kinds of internal constraints on selection.

We should thus expect the relationship between the first and second components of explanatory adaptationism to exhibit the same features as the relationship between functionalism as a stance and the particular theories about the nature of function and adaptation. And this is what we do find. For instance, my discussion above suggested that the functionalist stance is a part of philosophy, and the relevant theories part of science (interpreted broadly); Godfrey-Smith suggests that the first component of explanatory adaptationism is a philosophical claim (or at least, a mixture of science and philosophy), and the second component is a scientific hypothesis (although it might be more accurate to say that the second component combines the scientific hypothesis that selection explains adaptation with the philosophical claim that selection is *eo ipso* the most important evolutionary factor). Also, both relationships exhibit the features of non-entailment and indirect connection that characterise the relationship between stances and beliefs generally.

So, in summary, what pre-Darwinian and Darwinian functionalists/adaptationists have in common includes a commitment to a metaphysical stance defined in large part by a set of values and preferences[11] that identify adaptation and apparent design as the most interesting, striking

[11] van Fraassen is anxious to point out that, on his view, the values that partly constitute a stance are not just personal or subjective preferences (2004, 9, 17, 2007, 375–6), so we should avoid a simple equation of values with preferences here.

and problematic features of the natural world—the phenomena which above all others cry out for rational explanation.

What pre-Darwinian and Darwinian structuralists have in common includes a commitment to a different philosophical stance defined in terms of an interest in unity of type as the most interesting and striking feature of the natural world. Arguably something else that structuralists of different eras have had in common is a commitment to 'typological thinking'. Mayr ([1959] 1976) famously defined typological thinking as the approach which posits a small number of forms underlying biological diversity, and the view that it is those forms that are, in some sense, metaphysically 'real', while the variation we observe is secondary and ephemeral. It is opposed to 'population thinking', which sees variation as in some sense metaphysically primary (Lewens 2009b; see also Sober 1980). There is little doubt that typological thinking as characterised by Mayr underlay the views of the pre-Darwinian structuralists I have mentioned. According to Mayr, with the advent of Darwinism, typological thinking was rightly replaced by population thinking throughout much of biological thought. While this may have been true, in a recent paper Lewens (2009b) argues convincingly that a form of typological thinking (an acceptable form, he thinks) survives in modern evolutionary developmental biology. It isn't the case that all modern structuralists share this commitment—Gould himself, for instance, is a modern structuralist but has written an entire book critiquing typological thinking (1996). Typological thinking is plausibly thought of as a component part of the metaphysical stance of most or all pre-Darwinian structuralists, and that of *some* modern structuralists. (It is less clear that population thinking forms part of the functionalist stance in any very meaningful sense however.)

I said that according to Godfrey-Smith explanatory adaptationism and empirical adaptationism do not entail one another. But do they inductively support one another? Godfrey-Smith (*ibid.*, 336) notes that for explanatory adaptationists, selection can count as the most important evolutionary factor even if it is relatively rare. What matters is that it explains adaptation and apparent design *where and when* they occur. Even if adaptation and apparent design are not at all as common as empirical adaptationists suppose, and that therefore the field of activity of selection is highly circumscribed, so long as selection is the explanation for adaptation and design, it remains the most significant evolutionary force. This entails only that empirical adaptationism is not *necessary* for explanatory adaptationism however; the former may still *support* the latter, if true. But it seems to me that it would not. Whether adaptation and selection are rare, or more common, or ubiquitous, doesn't seem to be relevant to the question of their importance for explanatory adaptationists. That is, it would not strengthen their case if it turned out selection and adaptation were more rather than less common. It is the *existence* of adaptation and apparent design, and the special explanatory problems their existence raises, that matters for explanatory adaptationists; the plausibility of their arguments is not affected by empirical questions of objective relative frequency and causal dominance. Godfrey-Smith (*ibid.*, 342) agrees that it is 'hard to see an argument from empirical adaptationism to explanatory adaptationism.'

Conversely, and more obviously, explanatory adaptationism does not support empirical adaptationism. The empirical case in favour of selection and adaptation being dominant and ubiquitous phenomena in nature would not be strengthened by the truth of the claim that adaptation and apparent design are special problems to which selection is the special solution. Thus not only do explanatory and empirical adaptationism not entail one another; they do not support one another either.

So the negative side of the stance theorist's position—that stances and empirical scientific hypotheses neither entail one another, nor inductively support one another, in the sense of providing evidence for one another—is supported here. But there is a positive side to the stance theorist's position: that stances are capable of influencing empirical science, in the context of discovery, and that this influence can be profoundly important. The relationship between the functionalist/adaptationist stance and empirical science seems to bear this out also. Obviously, the identification of adaptation and apparent design as the central explanatory problems for biology acted as a powerful prod for Darwin's theory of natural selection, which purported to explain precisely these things. But it has likely acted as a stimulant for the development of any number of empirical hypotheses and research programs in functional biology. Empirical adaptationism itself, as a position and as a research program, as well as more specific empirical hypotheses within that research program, have undoubtedly felt the influence of the functionalist/adaptationist stance. (I said above that empirical adaptationism is not evidentially supported by the adaptationist stance; but I am talking here about non-evidential influence, in the sense of motivation or inspiration, in the context of discovery.) If a biologist adopts the position that functional adaptation is the most important and interesting feature of the natural world, clearly she is likely to be inspired to study and theorise about particular examples of functional adaptation. So the adaptationist stance has the effect of highlighting adaptation-related empirical problems. These problems are independent of the stance however. The general problem of explaining adaptation exists quite independently of the view that it is the *big* problem to be solved, as does Darwin's explanation of it. As Godfrey-Smith says, one could accept that it is *a* problem to be solved, and could accept selection as the solution, without holding it is *the biggest* problem. (And one can accept that selection explains adaptation without holding that that makes selection somehow special.)

Finally, if the adaptationist/functionalist stance does have a positive influence on empirical biology in the context of discovery, this can help to pragmatically justify it. This fits with van Fraassen's conception of the way that stances are justified.[12]

5 The Pragmatics of Explanation

I have argued that the 'question part' of explanatory adaptationism is relative to human interests, values and purposes. Explanatory adaptationists are particularly struck by adaptation and apparent design in nature, thus they regard it as especially in need of explanation. But, as Godfrey-Smith argues, this may well just reflect their psychology, values, and so forth, rather than any objective facts about the world. This assessment is very similar to what van Fraassen (1980) says about explanation in

[12] For convenience I have been focusing in this section on the adaptationist/functionalist stance, but it should be clear that what I have said applies *mutatis mutandis* to the structuralist stance, and its relationship to empirical scientific inquiry.

general: *all* explanations—all answers to 'why' questions—he argues (1980, 156–7), presuppose a specification of the phenomena to be explained, and rather than reflecting objective, context-independent truths, such specifications in fact reflect human values, interests, and purposes, and are radically context-relative.[13] Thus explanation possesses a crucial *pragmatic* dimension—an explanation is good, acceptable, illuminating, etc., only relative to specific human interests and purposes.

> Scientific explanation … is a use of science to satisfy certain of our desires; and these desires are quite specific in a specific context … The exact content of the desire, and the evaluation of how well it is satisfied, varies from context to context … there can be no question … of explanatory success as providing evidence for the truth of a theory that goes beyond any evidence we have for its providing an adequate description of the phenomena'. (*ibid.*, 156–7)

This view helps him to deny that the 'explanatory power' of a scientific theory constitutes an epistemic virtue of the theory (a factor contributing to the likelihood that the theory is true) over and above empirical adequacy and empirical strength. Thus the fact that some theory has great explanatory power is simply a pragmatic virtue of the theory, like simplicity or elegance. It may mean the theory helps us satisfy certain of our goals, given our interests and purposes (it helps us to answer 'why questions' that we find particular pressing, given our psychology) but it doesn't make the theory more likely to be true.

We can apply this analysis to evolutionary theory in the following way. Explanatory adaptationists regard the ability of the theory of natural selection to explain adaptation and apparent design as a great virtue of the theory. Of course the theory purports to explain a great many other things, but adaptationists regard its explanation of adaptation as especially important and impressive. According to van Fraassen, this may be a great virtue of the theory, but it is not an epistemic virtue. It is a pragmatic virtue, in that it merely records the fact that the theory provides a satisfying answer to a 'why question' that we (or some of us) find

[13] See Grimm (2008) for an interesting discussion of the kinds of factors that contribute to a phenomenon being interesting, striking or puzzling enough to demand an explanation.

particularly pressing, given our psychology. And this fits perfectly with Godfrey-Smith's assessment of explanatory adaptationism. I would suggest further that what lies behind particular, context-relative 'why questions' is sometimes precisely the value-laden metaphysical stances that van Fraassen discusses.

References

Agassiz, L. 1857. Essay on Classification. In *Contributions to the Natural History of the United States*, Vol. 1. Boston MA: Little Brown & Co.

Amundson, R. 2021. Form and Function in Evo-Devo. de la Rosa, L.N. & Müller, G.B. (eds.). *Evolutionary Developmental Biology.* Cham: Springer International. 457–67.

Appel, T. A. 1987. *The Cuvier-Geoffroy Debate: French Biology in the Decades Before Darwin*. New York: Oxford University Press.

Boucher, S.C. 2015. Functionalism and structuralism as philosophical stances: van Fraassen meets the philosophy of biology. *Biology and Philosophy*, 30 (3): 383–403.

——— 2020. Methodological naturalism in the sciences. *International Journal for Philosophy of Religion*, 88 (1): 57–80.

——— 2025. Functionalism and Structuralism in Biology: Stances or Explanatory Strategies? *Synthese*, 206 (1): 1–23.

Brown, R. 2023. Structuralism and Adaptationism: Friends? Or foes? *Seminars in Cell and Developmental Biology*, 145: 13–21.

Cronin, H. 1991. *The Ant and the Peacock*. Cambridge UK: Cambridge University Press.

Cuvier, G. 1800–1805. *Lecons d'anatomie comparee* de G. Cuvier, recueillies et publiees sous ses yeux par C. Dumeril. C. Dumeril, ed., vol. 1–2; G.-L. Duvernoy, ed., vol. 3–5, Paris.

Dawkins, R. 1976. *The Selfish Gene*. New York: Oxford University Press

——— 1982. *The Extended Phenotype*. New York: Oxford University Press.

——— 1987. *The Blind Watchmaker*. New York: W.W. Norton & Company.

——— 1996. *Climbing Mount Improbable*. New York: W. W. Norton & Company

Dennett, D.C. 1995. *Darwin's Dangerous Idea*. London: Penguin.

Donahue, M.K. 2024. Methodological Naturalism, Analyzed. *Erkenntnis:* Published online 29 February 2024.

Dowell, J. 2006. The Physical: Empirical, not Metaphysical. *Philosophical Studies* 131: 25–60.

Dupré, J. (ed.). 1987. *The Latest on the Best: Essays on Evolution and Optimality*. Cambridge, MA: MIT Press.

Forber, P. (ed.). 2009. Special issue of *Biology and Philosophy* on 30th Anniversary of Gould and Lewontin's *The Spandrels of San Marco and the Panglossian Paradigm: A Critique of the Adaptationist Program*.

Geoffroy Saint-Hilaire, E. 1818. *Philosophie Anatomique*. Vol. 1, *Des organs respiratoires sous le rapport de la determination et de l'identite de leurs pieces osseuses*. Paris: J.-B. Bailliere

Ghiselin, M.T. 1994. Darwin's Language may Seem Teleological, but his Thinking is Another Matter. *Biology and Philosophy*, 9 (4). 489–492

Godfrey-Smith, P. 1999. Adaptationism and the Power of Selection. *Biology and Philosophy*, 14 (2). 181–194.

——— 2001. Three Kinds of Adaptationism. S.H. Orzack and E. Sober (eds.). *Adaptationism and Optimality*. Cambridge: Cambridge University Press. 335–357.

Goethe, J. W. von 1790. *Versuch der Metamorphose der Pflanzen zu Erklaren*. Gotha: Etting.

Goodwin, B. 1994. *How the Leopard Changed its Spots: The Evolution of Complexity*. New York: Simon & Schuster.

Gould, S.J. 1983. Hyena Myths and Realities. Reprinted in S.J. Gould. *Hen's Teeth and Horse's Toes*. New York, NY: W.W. Norton & Company. 147–157.

——— 1996. *Life's Grandeur: The Spread of Excellence from Plato to Darwin*. London: Jonathon Cape.

——— 1997. The Exaptive Excellence of Spandrels as a Term and Prototype. *PNAS*, 94 (20). 10750–10755.

——— 2002. *The Structure of Evolutionary Theory*. Cambridge, MA: Harvard University Press

Gould, S.J. and R.C. Lewontin. 1979. The Spandrels of San Marco and the Panglossian Paradigm: a Critique of the Adaptationist Program. *Proc R Soc Lond B*, 205. 581–598.

Grene, M. and D. Depew, 2004. *The Philosophy of Biology: An Episodic History*. Cambridge: Cambridge University Press.

Grimm, S.R. 2008. Explanatory Inquiry and the Need for Explanation. *Brit. J. Phil. Sci*, 59 (3). 481–497.

Hempel, C.G. 1980. Comments on Goodman's *Ways of Worldmaking*. *Synthese*, 45 (2). 193–199.

Hull, D. and M. Ruse (eds.). 1998. *The Philosophy of Biology*. Oxford: Oxford University Press.

Kauffman, S. A. 1993. *The Origin of Order: Self-Organisation and Selection in Evolution*. Oxford: Oxford University Press.

Lennox, J.G. 1993. Darwin *was* a Teleologist. *Biology and Philosophy*, 8 (4). 409–421.

——— 1994. Teleology by Another Name: A Reply to Ghiselin. *Biology and Philosophy*, 9 (4). 493–495.

Lewens, T. 2009a. Seven Types of Adaptationism. *Biology and Philosophy*, 24 (2). 161–182.

——— 2009b. What is Wrong with Typological Thinking? *Philosophy of Science*, 76 (3). 355–371.

Maynard Smith, J. 1987. How to Model Evolution. In J. Dupré, (ed.). *The Latest on the Best: Essays on Evolution and Optimality*. Cambridge, MA: MIT Press. 119–131.

Mayr, E. ([1959] 1976). Typological versus Population Thinking. in *Evolution and the Diversity of Life*. Reprint. Originally published as "Darwin and the Evolutionary Theory in Biology", in J. Meggers (ed.), *Evolution and Anthropology: A Centennial Appraisal* (Washington, DC: Anthropological Society of Washington), 1–10. Cambridge, MA: Harvard University Press, 26–29.

Melnyk, A. 1997. How to Keep the 'Physical' in Physicalism. *The Journal of Philosophy*, 94 (12). 622–637.

Millikan, R. 1989. In Defense of Proper Functions. *Philosophy of Science*, 56. 288–302.

Neander, K. 1991. Functions as Selected Effects: The Conceptual Analyst's Defence. *Philosophy of Science*, 58. 168–184.

Novick, R. 2023. *Structure and Function*. Cambridge University Press.

Orzack, S.H. and E. Sober (eds.). 2001. *Adaptationism and Optimality*. Cambridge: Cambridge University Press.

Owen, R. 1848. *On the Archetype and Homologies of the Vertebrate Skeleton*. London: Voorst.

Paley, W. 1802. *Natural Theology*. London: R. Faulder.

Reiss, J. 2009. *Not by Design: Retiring Darwin's Watchmaker*. University of California Pres.

Resnik, D. 1997. Adaptationism: Hypothesis or Heuristic? *Biology and Philosophy*, 12. 38–50.

Ruse, M. 1979. *The Darwinian Revolution: Science Red in Tooth and Claw*. Chicago: The University of Chicago Press.

——— 2003. *Darwin and Design: Does Evolution have a Purpose?* Cambridge MA: Harvard University Press.

Russell, E. S. 1916. *Form and Function*. London: Murray.

Sober, E. 1980. Evolution, Population Thinking, and Essentialism. *Philosophy of Science*, 47. 350–383.

Sultan, S.E. 2015. *Organism and Environment: Ecological Development, Niche Construction, and Adaptation*. Oxford: Oxford University Press.

van Fraassen, B.C. 1980. *The Scientific Image*. Oxford: Clarendon Press.

——— 1996. Science, Materialism and False Consciousness. J.L. Kvanvig (ed.). *Warrant in Contemporary Epistemology: Essays in Honor of Plantinga's Theory of Knowledge*. Lanham, Md.: Rowman and Littlefield. 149–181

——— 2002. *The Empirical Stance*. New Haven, CT: Yale University Press

——— 2004. Replies to Discussion on *The Empirical Stance*. *Philosophical Studies*, 121 (2). 171–192

——— 2007. From a view of science to a new empiricism. In Monton, B. (ed.). *Images of empiricism: essays on science and stances, with a Reply from Bas C. van Fraassen*. Oxford, Oxford University Press, pp 337–383

Vicente, A. 2011. Current Physics and 'The Physical'. *Brit. J. Phil. Sci*, 62. 393–416

Williams, G. C. 1992. *Natural Selection: Domains, Levels and Challenges*. Oxford: Oxford University Press

Winther, R. G. 2015. Evo-Devo as a Trading Zone. In A.C. Love (ed.), *Conceptual Change in Biology: Scientific and Philosophical Perspectives on Evolution and Development*, 459–82.

Wright, L. 1973. Functions. *The Philosophical Review*, 82. 139–168

9

Punctuationism and Gradualism

1 Introduction

In this chapter I shall discuss punctuationism and gradualism. The argument for construing metaphysical gradualism and punctuationism as van Fraassian stances will be somewhat different from the one I presented in the previous chapter for so construing functionalism and structuralism. I shall not be arguing that the persistence of metaphysical gradualism and punctuationism across changes in theory means that they cannot be construed as factual beliefs. Although they have in fact so persisted, it is relatively straightforward, as we shall see, to articulate broad metaphysical theses—factual beliefs (even empirical beliefs) that may be true or false, justified or unjustified—with which we can, if we wish, identify metaphysical gradualism and punctuationism (which is not the case, I argued, for functionalism and structuralism understood as metaphysical positions). What I shall argue is that while we can define such broad factual theses, they are either inarguably false, or too vague to be of much interest. Hence, when we consider, and try to represent as rational, the influence that metaphysical gradualism and punctuationism have had on scientific theorising, a principle of charity constrains us to identify these

S. C. Boucher, *Empiricism, Naturalism and the Metaphysics of Biology*,
https://doi.org/10.1007/978-3-032-00555-7_9

views not with clearly false or uninteresting factual theses, but with more defensible, interesting, and fruitful intellectual constructs. The best candidate for these, I will suggest, is van Fraassian stances.

In broad terms then, the argumentative strategy *shall be* similar to van Fraassen's argument for construing materialism as a stance, that I adapted in the previous chapter. Van Fraassen suggests that any factual thesis with which we might identify materialism must be either clearly false or empty of content (given Hempel's dilemma), and thus uninteresting and unimportant. But the positive influence on science that materialism has had suggests that it is not uninteresting and unimportant; hence it is not a factual thesis. That is essentially what I shall be arguing, though in a somewhat different way, in the case of metaphysical gradualism and punctuationism.

2 Punctuationism[1]

Famously, Gould and Eldredge in the 1970s proposed the thesis of punctuated equilibrium (PE), which we encountered in Chap. 4, according to which the picture of evolutionary change that the fossil record appears to present—that most species change very little through the course of their existence—is not, as traditional Darwinians tended to argue, an illusion, but is in fact veridical: the majority of species are in 'stasis' for the great majority of their life-spans, rather than gradually and continually evolving, as the standard Darwinian 'phyletic gradualist' view has it (Gould and Eldredge 1972; Gould 2002). Evolutionary change, on this view, is largely a result of the processes associated with *speciation*. Thus speciation itself emerges as the crucial process responsible for the great bulk of genetic and phenotypic evolutionary change; species are seen to be discrete ontological *individuals*, rather than classes; and macroevolution (evolution above the species level), rather than being merely the summation of microevolutionary processes, emerges as a distinct and irreducible causal process, subject to its own laws.

[1] This section draws on my (2018) and (2019).

As stated, PE is a scientific theory that makes testable predictions concerning macroevolutionary patterns. It asserts that the punctuationist pattern is the *predominant* pattern for the life history of a species, not that it holds universally:

> [P]roponents of punctuated equilibrium have always recognised that the theory cannot be proven … from documentations, however rigorous and complete, of individual cases. As its primary claim, therefore, punctuated equilibrium must assert a dominant role for stasis within species and rapid cladogenesis between species in the construction of macroevolutionary patterns at the appropriate scale of speciation and trends across species within clades. This assertion requires that punctuated equilibrium maintain a *dominant relative frequency* in the origin of new paleospecies. Tests of the theory must therefore focus upon percentages of occurrence in exhaustive, or at least statistically definitive, surveys of particular taxa, faunas and times. (Gould 2002, 854)

The fact that PE is a claim about relative frequency makes it more difficult and complicated to empirically test than all-or-nothing positions. For instance, the validation of gradualist hypotheses regarding particular evolutionary histories will not undermine it, since it accepts that these hypotheses will sometimes be correct. Thus no 'crucial experiments' will be possible to test, once and for all, whether PE or phyletic gradualism is true (*ibid.*, 823). But this is, most theorists now accept, how things stand with most of the interesting debates in biology (*ibid.*). Few positions make all-or-nothing claims, and therefore crucial experiments of the classical kind are unlikely to be forthcoming. But it is recognised that this needn't be cause for despair, and recent work in the philosophy of biology has been devoted to showing how hypotheses about relative frequency and relative importance may be subjected to empirical test (see Beatty (1997), Kovaka and Novick (2024), Kovaka (2021), and Lloyd and Gould (1993)).

There are some more specific methodological issues surrounding the empirical testing of PE (*ibid.*, 784–874) but for present purposes we can accept that it is a respectable empirical, scientific hypothesis. However 'punctuationism' sometimes names a much more general, abstract and metaphysical position, according to which change over time in any natural or social system can be expected to instantiate the punctuationist rather than the gradualist pattern.[2] Gould and Eldredge have admitted that they are attracted by this general punctuational conception of change. In a television interview Gould (1984) contrasted the punctuationist with the gradualist conception of change:

> The issue at stake is the very nature of change itself. There is I think a pervasive bias in Western thought to see change as slow, steady, accumulative, gradual, to see change as the essence of nature. There's another view however, that stability and system and structure is more the essence, and the change, when it occurs, is difficult, that systems absorb stress, and try to maintain themselves, and that every once in a while the stress accumulates to a point where the system breaks and quickly reconstitutes in a new way so that change is not always continuously accumulating, but is rare and episodic, and that systems tend to sit at stable points as much as they can. It's a different way of viewing the world.

However Gould and Eldredge have, for the most part, been keen to sharply distinguish the general punctuational conception of change from the empirical hypothesis of PE (Gould 2002, 1011). The latter is a scientific theory, they argue, in support of which (or against which) empirical evidence may be marshalled. It is empirically verifiable and falsifiable. Punctuationism as a metaphysics of change on the other hand is a philosophical attitude that is not to be confused with a genuinely scientific hypothesis.[3]

But how precisely are we to understand this metaphysical position? In particular, should we think of it as a factual claim that one may believe or disbelieve, that can be true or false, and that can be supported or undermined by evidence and argument, i.e. a traditional metaphysical *thesis*? Or should we think of it as a *stance* in van Fraassen's sense? Gould (2002,

---

[2] Compare also the distinction between Darwin's theory of evolution—indeed any theory of biological evolution—and Herbert Spencer's 'Law of General Evolution', which was proposed as a general theory of change and development applicable to the whole of reality; see Medawar (1969).

[3] Gould has not however always clearly drawn this distinction. In the quote from the television interview above for instance, he suggests that 'the issue at stake' in the debate over PE is precisely the nature of change in general.

970) says that PE is a localised contribution to a 'broader punctuational paradigm about the nature of change—a worldview…' But 'paradigm' and 'worldview' could be interpreted either as broad factual claims or stances.

In an early paper in defence of PE (1977) Gould and Eldredge had conjectured that an interest in metaphysical punctuationism and its Marxist variant may have predisposed them towards considering punctuationist approaches in biology, thus influencing the development of PE. In his last book Gould (2002, 1018) responded to critics who have inferred from this that PE must represent an invalid attempt to impose a metaphysical or political agenda on empirical biology:

> We needed to say something about why we, rather than other paleontologists at other times, had developed the concept of punctuated equilibrium. We raised this point [about our interest in Marxist and punctuationist conceptions of change] as sociological commentary about the *origin* of ideas, not as a scientific argument for the *validity* of the same ideas. An identification of cultural or ontogenetic sources says nothing about truth value, an issue that can only be settled by standard scientific procedures of observation, experiment and empirical test.

Elsewhere he writes, '[t]he social and psychological contributions of a *Zeitgeist* to the origin of hypotheses bear no logical relationship to any subsequent scientific defense and validation of the same hypotheses' (*ibid.*, 31). Thus, metaphysical punctuationism influenced the development of PE, but it influenced it in the context of *discovery*, not the context of *justification*. Beginning with the metaphysical orientation made it easier to notice PE as a conceptual possibility—made it easier to 'find' it in conceptual space—but once developed, the empirical hypothesis left its metaphysical influences behind, to face the usual rigors of scientific testing.[4] The metaphysical origins of the hypothesis are strictly irrelevant to the question of its acceptability as a scientific hypothesis.

So, at least according to Gould, PE as an empirical hypothesis is distinct from, and independent of, punctuationism as a metaphysical

[4] 'As with all scientific theories, punctuated equilibrium will live or die by concrete and quantifiable evidence' (*ibid.*, 774).

position. (He suggests the same is true of gradualism as an empirical hypothesis and as a metaphysical position, as we shall see). The metaphysical positions may have influenced the development of the empirical hypotheses, but this influence took place in the context of discovery, not the context of justification. Thus reference to the metaphysical positions is pertinent when considering the historical (or 'sociological') question of the origins of the empirical hypotheses, but is not relevant when considering the empirical, scientific question of the rational acceptability of those same hypotheses. This accords with van Fraassen's claim, quoted above (1996, 157) that:

> …an empirical hypothesis introduced into science for reasons, and perhaps with putative credentials, coming from outside that science, will so to speak *forget its origins* … so that its fortunes within science will soon have little or nothing to do with those origins.

This suggests Gould may be thinking of metaphysical punctuationism along the lines of a van Fraassian stance. But we need to look more closely at this.

2.1 The Relation Between Science and Metaphysics: Van Fraassen, Dupré, and Gould

What I take to be the orthodox conception of the relation between metaphysics and science includes the view that metaphysical positions can enter into science as fundamental or foundational principles, underpinning and grounding the science. The idea, following Quine, is that they are 'in the same boat' as more obviously empirical hypotheses, differing from the latter only in being more abstract, general and conceptual—lying closer to the core of our web of belief. As van Fraassen (*ibid.*, 176) notes, this is 'the picture of science deeply imbued with metaphysics…'

The case of punctuationism and gradualism may controvert this picture however. If Gould is right, metaphysical punctuationism is not foundational with respect to PE. The metaphysical position does not underpin or ground the empirical hypothesis. The two need not go together; they are independent of one another. Acceptance of PE is thus consistent with the rejection of metaphysical punctuationism and vice versa.

Whether one's acceptance or rejection of the empirical hypothesis could or should have *any* bearing on one's attitude towards the metaphysical position is another matter however. It might be suggested that even if PE doesn't entail metaphysical punctuationism, it may still inductively support it. This conception of how metaphysical positions may be amenable to justification or refutation on the basis of empirical scientific evidence has been defended by Dupré (1993, 2): 'I … claim that empirical inquiry (which I do not limit to scientific inquiry) provides the evidence on which such [metaphysical] assumptions must ultimately rest. Thus I claim that founding metaphysical assumptions of modern Western science … have been shown, in large part by the results of that very science, to be untenable.' Dupré is suggesting that certain metaphysical presuppositions, in particular reductionism, determinism and essentialism, shaped modern Western science, explicitly and implicitly, from the beginning. Yet as science has developed, its theories, explanations and data have increasingly come into conflict with these very presuppositions: essentialism and reductionism have been undermined by developments in biology, and determinism by developments in modern physics, for example.[5] A similar case could be made for metaphysical gradualism, which I shall discuss in more detail below: it profoundly influenced sciences such as biology and geology in the eighteenth and nineteenth centuries, but has arguably been undermined by recent developments in evolutionary theory, such as PE. We could of course construe some of these inferences as deductive: 'gradualism holds of all systems' would be deductively refuted by the truth of 'gradualism is not true of system X' (although of course it could only be inductively supported by the truth of 'gradualism is true of system X'). But we are here considering weaker, inductive relations between science and metaphysics. On this picture, a general gradualist (or essentialist, or determinist) metaphysical view or perspective can be *somewhat* undermined, inductively, by the empirical failure of reductionism (essentialism, determinism) with respect to

[5] This is, of course, a kind of naturalised metaphysics—supporting or rejecting certain metaphysical claims on the basis of science—as discussed in Part I.

certain systems or phenomena, or somewhat supported, inductively, by empirical successes in science. Thus, the truth of PE (if it is true) may inductively support metaphysical punctuationism and inductively disconfirm metaphysical gradualism. Such, as least, is the Dupré picture as I am interpreting it.

As I suggested above, van Fraassen seems to reject these kinds of inferences from empirical science back to metaphysics—at least back to metaphysical stances. According to van Fraassen's view, as I am interpreting it, not only do empirical hypotheses not entail metaphysical positions of the sort we are considering in Part I, they do not so much as support them either, not least because metaphysical stances are not beliefs, and so not the kind of things that may be *inferred* at all. There may of course be *pragmatic* inferences in this direction. If metaphysical punctuationism inspired the development of PE, and the latter is true, this may give us a reason to endorse metaphysical punctuationism as useful, but it doesn't give us a reason to believe it is true.

As noted in the Introduction to Part II, van Fraassen does accept that there may be a degree of non-logical tension or dissonance—he calls it 'pragmatic incoherence'—involved in certain combinations of beliefs and metaphysical stances. It is plausible that there may be at least some pragmatic incoherence involved in, for instance, accepting metaphysical punctuationism while at the same time accepting empirical gradualism, or accepting metaphysical gradualism while accepting PE.

Is Gould closer to the van Fraassen conception or the Dupré conception? At least some of what he says is certainly compatible with, and suggestive of, the interpretation of them as stances in van Fraassen's sense. Firstly, he denies that they are 'high-order empirical observation[s], induced from the objective study of nature' (1977, quoted in Gould 2002, 1017).

Secondly, Gould is keen to stress the *values-driven* character of the metaphysical positions,[6] and their *nonrational* dimension (where 'rational' is to be understood as 'epistemically rational'). Thus he indicates that

[6] The focus here is on social, political and ideological values, but clearly the positions fit with certain epistemic values as well. Metaphysical punctuationism will appeal to those who value non-gradualist forms of explanation, for instance.

the basis of metaphysical gradualism, at least in Darwin's day, was his society's deep ideological commitment to gradual progress and improvement, in the natural, and especially the social, realm. Thus metaphysical gradualism is 'embedded in the modern history of Western cultures', and the influence on Darwin's science of this metaphysical position shows that 'even the greatest scientific achievements are rooted in their cultural contexts…' (2002, 1017).[7] He also stresses the 'cultural embeddedness of preferences for punctuational change' (*ibid.*). (The term 'preference' itself suggests a nonrational psychological inclination.) One example he gives of this (1980a, 184–5) is the fact that in the Soviet Union, scientists were trained in the official state ideology of Marxist dialectics, which is explicitly punctuational in character (for instance in the dialectical law of the transformation of quantity into quality (Engels 1972)). This made them more naturally receptive to theories like PE than scientists in the gradualist-dominated West. This influence of cultural, political and ideological factors on scientific thinking in the Soviet Union was clear and explicit, Gould notes, whereas in the West these factors tend to be hidden, or even denied; but they are no less real and important.

Thirdly, Gould defends *pluralism* about metaphysical positions, at least with respect to their influence on biology (Gould and Eldredge 1977, quoted in Gould 2002, 1018; Gould 1980a, 185). We should, he argues, draw on more than one general philosophy of change in trying to understand the natural world; only having gradualism to draw on in the past has been limiting. He doesn't argue for the correctness of punctuationism ('We emphatically do not assert the 'truth' of this alternative metaphysic of punctuational change' (Gould and Eldredge 1977, *ibid.*)); he just thinks it should be added to gradualism to enrich the philosophical influences on empirical biology.[8] So he admits gradualist hypotheses are sometimes true, and that gradualism as a philosophy has often been useful and encouraged good work. Pluralism is, *prima facie*, inconsistent with the interpretation of the metaphysical positions as factual beliefs that may be supported by empirical evidence from science. Dupré, for instance, is not

[7] I discuss this in more detail in Sect. 3.

[8] This is reminiscent of Feyerabend's plea for pluralism with respect to metaphysical influences on science. See van Fraassen (1996).

a pluralist in this sense (although he is in other senses). He regards the positions he is arguing against, such as essentialism, reductionism, and determinism, as *wrong*, and the opposing positions as *right*. Gould on the other hand is not interested in questions of 'right' and 'wrong', 'true' and 'false'; his focus is on the *usefulness* the different positions may have for empirical work. So we do not, for the most part, find him arguing that the truth of PE gives us a reason to think that metaphysical punctuationism is true and metaphysical gradualism is false (although see interview quote above). But he would presumably endorse the pragmatic inference that the truth of PE gives us a reason to accept metaphysical punctuationism as useful. I leave aside for now the plausibility of the further, abductive, inference that the best explanation of the usefulness of metaphysical punctuationism is that it's true.

This pluralist/pragmatist orientation is reminiscent of pluralism about the units of selection, as defended by Dawkins among others, and discussed in Chap. 3. Dawkins is of course best known for defending genic selectionism (2006), the basic idea of which is that most or all traits evolve because they benefit the genes that are causally responsible for them. But in his (1982), Dawkins defends the gene's eye view as the most illuminating heuristic or perspective on evolution, not as factually correct. 'The vision of life that I advocate … is not provably more correct than the orthodox view … They are equally correct' (*ibid.*, 1). He suggests that the gene's eye view cannot be empirically verified or falsified: 'What I am advocating is not a … hypothesis which … can be judged by its predictions. What I am advocating is a point of view, a way of looking at familiar facts and ideas, and a way of asking new questions about them' (*ibid.*).[9] This viewpoint 'has made me see animals and their behaviour differently, and I think I understand them better for it. [The stance] does not constitute a testable hypothesis in itself, but it so far changes the way we see animals and plants that it may cause us to think of testable hypotheses that we would otherwise never have dreamed of' (*ibid.*, 1–2). '[I]t is

[9] He quotes Bonner: 'I do not propose to say anything new or original in these lectures. But I am a great believer in saying familiar, well-known things backwards and inside out, hoping that from some new vantage point the old facts will take on a deeper significance' (*ibid.*, 7).

not a factual position I am advocating, but a way of seeing facts…' (*ibid.*, vi).

Thus, for Dawkins, the gene's eye view and the orthodox individual-level perspective are equally correct perspectives on the evolutionary process—they are different ways of representing the same facts. He favours the gene's eye view—and argues strongly for it—because he finds it better illuminates those facts than the orthodox view. But this is an entirely pragmatic question,[10] and he accepts that for others, the orthodox view may be pragmatically preferable.[11]

The analogy with Dawkins' view can help to dissolve the apparent conflict involved in Gould, on the one hand, defending the punctuationist stance as superior to others, while on the other endorsing pluralism about metaphysical positions. According to Dawkins, there is no fact of the matter about whether the gene's eye view or the individual-level perspective is true, yet he favours the former for pragmatic and heuristic reasons. It is the pragmatic (nonempirical) character of these reasons that licenses a pluralistic and tolerant attitude towards the individualist perspective. If biologists find the latter pragmatically or heuristically preferable, he argues, they should feel free to adopt it. Similarly, so long as Gould's preference for metaphysical punctuationism is understood in pragmatic terms, there will be nothing inconsistent in his adopting a pluralistic and tolerant attitude towards other positions (such as metaphysical gradualism). And, like Dawkins, Gould argues for a minimal claim and a stronger claim. The minimal claim is that punctuationism should be included with gradualism among the philosophical positions biologists may draw on. This is the pluralist claim. The stronger claim is that, in the context of this pluralism, punctuationism will be found to be generally superior to gradualism, not in the sense of more objectively correct, but in the sense of more useful, fruitful and so on. It is the minimal, pluralist claim that he is keen to establish; the stronger claim is, he accepts, more uncertain and conjectural, and he is less committed to it (Gould and Eldredge 1977, quoted in Gould 2002, 1018). Dawkins, in

[10] The gene's eye view is 'not demonstrably more correct' than the orthodox view, but it 'is more elegant and parsimonious' (*ibid.*, 232).

[11] '[T]he biologist should try both ways of thinking, and choose the one he or she prefers' (*ibid.*, 7).

the same vein, notes that his '*minimum*' claim, of which he is 'pretty confident', is that the gene's eye view is 'at least as satisfactory' as the orthodox individualist view, while his 'wildest daydream' is that the gene's eye view will prove more illuminating than the orthodox view throughout 'whole areas of biology' (Dawkins 1982, 7).

Finally, in a passage from a 1982 paper (quoted in Gould 2002, 1012), Gould makes three points that are very much in keeping with van Fraassen's stance approach. Firstly, he argues that punctuationism is a 'style of thinking' that is characterised by a focus 'upon the stability of structure, the difficulty of its transformation, and the idea of change between stable states' (see also quote from television interview above). In other words, it is not a particular factual belief about how the world is; it is rather an orientation, or perspective: a tendency or inclination to focus on and emphasise certain aspects of reality. It is a style of thinking, not a particular thought.[12] It is likely that Gould opts for this stance-like characterisation of punctuationism because candidate beliefs—for example, the belief that punctuational change characterises all or most systems—do not adequately capture the punctuationist's position. It seems clear in fact that one may adopt the punctuationist stance—that is, may regard as fruitful the 'style of thinking' that encourages a focus on punctuational change—without actually believing that punctuational change characterises all or most systems. Another way of thinking of the punctuationist stance is that it involves an *expectation* of finding punctuational patterns in phenomena, rather than a *belief* that punctuational change characterises all or most systems. Of course the expectation may in some sense *result* from the factual belief; but there doesn't seem to be any reason why logically it could not exist independently of it, or in the absence of it.

Secondly, he notes that this style of thinking has several manifestations within biology. But these manifestations are independent of one another. 'Any manifestation may be true or false, or of high or low relative frequency, without affecting the prospects of any other.' The different manifestations of metaphysical gradualism, such as Darwin's theory in biology and Lyell's in geology, are also independent of one another. The particular manifestations—assuming the systems they describe are relatively

[12] Compare Rowbottom and Bueno (2011) on stances as 'styles of reasoning'.

causally isolated from one another—neither entail nor inductively support one another; yet they are united by their shared connection with the overarching metaphysical stance/orientation. Certainly, if metaphysical punctuationism was a factual thesis, the local manifestations may still be logically independent of *one another*: if it was true that all change in all systems was punctuational, it would follow that change in system A and in system B must be punctuational; but from the mere fact that change in system A was punctuational it wouldn't logically follow that change in system B must be punctuational. So the logical independence of the local manifestations doesn't support the stance view over the factual thesis view. Also, on the factual thesis view, the local manifestations presumably would not *directly* inductively support one another. The fact that change in system A was punctuational would not on its own, presumably, increase the likelihood that change in system B was punctuational. But if the factual thesis view is correct, the truth of the local claims may *indirectly* inductively support one another, via inductively supporting the overarching metaphysical claim. If it turned out change in system A was punctuational, this could offer some inductive support (even if rather minimal support) to the claim that all change in all systems is punctuational, which then in turn would support (indeed entail) that change in system B was punctuational. (If the overarching factual thesis states rather that most change in most systems is punctuational, then if change in system A is punctuational, this can inductively support the overarching claim, which then would inductively support, but not entail, the claim that change in system B is punctuational.) Thus if, as Gould suggests, *no* relations of inductive support hold, either directly or indirectly, between the local claims, this does support the stance view over the factual thesis view—assuming we want to connect the local claims with some overarching metaphysical position or other—since stances are not the sorts of things that can receive inductive support.

Thirdly, Gould suggests that the punctuationist style of thought is valuable as 'a fruitful source for hypotheses.' This is in line with van Fraassen's conception of how metaphysical positions may be (pragmatically) justified.

2.2 Punctuationism as Factual Thesis and as Stance

Elsewhere however, Gould appears to be tempted by the factual thesis interpretation. I shall argue that he ought to have resisted this temptation.

In the section of his (2002) entitled 'Punctuational Models in Other Disciplines: Towards a General Theory of Change', he discusses several contemporary applications of punctuational thinking in disciplines other than biology, and considers what implications can be drawn from the success of these approaches. It's true that in these other examples it is PE itself (the empirical claim about evolution at the species level) that is explicitly cited as the inspiration, and the source of the hypotheses, not metaphysical punctuationism as such. But we can construe this in terms of both PE and the views in the other areas being implicitly inspired by the punctuationist stance.

In this section he distinguishes between punctuational approaches that are only metaphorically analogous to PE, and applications that are 'homologous' to PE in that they possess a number of the conceptual or structural features that PE possesses beyond the bare notion of jerkiness (stasis and punctuational change). The metaphorical applications he regards as useful, but he's more interested in the homologous ones.

The title of the section includes 'Towards a General Theory of Change', and he indicates that we may be able to find empirical support in such examples for the 'overarching hypothesis that punctuation records something quite general about the nature of change itself…' (*ibid.*, 929). This clearly suggests the factual-thesis interpretation of metaphysical punctuationism. Such an interpretation seems in tension with his claim elsewhere (see above) that he 'emphatically does not assert the 'truth' of this alternative metaphysic of punctuational change.' There are, indeed, good reasons for refraining from endorsing such a view understood as a factual claim. If the claim is that *all* change takes place in the punctuational mode, that *all* change and alteration in nature and society is episodic or discontinuous ('nature always makes leaps', perhaps) it is indisputably false. Gould himself acknowledges that even in the evolutionary domain in which he thinks PE generally holds true, it would be foolish to deny that at least some of the time change takes place gradually and incrementally. (He also accepts that in geology 'gradualism does…maintain at least a respectable relative frequency among patterns of geological change' 2002, 481.) But if the claim is softened to the claim that '*most* change takes place in the punctuational mode', it seemingly becomes too vague to be of much interest (how much is 'most'? 90 percent? 70 percent? 51 percent?), or for us to have any notion how such a view could be established. It may be possible to establish that in a particular domain, such as biological evolution at the species level, the majority of change satisfies the punctuational model—though even this modest claim, as Gould shows, requires an enormous amount of empirical work. The idea that we could hope to justify the claim that the majority of change *of all kinds* takes place in the punctuational mode is ambitious in the extreme. Thus in his (1980a, 185) he 'confesses to a personal belief' that punctuationism is true more often than gradualism across the board. But he is careful to say that this is just a personal belief; he is not defending it, or claiming there is compelling evidence that makes the belief justified. If metaphysical punctuationism is to be a respectable position, it can't, seemingly, be as a factual thesis about the nature of change in general.

And indeed Gould, elsewhere in this section, suggests that the successful applications of punctuational thinking he discusses support a more modest claim, that punctuational approaches (when homologous with PE) are useful in explaining a number of other systems (e.g. *ibid.*, 952: that PE can 'provide causal insights into other styles of change [aside from biological evolution]'). And in his discussions of the different examples, it is again the stance conception, and this more modest claim, that appear to be to the fore. Thus he talks about how, when it comes to explaining change in cave art, gradualist assumptions can be limiting, and 'punctuated equilibrium can play a salutary role as a potential corrective, or at least as a source of novel hypotheses…' (*ibid.*, 953). (We find similar formulations for the other examples.) This is, again, strongly suggestive of the stance view. Note, firstly, the pluralism again: we should add punctuationist approaches to gradualist ones to enrich the set of general perspectives or models we may draw on. If punctuationism was 'right' and gradualism 'wrong' it would be misplaced to talk of the former as a potential 'corrective' to the latter; it should rather replace the latter. Darwin's theory was not a 'corrective' to creationism; Heliocentrism was not a 'corrective' to Geocentrism. The idea here is that one 'style of thinking' (stance) has dominated historically, and has been, not so much 'wrong', as limiting in its unchallenged dominance. We should add to it (rather than replace it with) a different 'style of thinking' (stance). Secondly, note punctuationism as a 'source of novel hypotheses'. As we've seen, this is a key function of a stance, in my view. It may be objected that broad factual claims may also be a source of novel hypotheses, which is true, but there is a sense in which stances are 'gentler', less rigid or dogmatic, and thus more palatable and productive sources of empirical hypotheses than broad, strong factual claims. If we accept the broad factual claims, we can after all *deduce* the answers to the specific empirical claims: since it is 'true' that change in all systems exemplifies punctuationism, it follows that this particular system must exemplify it! Whereas the punctuationist stance doesn't deductively entail anything, all it does is, as it were, gently suggest possible hypotheses in the punctuational mode that then have to be tested by standard empirical means.

The upshot of Gould's discussion in this section should, I claim, be that there is, on the one hand, the metaphysical punctuationist stance,[13] and, on the other, the specific punctuationist hypotheses about particular systems, that may be true or false, justified or unjustified; and that is all. There is not, in addition to these, a General Punctuational Theory of Change (or if there is, we should not believe it, or worry about debating it).

3 Gradualism

As with punctuationism, gradualism comes in both a metaphysical form (which I have alluded to above), and more empirical, scientific forms. Darwin's theory of evolution by natural selection is gradualistic, in more than one sense. First, evolution and speciation, he claims, are the result of the selective retention, and gradual accumulation, over great lengths of time, of slight phenotypic modifications, each advantageous to its bearer

[13] The punctuational stance may be associated with other stances, or notions. At the end of this chapter Gould notes that PE was part of a 'Zeitgeist' that favoured punctuation, contingency, chaos and indeterminacy over the traditional notions of orderly, predictable, gradual determinacy (*ibid.*, 971–2). See my Coda to Part II.

in the struggle for life. This sense of gradualism is opposed to *saltationism*, the view (quite popular both before and after the *Origin*), that major evolutionary change results primarily from occasional, sudden, large macromutations (saltations). This issue has to do with the *mechanism* or *process* of evolutionary change. The second sense in which Darwin's theory is gradualistic is in its commitment to *phyletic gradualism*, the view that most lineages most of the time are evolving in a slow and steady fashion and that there is no qualitative difference between the tempo and mode of evolutionary change within evolving species, and that associated with the transformation of one species into another. This is the sense of gradualism that is opposed to PE, and it has to do mainly with the *pattern* of evolutionary change.

While clearly both of these gradualistic features of Darwin's theory were shaped by his metaphysical gradualism, in this chapter it is his phyletic gradualism, and the influence on it of the gradualist stance, in which I am most interested (and hence why this chapter is framed in terms of the opposition between gradualism and punctuationism[14]). The gradualism that is opposed to saltationism is really a view about the power of selection as the primary agent of evolutionary change (with variation providing merely the raw material for selection to act upon), opposed to internalist theories (such as saltationism and orthogenesis), that cite non-selectionist, internal factors as the primary evolutionary agents. Thus this dispute relates more to the topic of the previous chapter, functionalism vs. structuralism. 'Gradualists' in this sense are functionalists (stressing the power of selection), and saltationists are structuralists or formalists of a certain sort. Defenders of PE have always stressed that the theory they reject is not gradualism in the sense of anti-saltationism. PE is the view that evolution is punctuated *not* in the sense that it takes place via saltations,[15] but in the sense that typically the majority of evolutionary

[14] Thank you to an anonymous referee for encouraging me to clarify this point, by asking why it is gradualism vs. punctuationism, rather than gradualism vs. saltationism, that is the theme of this chapter.

[15] It is true that Gould, notoriously, flirted with something like saltationism at one point in his career (1980b). But, firstly, the version he entertained was quite moderate and, in his view, compatible with standard Darwinian selectionist processes; and secondly, there was no suggestion that such a view was *logically required* by PE, even if it would, if true, account for the type of punctuational change we find reflected in the fossil record.

change in a lineage is concentrated into the brief period during which speciation takes place, and that once established, most species change very little in the course of their existence. This claim is intended to be consistent with gradualism in the sense of anti-saltationism and pro-selectionism. PE, in other words, is neutral when it comes to the functionalism vs. structuralism dispute. Thus, functionalist anti-saltationism does not imply phyletic gradualism. One could consistently be a functionalist anti-saltationist and accept PE.[16]

With this is mind, let us turn to consider metaphysical gradualism and its influence on biological science. There appear to have been three key sources of Darwin's commitment to metaphysical gradualism: 1. His acceptance of gradualism as a metaphysical picture of the world in a tradition going back at least to Leibniz and Linnaeus, and the latter's famous phrase *Natura non facit saltum* ('nature does not make leaps'). 2. Lyell's notion that gradualism is a necessary presupposition of empirical historical science; and 3. The cultural context of Victorian England, and bourgeois liberalism more generally, with its deep commitment to the values of gradual progress and improvement. I shall discuss the first two in Sects. 3.1 and 3.2, and the third in Sect. 3.4.

3.1 *Natura non facit saltum*

Darwin followed Lyell in professing a deep attachment to the principle defended by Leibniz among others of *Natura non facit saltum*—nature does not make leaps.

> As natural selection acts only by accumulating slight, successive, favourable variations, it can produce no great or sudden modification; it can act only by very short and slow steps. Hence the canon of 'Natura non facit saltum,' which every fresh addition to our knowledge tends to make more strictly correct, is on this theory simply intelligible. (Darwin 1859, 471)

[16] It is less clear that saltationism is consistent with phyletic gradualism. It may be that anti-saltationism is necessary, but not sufficient, for phyletic gradualism.

Natura non facit saltum is mentioned seven times in the First Edition of the *Origin* (Fishburn 2004). Fishburn brings out the way in which Darwin transformed the principle in an evolutionary context:

> The 'old canon' served for most of its career—whether with Leibniz or Linnæus—as an expression of divinely-ordered progression, but never movement, in a fixed hierarchy of creation. Darwin, under cover of its ancient authority, used it to inject change into the system, thus ending stasis and, ultimately, the implicit ranking implied by the old hierarchical order of things. (*ibid.*, 64)

The 'old canon' is, at least on the face of it, a factual claim about the nature of reality. It (or at least Darwin's version of it) asserts that all change in nature takes place gradually, through small, incremental steps, rather than through sudden, rapid shifts or jumps. To the extent that a large change in a system takes place from state A to state B, these must be connected by a series of smaller, intermediate, transitional states, whether observed or not. Given we are interested in the nature of change, we are focusing on the diachronic version of the principle, but it's worth noting it has a synchronic version, to wit: between all apparent discontinuities of type at a time, there exist (or could exist) intermediate forms connecting those types by continuous gradations.

There are two ways of understanding the 'old canon'. The first is as a broadly empirical, speculative generalisation about nature, subject in principle to empirical test. What kinds of evidence would count against it depends on how loosely or strictly it is interpreted; after all, talk of 'gradual' vs 'rapid' and 'large' vs 'small' change is plainly vague. Changes that at the macro-level, and relative to our ordinary frame of reference, appear large and sudden, may resolve at the micro-level into relatively gradual and incremental change, and so may not count against the principle. On a weak enough interpretation (that we might call micro-gradualism), it seems that the only rapid change that could really count against it would be genuinely discontinuous change at the micro-level, such as the kind of discrete transitions between states posited by quantum mechanics on some interpretations. (Schrödinger 1952)

We need not resolve this interpretive issue, since what matters for our purposes is the way Darwin (following Lyell) interpreted the principle, and he clearly had in mind the strong version that is inconsistent with macro-level change of a relatively rapid, discontinuous or episodic character, and which, as he saw it, mandated the kind of gradualist and extrapolationist conception of evolutionary change and the history of life that he advocated. He was committed, that is, to macro-gradualism. The kind of change that, in his view (and following Lyell again), would violate the principle was exemplified by so-called catastrophist views in geology, and saltationist views about the origin of species, that I mentioned above, where species may arise fully-formed in a single generation. PE, even if it is only true of a small number of lineages, would also be ruled out by the type of gradualism he adhered to, despite the fact that it is compatible with a weaker form of gradualism. As I noted above, PE does not conflict with gradualism understood as anti-saltationism. Gould distinguishes evolutionary gradualism as a claim about 'insensible intermediacy' from gradualism as a claim about 'slowness and smoothness of (evolutionary) rate' (2002, 755–6). He notes that in rejecting the latter, PE need not reject the former. That is, it need not reject the claim that 'in going from A to a substantially different B, evolution must pass through a long and insensible sequence of intermediate steps' (*ibid.*, 150). Darwin was committed to the stronger claim about slowness and smoothness of rate.

On such an interpretation, the gradualist view is no more plausible, and seems just as clearly to be amply refuted by the evidence, as the corresponding strong punctuationist claim (that all change in all systems is punctuational, rapid, episodic). Change at the macro-level is, at least sometimes, rapid and sudden. To give just one example, the asteroid impact that caused the K-Pg mass extinction was a very sudden event at the macro-level, and a mere nanosecond in the context of geological time.[17] Understood in this way, the principle of *Natura non facit saltum* joins a long list of metaphysical, factual claims that have not been borne out by the scientific evidence, like Aristotle's claim that time must be infinite, and Kant's claim that space must be Euclidian. Note that the weakened version of the punctuationist factual thesis that I mentioned

[17] And thus was rejected out of hand by those still committed to gradualist explanations.

(and rejected) above—that punctuationism is true of *most* forms of change—doesn't appear to be an option here. Nobody is arguing that nature *mostly* doesn't make leaps. It seems the principle must be held to apply universally or not at all. Thus Darwin talks of it being 'strictly true' (1859, 206). Whether more local forms of (weakened) empirical gradualism can be articulated, for instance the claim that with respect to evolutionary change phyletic gradualism holds in the majority of cases, is of course a further question.

A different interpretation of the gradualist view however is as a strongly metaphysical and a priori claim about what reality is fundamentally like, behind the appearances. On this interpretation all apparent counter-evidence can be dismissed as illusory, meaning that in practice the view is not empirically falsifiable. Darwin might seem to be advocating such a view in the biological context when he dismisses the 'evidence' of the fossils, which on the face of it doesn't support gradualism, as an artefact of the imperfection of the fossil record: '…Darwin had to use Lyell's most characteristic method of argument—he had to reject literal appearance and common sense for an underlying "reality"' (Gould 1980a, 180–1). But I have in mind a much stronger claim than this; after all, dismissing as illusory or defective, or attempting to explain away, certain types of evidence that conflict with one's favoured theory is a standard strategy employed by scientists. On the view I am considering *no* amount of empirical evidence is capable of counting against the theory, understood as an a priori-justified metaphysical claim concerning ultimate reality lying behind the phenomena. It would thus be analogous to metaphysical monism—more precisely what Schaffer (2010) calls 'existence monism'—the view that reality is fundamentally one in the strong sense that only one thing or substance (the cosmos, or pure being) exists, complete, all-embracing, unchanging (see Della Rocca 2020 for a defence). On such a view all our evidence for the existence of a plurality of things and properties, for change, and for distinctions between things, is illusory.

I think it is clear that the empirical-generalisation interpretation of *Natura non facit saltum* is preferable to this latter interpretation—probably for the majority of its adherents, but certainly for Darwin (and Lyell). They clearly believe it holds of the natural world *as we find it* and *as science describes it*. While there may be some instances (such as the fossil record) where the evidence seems *prima facie* to contradict it, and thus wants some explaining away, *on the whole* it saves the phenomena. It is not like existence monism, by its very nature flying in the face of all the evidence of science, perception and common sense.[18] The phenomena are in keeping with gradualism, they think. Now we can of course object, along with Gould, that this was self-deceptive; that much of the crucial phenomena of natural history and geology was not and is not in accord with gradualism, meaning that their preference for gradualism 'had nothing to do with superior perception of empirical information' (1980a, 180).[19] It might be that, as Gould suggests, metaphysical gradualism couldn't be validly inferred from the evidence, so the only explanation for Darwin and Lyell's adherence to it is that it was a prior assumption or prejudice, rooted in values, culture or ideology, that they and their followers imposed on, rather than discovered in, nature. This is of course more in keeping with the view I am defending that metaphysical gradualism is best construed as a stance rather than a factual belief. But we are concerned here with what Darwin *thought* he was committed to, and there seems no reason not to attribute to him the view that what he was committed to was a factual claim understood as an empirical generalisation about nature. The problem, as we've seen, is that this empirical generalisation is false. Thus we can claim that Darwin would have done better to have adopted explicitly what Gould argues he adopted implicitly: metaphysical gradualism construed as a value-laden metaphysical stance. I shall say more about this in Sect. 3.3.

3.2 Gradualism as Methodological Presupposition

Another significant influence on Darwin's metaphysical gradualism was Lyell's dictum that with respect to geology and other historical sciences,

[18] While there are a priori arguments for strong monism (Della Rocca 2020) it's not clear what a priori arguments could be given for metaphysical gradualism.

[19] Compare van Fraassen on the 'false consciousness' involved in treating as factual beliefs or theses what are in fact stances (1996).

gradualism was a necessary methodological assumption.[20] Only gradualist hypotheses could be genuinely scientific, as only gradualist hypotheses (invoking causes that could be observed to be operative in the here and now) could be testable empirically. This had the convenient implication that the anti-gradualist 'catastrophist' theories advanced by his intellectual opponents (invoking cataclysmic events that for the most part were not observable in the present) were to be ruled out a priori as unscientific, rather than treated as potential scientific rivals to his own gradualist geological theories. On this view metaphysical gradualism is not a factual claim or empirical generalisation. Neither is it a stance in our sense. It is a kind of metaphysical constraint on science: an a priori presupposition of historical science, a precondition for the scientific method to be successfully applied in these areas. Gould (2002, 481–2) suggests Lyell accomplished this 'victory' through a kind of sleight-of-hand whereby he gave the same name, 'uniformity' (closely related to gradualism) to both a set of methods that do arguably partly define science (uniformity of law etc.) and to controversial and testable empirical claims about processes on the earth, thus making the latter out to be a priori assumptions of science as well as the former. Thus gradualism becomes metaphysical, a priori, quite explicitly not inferred from the facts but assumed in order to study the facts scientifically.

The two versions of gradualism, as factual thesis and as methodological constraint, are logically independent of one another. One could hold that the factual thesis is true without regarding empirical claims that violated it as non-scientific; and one could hold that gradualism is a necessary assumption of scientific method on entirely methodological grounds, i.e. without committing to the truth of the factual metaphysical claim. Compare the logical independence of metaphysical naturalism (the view that only the natural world exists) and methodological naturalism (the view that naturalism is a necessary methodological assumption of science); see Boucher (2020) for discussion.

[20] Gould also notes (*ibid.*, 480) that Darwin needed (or felt he needed) Lyell's theory to be true because he thought that geology had to satisfy Lyellian gradualism in order for natural selection to be able to do its work the way Darwin needed it to. But this is really a matter of Darwin's *empirical* gradualist evolutionary theory requiring Lyell's *empirical* gradualist geological theory, which need have nothing to do with *metaphysical* gradualism.

John Searle defends a similar methodological view with regard to metaphysical realism. The latter is, he claims, is a fundamental metaphysical presupposition of all science—a necessary condition of the possibility of doing science, we may say. Theories have to be consistent with realism, or be capable of being interpreted realistically, in order to count as scientific: '…external realism is not a theory… It is rather the framework that is necessary for it to be even possible to hold opinions or theories about such things as planetary movements… you have to take it for granted that there is a way that things really are. Otherwise, the debate can't get started. External realism is not a claim about the existence of this or that object, but rather a presupposition of the way we understand such claims' (Searle 1999, 32).

Most today would regard the view that gradualism is a necessary methodological presupposition of scientific method as incorrect. Non-gradualist theories, such as PE, may be testable and certainly count as scientific. We don't have to assume gradualism is true to do proper historical science. And indeed, as Gould shows (1980a, 2002), this was clear enough in Darwin's and Lyell's day. There was no reason, even then, to regard the 'catastrophist' theories in geology that Lyell rejected as non-scientific, no reason to regard their proponents as violating the canons of proper scientific method.

3.3 Gradualism as a Stance

I have discussed metaphysical gradualism as a factual hypothesis about nature, and as a necessary methodological assumption of historical science. Both, I argued, should be rejected, and I follow Gould in claiming that they should have been rejected in Lyell and Darwin's day. This result is quite negative for Darwin (and Lyell). It implies they were unjustified in accepting gradualism understood in either of these two senses. Yet as Gould admits, it would be foolish to deny that metaphysical gradualism has had a positive influence on scientific inquiry. Lyell's gradualist empirical theories in geology, and Darwin's gradualist empirical theory of evolution, were of course extremely important and influential, and count as successful science on any measure. Thus, I suggest, in order to do justice to this positive role that metaphysical gradualism has played, we ought to reinterpret it as a stance. As I noted above, it is permissible to regard Darwin as *implicitly* treating gradualism as a stance, even if this was not his official view. In keeping with our discussion of the punctuationist stance above, the gradualist stance would have the following features:

- It would be logically distinct from the various empirical forms of gradualism, such that accepting it would be neither necessary nor sufficient for accepting any particular empirical claims.
- It would be understood as influencing science in the context of discovery, not the context of justification. In their (1977), Gould and Eldredge suggest that Darwin's commitment to metaphysical gradualism (reflecting the Victorian ideology of gradual progress) influenced his gradualist conception of evolution, but that this exposure of the philosophical influences on Darwin's science—belonging to the context of discovery—says nothing about the relative merit of the science itself—belonging to the context of justification. 'The general preference that so many of us hold for gradualism is a metaphysical stance embedded in the modern history of Western cultures; it is not a high-order empirical observation, induced from the objective study of nature … We mention this not to discredit Darwin in any way, but merely to point out that even the greatest scientific achievements are rooted in their cultural contexts…' (quoted in Gould 2002, 1017).
- It would be value-laden. We have seen that Darwin's gradualism was imbued with the social, cultural and political values of his liberal bourgeois society. The preference for constant, gradual, orderly, i.e. not sudden or 'revolutionary' (hence dangerous and unpredictable) change and progress in human affairs was a value deeply rooted in his social and cultural context, and his gradualist theory reflected that. I shall say more about this in the following section.
- It would allow for *pluralism* with respect to its relation to other metaphysical views of change. Gradualism as a stance may sit happily alongside punctuationism as a stance as part of the legitimate set of metaphysical influences on science, in a way that would be ruled out if these were understood as incompatible factual claims about nature.

- It would be conceived as a *style of thinking*, rather than a particular thought.
- Its various empirical manifestations within science would (so long as they were dealing with causally isolated systems) neither entail one another, nor inductively support one another, directly or indirectly.
- Its value for science would be as a fruitful source for empirical hypotheses, where the latter are 'gently' suggested by the stance, but then have to be tested by standard empirical means, independently of the stance. To the extent that such hypotheses are borne out, the stance will be pragmatically justified.

3.4 Cultural Contexts

The third important source of Darwin's commitment to metaphysical gradualism was the social, political and cultural context in which he lived and worked. This had a number of elements relevant to Darwin's intellectual development, perhaps most famously the capitalist social relations, liberal economic theory and free market and Malthusian ideology that so patently influenced his theorising, and which have been the focus of much recent work on Darwin and his cultural context. As Marx famously stated, 'It is remarkable how Darwin recognises among beasts and plants his English society with its division of labour, competition, opening up of new markets, 'inventions', and the Malthusian 'struggle for existence" (Marx, quoted in Radick 2009, 147). But it also clearly included as a central element his Victorian society's commitment to an ideology of gradual progress and improvement, as alluded to above. This was, Gould notes, the culmination of the post-enlightenment liberal commitment to the virtues of 'slow and orderly change' in the social realm, applied then to nature (1980a, 180). Natural cataclysm was threatening the way social revolutions were.

Radick (2009) distinguishes two answers to the question, 'Is the theory of natural selection independent of its history?'[21] The *independence* thesis

[21] I shall assume, I think not unreasonably, (1) that Darwin's theory of natural selection is essentially gradualistic, or at least that gradualism is a core component of the theory, and (2) that the ideology of gradual progress was a fundamental and ineliminable component of the cultural context. If

holds that Darwin was influenced by the culture and ideology of his day (including the ideology of gradual progress) but the theory was so intellectually compelling and well-supported by the evidence it would've been developed by somebody, somewhere, anyway, with the 'external' cultural and societal influences just speeding up the process. If Darwin hadn't been the one to think of it someone else, perhaps in a different culture and subject to quite different influences, no doubt would have or could have, at some point or other. The *inseparability* thesis, on the other hand, holds that the theory of natural selection was 'a contingent result of a unique social history' (Radick *ibid.*, 148). Only Darwin, or someone very much like him, at that precise juncture in history, subject to the particular confluence of cultural, social, economic and political factors characteristic of nineteenth century English bourgeois society, could have come up with the particular theory Darwin proposed.

It seems to me however that both of these theses may come in stronger and weaker forms, and that the weaker forms may be reconcilable. One way to think about this is to distinguish 'in practice' from 'in principle' claims. It might be, as a contingent fact about the actual history of ideas among human beings, and close possible histories, that only Darwin's specific context would or could have given birth to that theory. As Gould writes, '[p]eople of equal intelligence could neither have formulated nor owned such a concept [as Darwinian evolution] in Newton's generation, even if some hypothetical Darwin had then advanced such a claim…' (2002, 29). So in practice, as it were, the inseparability claim holds. But we might want to allow the *in-principle* possibility that the theory could be or could have been developed by different people in a different time period or different cultural context. So in principle the independence thesis holds. If this is right then there is truth in both the independence view and the inseparability view, and we don't need to choose between

either of these claims were denied, the relevance of the independence vs. inseparability debate to my concerns in this chapter would be greatly reduced. If the debate is really about whether a theory which has at its heart the notions of competition, adaptation, innovation, and individualism, could have arisen in a context other than one in which these notions were as embedded in the intellectual and cultural landscape as they were in Victorian England, then the debate will be quite removed from the question of the nature and influence of metaphysical gradualism. I will also set aside, for simplicity's sake, the possibility that Darwin's theory may be independent of some features of the cultural context (which may or may not include the gradualist stance) but not others.

them. Thus Radick says the issue is about whether someone other than Darwin in his context was 'likely' to have discovered the theory. But we can accept it wasn't at all 'likely'—the in-practice inseparability claim—but was eminently possible in-principle—the in-principle independence claim.

These points would apply *mutatis mutandis* to punctuationism: perhaps in practice only a context, such as the 1960s and 1970s, in which metaphysical punctuationism exerted the sort of influence it did, and related notions such as contingency, chaos, indeterminacy, and so on, characterised the intellectual and cultural 'zeitgeist' (Gould 2002, 971–2) could have given rise to a theory such as PE, but it is plausible that it might in principle have been proposed and defended in a quite different context.

If we qualify the two claims in this way—inseparability as an in-practice claim, and independence as an in-principle claim—they are compatible. They only become incompatible if they are interpreted in stronger ways, as both making claims about *both* 'in-practice' and 'in-principle': the strong independence claim saying not just that in-principle someone else could have come up with the theory, but in practice, in the actual history of ideas or closely related histories, someone else would have or could have; and the strong inseparability claim saying not just that in practice it is unlikely anyone else would have come up with it, but in principle it was effectively impossible that anyone else in any other context would or could have. We can perhaps cash this in terms of possible worlds. On the strong independence view, the possible worlds in which someone other than Darwin, in a different context, discovers the theory, include possible worlds quite close to the actual world (as well as ones more distant). And there are many such worlds. On the strong inseparability view, the only non-actual worlds in which someone in a different context discovers the theory are so distant from our world that we can for all intents and purposes regard the scenario as impossible (it would clearly be too strong to regard this scenario as literally logically or metaphysically impossible).

Of course we can define the two strong views, and there may well be people who endorse them so understood. But they may be too strong to be very plausible, and the compromise view, which combines the qualified versions of each, may be more defensible, uniting what is plausible in each and discarding what is implausible. On the compromise view, there are no very close possible worlds in which someone in a different context discovers the theory, but the possible worlds in which this does happen are not nearly so distant or rare as the strong inseparability view suggests—some are worlds recognisably like our own. We have to travel *some* distance from the actual world to find this (so to speak), but not *too* far.

How do these views relate to the stance interpretation of metaphysical gradualism? We can say that the strong version of the inseparability thesis is incompatible with the stance view. Here the relevant version of the inseparability thesis states that it would have been near-impossible for anyone who had not adopted the gradualist stance to come up with the theory of natural selection, thought of as a form of empirical gradualism. This is clearly inconsistent with the theme I have emphasised throughout Part II, of the relative independence of scientific theories from the metaphysical stances that may influence them in the context of discovery. On the stance view it may have been unlikely, but it was far from impossible, that a context that did not include the gradualist stance could have given birth to Darwin's theory.

What about the strong version of the independence view? This appears to be *compatible* with the stance view—it doesn't seem to be *ruled out* by it. But while the views are consistent, the stance view is seemingly also consistent with the denial of the strong independence thesis. The processes within the context of justification can't get to work on a theory if the theory hasn't been proposed, and it might be that at least in practice, in histories of science not far removed from the actual history, certain theories wouldn't exist without the quite specific cultural, philosophical, ideological, and social contexts in which they were actually developed. Perhaps in such close possible histories, there is no theory of natural selection without the gradualist stance. The stance view doesn't rule this out, it seems.

So there are three theses:

1. The strong inseparability thesis. This is inconsistent with the stance view.

2. The strong independence thesis. This is consistent with stance view, but the stance view is also consistent with its denial.
3. The compromise view incorporating the softened version of both theses (inseparability as an in-practice claim, independence as an in-principle claim). The stance view is consistent with this, without entailing it or being entailed by it.

For my part, as I've intimated, I find (3) to be the most plausible when these positions are considered in themselves—though I will not argue in detail for that claim here—so I would be inclined to combine (3) with the stance view. But the stance view is compatible with (2), so if (2) turned out to be correct that would not undermine the stance view.

* * *

To sum up the central argument of Sects. 2 and 3. We have seen that both metaphysical gradualism and punctuationism can be construed as broad factual theses about nature, that are subject to empirical test. We have also seen that they can be construed as van Fraassian stances. The problem with the former is that the broad factual theses in question are either clearly false, overly vague or unjustified. There are no rational grounds for endorsing them (and weren't in Darwin's day). Construed as stances however, they can be much more positively evaluated. Both the gradualist and the punctuationist stance have seemingly been fruitful prods for scientific research and theorising, in different contexts and at different times, and thus their adoption can be seen as rationally justified in the sense in which value-laden metaphysical stances can be rationally justified: as being useful in inspiring and motivating good scientific work. Gould, as we have seen, is explicitly aware of this. Although at times he is tempted by the factual thesis interpretation, he is circumspect enough to recognise that 'any attempt to support the exclusive validity of such a grandiose notion [as metaphysical punctuationism] would border on nonsensical' (1980a, 185). Thus he recommends interpreting metaphysical punctuationism in a way that is highly congenial to the stance construal.

Darwin was less cautious. There is no reason not to interpret him, at least officially and explicitly, as endorsing metaphysical gradualism as a factual claim about nature. Given the falsity of such a claim however, we can say at least that he would have been *better off* conceiving of his gradualism the way Gould conceived of punctuationism, as a stance that *qua* stance would not be undermined by any amount of empirical evidence. Indeed, as Gould suggests, we can interpret Darwin as *implicitly* so conceiving of it. If *Natura non facit saltum* as an empirical generalisation about nature fitted the facts as poorly as Gould suggests (and I think he is right about this), then we need to explain why intelligent scientists such as Darwin (and Lyell) continued to adhere to it so rigidly in the face of countervailing evidence. It is at least a plausible explanation that its importance and role for Darwin was not in fact as an empirical generalisation at all, but rather as a value-laden metaphysical stance with a basis much more firmly in ideology, social and cultural prejudices and a priori assumptions, than in facts and evidence. That this was not explicitly recognised by Darwin can then be attributed to the kind of false consciousness that van Fraassen claims philosophers are characteristically subject to when reflecting on the nature of philosophical positions and disputes. The key move however, and one that Gould makes in this context[22] (and van Fraassen makes in others), is to regard this interpretation of the nature and role of metaphysical gradualism as in no way a criticism. Quite the contrary. It is when interpreted as a factual claim that metaphysical gradualism should be dismissed as rationally unjustified. As a stance, its value and importance can be validated.

It may be objected that the positive role and influence of views like metaphysical gradualism for science doesn't depend on them being interpreted as stances. Even if metaphysical gradualism as a factual thesis is false, cannot it still play a useful motivating role in science, and indeed has it not done so in Darwin's case? Huw Price (2002, 4) notes that despite being regarded as false by philosophers, verificationism 'remains influential and useful as a kind of pocket philosophy for practising scientists, where its effect is to focus the minds of theorists on the question as to how their proposals might be tested.' False theories can be useful for

[22] As we have seen, he insists that his and Eldredge's claim that Darwin's gradualism had a social and cultural rather than an empirical basis was not meant 'to discredit Darwin in any way' (1977, quoted in Gould 2002, 1017).

scientists. In response I would make three points. Firstly, since beliefs aim at truth, it is plausible that we should not believe theories or claims that are false—certainly that we should not believe those that are not epistemically justified—even if they do have positive effects. If there are convincing reasons for thinking that verificationism (or metaphysical gradualism) is false, scientists should not believe it, even if such belief has good practical (and indeed epistemic) consequences. The kind of rampant pragmatism according to which either pragmatic usefulness is what truth consists in, or that we are rationally justified in believing anything, even a falsehood that is known to be falsehood, if it has sufficiently significant practical or epistemic benefits, is rightly unpopular among contemporary philosophers, and we may wonder whether it is even intelligible. Secondly, I have suggested that stances are much less rigid or commanding as sources of empirical hypotheses than dogmatically-held factual beliefs, and thus more conducive to the values of open inquiry. If one believes that *all Fs are G*, then one can *deduce* that *this F is a G*, without needing to test the claim through observation and experiment. (This is of course especially problematic when we are considering cases in which *all Fs are G* is false.) But if one accepts a stance that tends to encourage a focus on the G-ness of Fs, then the hypothesis that *this F is a G* will be suggested, gently as it were, but the hypothesis will then need to be rigorously tested to discover whether it is true or justified. It might be objected that in practice a scientist such as Darwin, even if he accepts metaphysical gradualism as a factual thesis, is unlikely to reason, 'metaphysical gradualism is true, therefore my gradualist theory of biological evolution must be true' and leave it at that. In practice Darwin still tried to establish the truth of the empirical theory through observation and experiment. But, while this might be true of his theory as a whole—which of course involved much more than a claim about gradualism—it's not at all clear that commitment to metaphysical gradualism didn't entitle him to reason: 'metaphysical gradualism is true, therefore *whatever precise form the correct theory of biological evolution will take, we know it will have to be a gradualist theory*'.

Thirdly, to the extent that we think that pluralism with respect to metaphysical influences on science is generally a good thing, we will be inclined to favour stances over factual beliefs, since pluralism about stances is possible and rational in a way that pluralism about conflicting factual beliefs is not. Metaphysical gradualism and punctuationism as factual beliefs cannot both be true. But the gradualist and punctuationist stances can both be, and arguably both have been, useful as stances. Gould's 'plea for pluralism' with respect to the metaphysical influences on science can only be heeded if the metaphysics comes in the form of stances rather than factual beliefs.[23]

4 Kuhn and Theory-ladenness

Gould and Eldredge's original paper presenting the theory of PE was strongly influenced by Thomas Kuhn's *The Structure of Scientific Revolutions* (1996). Kuhn (along with others such as Feyerabend (1975) and Hanson (1965)) argued for the theory-ladenness of observation: what we observe depends on the theories we accept. Our theories 'colour' our observations of the world. There is no theory-neutral observation, or theory-neutral observation language. Thus when we look at the fossil record from gradualist standpoint, we tend to 'see' gradualism; the evidence against it (evidence for stasis and rapid transformation) is not seen as such. But when we look from the punctuationist standpoint, we 'see' punctuationism in the data. We cannot get outside our frameworks to observe the objective facts in a neutral way.

But if that's true, how can the theories be tested at all (Turner 2011)? To test a theory we need to compare it to the facts (the evidence). But if

[23] One disanalogy between the punctuationist and gradualist stances on the one hand and the functionalist and structuralist stances, discussed in Chap. 8, on the other, is that the former do not seem to involve the kinds of *judgements of importance* that the latter do (see Novick 2023). They do not concern the most important phenomena to be explained, the way I argued functionalism and structuralism do. In Boucher (2025) I distinguish (following a suggestion from Novick 2023) between Stances Involving Judgements of Importance (SIJIs) and stances that don't involve such judgements. The punctuationist and gradualist stances are not SIJIs. It does seem correct however to understand them as involving the 'tendency or inclination to focus on and emphasise certain aspects of reality'. If, as Gould argues, gradualism is not 'wrong' and has often been useful, but is 'limiting', such that punctuationism is needed as a 'corrective', with pluralism being the best approach, it doesn't seem incorrect to say that those who adopt the punctuationst stance highlight or foreground or emphasise or draw attention to aspects or features of reality that are neglected or sidelined by gradualist approaches.

the facts we see depend on the theory we accept, it seems we can't test the theory against the facts. How then can we use empirical evidence to decide between gradualism and punctuationism? It seems we have just two general frameworks; we can adopt one or the other, but we can't do so on the basis of the evidence (at least in a non-circular way), since what we take to be the evidence depends on the theory we've already adopted: 'Science progresses more by the introduction of new world-views or 'pictures' than by the steady accumulation of information' (Gould and Eldredge 1972, 86). These worldviews are like Kuhn's 'paradigms': different paradigms are incommensurable; accepting a paradigm is like undergoing a religious conversion. The evidence can't decide between them, as all evidence is 'internal' to a paradigm:

> The idea of *punctuated equilibria* is just as much a preconceived picture as that of phyletic gradualism. We readily admit our bias towards it and urge readers, in the ensuing discussion, to remember that our interpretations are as colored by our preconceptions as are the claims of the champions of phyletic gradualism by theirs. We merely reiterate: (1) that one must have some picture of speciation in mind, (2) that the data of paleontology cannot decide which picture is more adequate, and (3) that the picture of punctuated equilibria is more in accord with the process of speciation as understood by modern evolutionists. (Gould and Eldredge 1972, pp. 98–99)

Despite some superficial similarities, I wish conclude this chapter by distinguishing this view from the stance view I am defending. On the strong Kuhnian view, the assumptions of a theory or paradigm, including its metaphysical claims, so thoroughly permeate and infect the process of empirical testing in the context of justification as to make it unclear how a theory can be tested against the data at all, at least in a non-circular way. On the stance view however, a metaphysical stance influences scientific theorising in the context of discovery only. Theories tested in the context of justification then 'leave behind' these metaphysical influences, for the most part, so that whatever role the metaphysical stances play, they certainly do not compromise the ability of scientists to test a theory against the data, through the ordinary methods of observation and experiment.[24] In this respect the stance view is relatively conservative and conventional. The question of the nature and extent of the theory-ladenness of observation, and its implications for the objectivity and rationality of science, is of course a long-standing and difficult problem in the philosophy of science, and one on which the stance view—having to do primarily with the relation between *metaphysical stances* and science—has nothing very original to say. But Gould at least appears to have softened his Kuhnian views, since, as I noted above, he later came to regard the issue of PE vs. phyletic gradualism as eminently empirically decidable, with the evidence pointing incontrovertibly, in his view, towards the correctness of PE understood as a claim about relative frequency. Thus he presumably came to accept that the evidence concerning evolutionary rates can be (and has been) identified and put to use in adjudicating between PE and gradualism without presupposing the truth of either. The 'exhaustive, or at least statistically definitive, surveys of particular taxa, faunas and times' (Gould 2002, 854) that he suggests are the basis on which to test PE, must be capable of being carried out without assuming that PE is true. Strong theory-ladenness must not hold.

References

Beatty, J. 1997. Why Do Biologists Argue like They Do? Philosophy of Science 64: S432–43

Boucher, S.C. 2014. What is a philosophical stance? Paradigms, policies and perspectives. *Synthese*, 191(10): 2315–2332

———. 2018. Stances and epistemology: values, pragmatics and rationality. *Metaphilosophy* 49 (4): 521–547

———. 2019. An Empiricist Conception of the Relation between Metaphysics and Science. *Philosophia,* 47 (5): 1355–1378

———. 2020. Methodological naturalism in the sciences. *International Journal for Philosophy of Religion*, 88 (1): 57–80

[24] See Boucher (2014), in which I argue against Rowbottom's suggestion (2011) that stances can be understood as Kuhnian paradigms.

———. 2025. Functionalism and Structuralism in Biology: Stances or Explanatory Strategies? *Synthese*, 206 (1): 1–23.

Dawkins, R. 1982. *The Extended Phenotype*. Oxford: Oxford University Press

———. 2006. *The Selfish Gene, 30th Anniversary edition*. Oxford: Oxford University Press

Darwin, C. 1859. *On the Origin of Species by Means of Natural Selection*. London: John Murray.

Della Rocca, M. 2020. *The Parmenidean Ascent*. Oxford University Press.

Dupré, J. 1993. *The Disorder of Things: Metaphysical Foundations for the Disunity of Science*. Cambridge MA: Harvard University Press

Engels, F. 1972. *Dialectics of Nature. 5th edition*. Moscow: Progress Publishers

Feyerabend, P. K. 1975. *Against Method: Outline of an Anarchistic Theory of Knowledge*. London: New Left Books

Fishburn, G. 2004. *Natura non facit saltum* in Alfred Marshall (and Charles Darwin). *History of Economics Review*, 40:1, 59–68

Gould, S.J. 1980a. The Episodic Nature of Evolutionary Change. Reprinted in S.J. Gould. *The Panda's Thumb*. New York, NY: W.W. Norton & Company. 179–185

———. 1980b. Return of the Hopeful Monster. Reprinted in S.J. Gould. *The Panda's Thumb*. New York, NY: W.W. Norton & Company.186–193

———. 1984. *Stephen Jay Gould: This View of Life*. NOVA television portrait/ interview. Original broadcast date: 18 Dec. 1984. Youtube video accessed 15 June 2010. <http://www.youtube.com/watch?v=vJpyOdwBUCQ>

———. 2002. *The Structure of Evolutionary Theory*. Cambridge, MA: Harvard University Press

Gould, S.J. and N. Eldredge. 1972. Punctuated Equilibria: an Alternative to Phyletic Gradualism. T.J.M. Schopf (ed.). *Models in Paleobiology*. San Francisco: Freeman, Cooper & Co. 82–115

———. 1977. Evolutionary Models and Biostratigraphic Strategies. E.G. Kauffman and J.E. Hazel (eds.). *Concepts and Methods of Biostratigraphy*. Stroudsburg PA: Dowden, Hutchinson & Ross. 25–50

Hanson, N.R. 1965. *Patterns of Discovery*. Cambridge: Cambridge University Press

Kovaka, K. 2021. Fighting about Frequency. *Synthese* 199 (3): 7777–97.

Kovaka, K. & Novick, R. 2024. Relative Frequency Controversies and the Growth of Biological Knowledge. *Philos Theor Pract Biol* 16 (1): 1–19

Kuhn, T. 1996. *The Structure of Scientific Revolutions*. Chicago: The University of Chicago Press

Lloyd, E. A., and S. J. Gould. 1993. Species Selection on Variability. *Proceedings of the National Academy of Sciences of the United States of America* 90 (2): 595–99

Medawar, P.B. 1969. Herbert Spencer and the Law of General Evolution. Reprinted in P.B. Medawar. *The Art of the Soluble*. London: Penguin. 45–67

Novick, R. 2023. *Structure and Function*. Cambridge, Cambridge University Press

Price, H. 2002. Starving the Theological Cuckoo. Unpublished MS.

Radick, G. 2009. Is the theory of natural selection independent of its history? In Hodge, J & Radick, G. (eds.). *The Cambridge Companion to Darwin*. Cambridge: Cambridge University Press. 147–172

Rowbottom, D.P. 2011. Stances and Paradigms: A Reflection. *Synthese*, 178 (1).111–119

Rowbottom, D.P. and O. Bueno. 2011. How to Change It: Modes of Engagement, Rationality, and Stance Voluntarism. *Synthese*, 178 (1). 7–17

Schaffer, J. 2010. Monism: The Priority of the Whole. *Philosophical Review*, Vol. 119, No. 1: 31–76

Searle, J. 1999. *Mind, Language and Society: Philosophy in the Real World*. Basic Books

Schrödinger, E. 1952. Are there quantum jumps? Part I. *The British Journal for the Philosophy of Science*. 3 (10): 109–123

Turner, D. 2011. *Paleontology: A Philosophical Introduction*. Cambridge: Cambridge University Press

van Fraassen, B.C. 1996. Science, Materialism and False Consciousness. J.L. Kvanvig (ed.). *Warrant in Contemporary Epistemology: Essays in Honor of Plantinga's Theory of Knowledge*. Lanham, Md.: Rowman and Littlefield. 149–181

10

Holism and Reductionism

1 Introduction

It is not unusual in discussions of certain concepts, doctrines or 'ism's for authors to complain about the existence of multiple, not always consistent uses of the relevant terms, but this seems to be especially true for reductionism and holism. Hull complains that with respect to these concepts and related ones like organicism, mechanism and vitalism, '[n]o two people seem to use them in the same way' (1974, 125). Okasha notes that even within the units of selection literature, 'holism' and 'reductionism' seem to be used differently by different people (2006, loc. 1816–1821).[1]

Rosenberg suggests that reductionism in biology (and thus, presumably, also holism) comes in three forms: a metaphysical thesis, an

[1] This in itself can provide an argument for the stance view (though not one I am laying primary stress on in this chapter): many different things have been meant by 'holism' and 'reductionism', such that identifying them with particular factual beliefs is forlorn; but the different versions of each view do seem to be united as different expressions or manifestations of the reductionist or holist spirit, stance or attitude. In its historical version this of course is the van Fraassen-type argument for stances I employed in the chapter on functionalism and structuralism; see below.

S. C. Boucher, *Empiricism, Naturalism and the Metaphysics of Biology*,
https://doi.org/10.1007/978-3-032-00555-7_10

explanatory thesis, and a research program (2020, 1). The metaphysical thesis of reductionism is, he says, physicalism: the view that the physical facts fix all the facts. At times (*ibid.*, 2–3) he seems to imply that physicalism *is* reductionism (in its metaphysical dimension), but this clearly can't be the case, since, as he himself notes, most antireductionists today are also physicalists. At most the truth of physicalism might give us a *reason to believe* reductionism (though physicalist anti-reductionists obviously reject this inference).[2] Other candidate metaphysical theses with which to identity reductionism and holism might be: with respect to some particular system, or all systems, 'the whole is (or is not) more than the sum of the parts' or 'wholes are (or are not) prior to parts'. These claims might however (like the functionalist claim that 'function is prior to structure'; Chap. 8) be too vague to be very helpful, at least without a clearer understanding of notions like 'more than', and 'prior to', than we seem to possess. The notion of 'emergence' might provide us with a more definite metaphysical doctrine: holists affirm, while reductionists deny, the existence of emergent properties. 'Emergence' and 'emergent properties' are sometimes used rather loosely by biologists (e.g. Vermeij 2010); philosophers have tried to define these concepts more precisely (Kim 1999). The problem is that holism doesn't seem to *require* emergence. I shall below give examples of holist positions that are not emergentist positions, suggesting that emergentism is at most sufficient, not necessary, for holism.

The explanatory (i.e. epistemological) doctrine Rosenberg associates with reductionism is roughly that reductionist explanations provide a model for good explanations in biology (*ibid.*). Clearly this isn't very informative until we are told what 'reductionist explanations' are. It is clear from the rest of his discussion that this has mainly to do with the relation between functional biology and molecular biology.[3] Reductionists hold that 'the [molecular] explanations they favor enhance the depth and detail of higher-level biological explanations – improving them and providing something more, something that corrects them, completes them,

---

[2] Of course, if van Fraassen is right physicalism is a stance, not a factual belief. Thus Rosenberg should be interpreted as highlighting the connection between the physicalist stance and the reductionist stance.

[3] Though as Rosenberg notes, this is an unfortunate way of putting it, since the molecular level is also shot through with functions (*ibid.*).

or fills a lacuna in these explanations, something that makes them better' (*ibid.*, 30). Anti-reductionists deny this: they hold that there are explanations at the higher, functional level that are complete and adequate, and there are no superior explanations of the relevant phenomena to be had a deeper (molecular) level. His almost exclusive focus on the relation between functional and molecular biology reflects the fact that this has been the area where some of the most prominent debates between reductionists and holists in biology have been carried out in recent decades, but it is nonetheless unfortunate, since it overlooks holism-reductionism disputes in other areas of biology, such as in ecology, where this has had nothing to do with the relation between functional and molecular biology.[4] In ecology the debate has centred on issues such as whether ecosystems possess their own higher-level properties and capacities that are irreducible to the properties and capacities of, and interactions between, the species and organisms that make them up (Lefkaditou and Stamou 2006).[5] Still, I think Rosenberg is right to see the question of explanation as central to most or all holism-reductionism disputes. In ecology, for instance, it is natural to interpret the reductionist as affirming, and the holist denying, that the properties and capacities of ecosystems can be explained by reference to the properties and capacities of, and interactions between, the species and organisms that make them up.

Reductionism as a research program is the attempt to, so to speak, put the explanatory doctrine into practice within biological science. *Qua* program for research it is not itself a factual claim that might be true or false, but it could count as successful or unsuccessful ('progressive' or 'degenerating' in Lakatos' terms (1970)). It is clearly associated with methodological versions of reductionism: claims to the effect that in biology one

[4] There are a number of other examples. Darwin's version of evolutionary theory, for instance, is often deemed reductionist, yet obviously had nothing to do with reducing functional to molecular biology.

[5] This issue has played out in the shift in recent years towards a 'systems theory' approach in ecology and environmental science—for instance 'Earth Systems Science' (ESS)—with particular focus on its implications for understanding the possible effects of climate change. ESS is strongly holistic, with its emphasis on the emergent high-level properties of, and complex relations both within and between, environmental systems conceived as 'more than the sum of their parts', and it explicitly rejects the more traditional 'fragmentary reductionist' approach; see Steffen et al. (2020), Ernst (2000). Thank you to Fiona Utley for helpful discussion on this issue.

ought to pursue reductionist methods and forms of explanation, that these are likely to succeed or lead to progress. The holist can be understood as claiming that the success of the reductionist research program is likely to be limited at best. Whether there has even been anything that might be called a holist research program in biology (outside, perhaps, of ecology), let alone a successful one, may be questionable. Part of what holists are typically doing is calling for such a research program to be initiated and carried through.

Let us work with an initial understanding of reductionism and holism as any views to the effect that, locally or globally, the properties of wholes can (or cannot) be explained in terms of the properties of their parts, or that the properties, laws, and processes at one level can (or cannot) be explained by reference to the properties, laws, and processes at one or more lower levels.

I shall also consider methodological versions of reductionism and holism, that state that certain methods or approaches ought to be pursued in seeking biological explanations.

Historically the issue of reductionism in biology has concerned the question of whether biology is reducible to, or is autonomous from, physics and chemistry. While this debate still continues (see e.g. Fang 2022) it is not very relevant to most interesting contemporary reductionism debates, since they are about whether certain parts of biology can be reduced to other parts of biology. So I shall set it aside. Even if biology is an autonomous science (which is the near-consensus view today), this doesn't settle whether, say, functional biology can be reduced to molecular biology,[6] whether all selection is selection on genes, whether ecosystems have emergent properties, and other live questions concerning reductionism and anti-reductionism in biology.

It will be seen from the above that under 'holism' I am including views about the irreducibility of higher levels to lower levels (e.g. the irreducibility of the functional to the molecular level), despite the fact that these views may or may not concern the relation between wholes and their parts. Take genic selectionism. Although genes are parts of organisms, it's not really the case that genic selection itself is 'part of' organisms (though

[6] Assuming that molecular biology is part of biology, not part of chemistry.

of course it can take place within organisms), or 'part of' organismic or higher-level selection. The reductionist claim that all evolutionary phenomena can be explained as the result of selection acting on genes is not best understood as an attempt to explain wholes in terms of their parts, in my view.

It seems reasonable, and historically not inaccurate, to classify such views as holist views (Hull 1974). A concomitant of this is that I will apply the label of 'holist' to what might more commonly be deemed *hierarchical* views, such as multi-level selection theory, that postulate a hierarchy of autonomous levels to reality, each possessing their own (irreducible) features, capacities, etc. If it be objected that holism by definition can only concern the relation between wholes and their parts (with wholes being 'more than the sum of their parts'), allow me to stipulate a meaning of 'holism' such that it includes claims about levels as well.[7]

There is of course a lot to be said about the history of the conflict between reductionism and holism in biology; my excuse for not engaging with much of this is that I believe there are quite simple and persuasive arguments that demonstrate that at the level of factual belief and scientific method, the reductionism-holism debate, at least at a general, across-the-board level, is futile: both sides, with their strong, universal claims about the correctness of holism or reductionism as factual claims, or the fruitfulness of reductionist or holist methods, are wrong.

We came across reductionist and holist (emergentist) positions in Part I, in the chapter on functions and teleology. My focus there was quite different. I was interested there in questions of ontology. I wasn't interested in reductionism and holism/emergentism *as such*, but only in the question of what *types of entities* certain reductionist and emergentist positions committed one to (e.g. reductionists about functions were committed to selected effects, while emergentists were committed to *sui generis* teleology). The relevant epistemic stances, standing above these positions so to speak, dictated the sorts of entities permitted. In this chapter I will consider the nature of reductionism and holism/emergentism themselves, in particular whether they should be understood as

[7] I use the term 'holism' rather than the apparently weaker 'anti-reductionism' since the latter is entirely negative, while the former captures the more positive dimension of the view.

metaphysical stances, in the sense of Part II. Our discussion in the chapter on functions was neutral on this question. Even if emergentism and reductionism about functions are factual beliefs, it doesn't follow that holism/emergentism or reductionism in general are factual beliefs. A stance can issue in or be associated with factual beliefs without itself being a factual belief. Having said that, there may be interesting relations between the stances discussed in Parts I and II. Certain metaphysical stances and certain epistemic stances tend to fit together. Reductionist positions, for instance, tend to be closer to the empiricist end of Chakravartty's continuum, and holist views closer to metaphysics end. There is a rough affinity between reductionism and empiricism,[8] and between holism and strong metaphysics of science or armchair metaphysics. I shall say more about the relation between epistemic and metaphysical stances in the Conclusion.

2 Two Arguments for Interpreting Views as Stances

We have encountered, in the previous two chapters, two related but distinct arguments for construing the positions in question as stances. The historical argument I offered in the chapter on functionalism and structuralism considers the way the factual beliefs associated with the positions change over time, and the way the views seem to be capable of surviving such changes. The argument in the chapter on punctuationism and gradualism focuses not on the way the factual beliefs connected with the view have changed over time, but rather on what we have reason to believe *now*. It suggests that the prevalence of factual or empirical versions or instances of rival metaphysical positions—versions or instances that we have good reasons today to think are true—means that any broad factual thesis we may identify with the metaphysical position in question is very unlikely to be correct. For example, the prevalence of gradualist

[8] With the caveat that to the extent that (as I have suggested in this chapter) reductionists tend characteristically to seek reductionist *explanations*, this might fit ill with the empiricist rejection of demands for explanation. As we saw in Part I, some reductionists (e.g. about teleology) adopt the MoS, rather than the empiricist, stance.

theories or claims that we have good reason to suppose are true means that punctuationism understood as a broad factual claim about change in all or most systems is very likely to be false. Given this, the principle of charity, along with the evident fruitfulness of the metaphysical position in question for science, means we should identify the position not with such a false factual claim, but with a van Fraassian stance. Or so I argued.[9]

The arguments are independent of one another. Even if, say, there was a version of functionalism understood as a factual claim that we had good reason for thinking was true today, the fact that it has survived radical changes in belief and theory over the centuries would give us reason to construe it as a stance as well. For instance, with respect to physicalism, even if a view that we might call 'present physics-alism', according to which all that exists is, or reduces to, the entities of present-day physics, was true, the fact that *in the past* physicalism/materialism has been associated with quite different physical ontologies, gives us a reason to not define physicalism as present physics-alism. This is known as the backwards-looking continuity objection to currentism about physicalism (see my (2020) for discussion). Conversely, even if the positions didn't form a historical lineage that has persisted through multiple changes in theory and belief, the falsity of broad factual theses that may be candidates for the position in question could (given various other assumptions) give us grounds for construing it as a stance.

Both of these arguments can be applied to reductionism and holism. As Hull notes (1974, ch. 5), holism and reductionism in biology have long been in conflict, and have changed their content as scientific theory has changed. The form that holism used to characteristically take was vitalism. But when vitalism was refuted, holism endured. Holists today are not vitalists; indeed, most of them are physicalists. Holism today in its sophisticated versions is (like functionalism and structuralism) framed so as to be consistent with Darwinian evolutionary theory[10]; this wasn't the case, obviously, prior to Darwin. Similarly with reductionism. In the

[9] Novick (2023) in fact applies precisely this sort of argument to functionalism and structuralism, but stops at the first step: she claims that any broad factual version of either view must be false given functionalism is true of some domains and structuralism of others, but does not infer from this—indeed explicitly rejects the inference—that the positions are stances.

[10] Though not necessarily Modern Synthesis neo-Darwinian theory.

Seventeenth and Eighteenth Centuries reductionists tended to be atomists and mechanists as atomism and mechanism were understood at the time. When these doctrines were rejected or modified with the advance of science, reductionism survived. Contemporary reductionists are not classical atomists, and are not mechanists of *this* sort (though some are mechanists in the different, modern sense).

An even more explicit attempt to apply the historical argument to reductionism and holism can be found in Hein (1969). She suggests that the same fundamental issues that animated the dispute between vitalism and mechanism in a previous age are found to recur in the very different theoretical context of the (then) contemporary dispute between molecular biologists and organicist biologists, and that this gives us reason to construe holism and reductionism as 'meta-theoretical commitments', which are much like metaphysical stances in our sense. Modern molecular biologists and earlier mechanists, and modern organicists and earlier vitalists, have very different theoretical commitments and beliefs about the natural world. But there is clearly also significant philosophical continuity—central philosophical commitments are shared by the old and the new reductionists, and the old and the new holists.[11] '(T)he controversy between organicism and molecular biology is not an isolated historical dispute; but is a manifestation of a perennial disagreement between two types of meta-theoretical commitment' (*ibid.*, 250).[12]

While acknowledging the force of Hein's historical argument, in this chapter it is the second, more epistemic-type of argument, that I shall be pursuing. I shall argue that we have good reason for thinking that candidate global factual or methodological claims or theses with which reductionism or holism in biology may be identified are false. I shall argue further that it is plausible that there are nonetheless general reductionist

[11] As well as identifying shared commitments among the older and newer thinkers, Hein bases her argument on the fact that the disputes in question exhibit the characteristic features of 'meta-theoretical disputes': (1) They are strongly polemical; (2) They largely concern methodological matters; (3) They in part concern the question of what counts as evidence; (4) The views in question are defended in large part in terms of their heuristic value (*ibid.*, 239–40).

[12] The argument could be extended by noting that candidate *factual* claims that all reductionists (holists), in all eras, might be supposed to agree on, such as 'wholes are (or are not) more than the sum of their parts', or 'wholes are (or are not) prior to their parts', are too vague or indeterminate to give sufficient content to the views.

and holist positions that can play and have played a fruitful role in scientific theorising. These positions should be identified not with false factual or methodological claims but with philosophical stances. But the epistemic argument I offer in this chapter, and the historical argument offered by Hein, are complementary, leading by different paths to the same conclusion.

3 Pluralism[13]

For the moment, let us consider versions of reductionism and holism as *universal* factual or methodological claims about biology, as holding without exception. Later we shall consider weaker versions according to which they apply to the *majority* of biological phenomena.

There are two ways one might try to establish that both reductionism and holism construed as global factual or methodological claims are false. The first is to show that, as a matter of objective fact, reductionism is true with respect to some domains, contexts, or phenomena while holism is true with respect to other domains, contexts, or phenomena. A related but somewhat different approach would be to claim that *arguments* for global reductionism or holism are unpersuasive, because, for example, they involve inferring inductively from a relatively small number of unambiguous reductionist or holist explanatory successes to the truth of reductionism or holism across the board: a weak inductive inference (Rosenberg makes precisely this criticism of arguments for global reductionism; *ibid.*, 1).

The second tries to show that for some, many or all contexts or phenomena, there is no objective fact of the matter concerning whether reductionism or holism holds, rather this is a question of our interests, values, explanatory priorities, tastes, and so on. These approaches may seem to be in conflict—and if they are applied across the board they

[13] There is of course a more general pluralist position in the philosophy of science, associated with the 'Stanford School' (see Dupré 1993, Mitchell 1992, 2003, Longino 1990), and with which I am quite sympathetic, but here I defend a more limited pluralism concerning the legitimacy of both holist and reductionist views and methods.

are—but I shall pursue a combination of the two strategies, arguing that each holds for certain domains and contexts. I begin with the former.

A recent notable defense of holism (in the form of emergentism) in biology is Walsh (2015). Walsh is part of a recent trend in biology calling for a return to organicism, i.e. a re-centreing of the organism in evolutionary theory, and a move away from the gene-centrism of orthodox Modern Synthesis-based neo-Darwinism (see also Nicholson (2014), Nicholson and Dupré (2018), Dupré (2021), Huneman (2010), Reiss and Ruse (2023), Mossio (2024)). On this view organisms are complex, integrated, cohesive, functional units—agents—with emergent properties and behaviour—they are more than the sum of their parts. They are the locus of irreducible teleology (goal directedness; see Chap. 6). They play an active and irreducible role in evolutionary processes. And they require a holistic approach to understand and explain their nature and activities as 'dynamic, self-organising, adaptive entities', rather than 'mere aggregates of mechanical parts' (Walsh *ibid.*, 41). 'The properties and behaviour of the parts are partially determined by the organization and activity of the organism as a whole' (Nicholson *ibid.*, 353). Though his focus is primarily on organisms, Walsh suggests that sub-organismal systems also instantiate holistic emergence, such as gene regulatory networks (*ibid.*, 39): 'In systems like this there is a sort of causal reciprocity between part and whole. The way that a part affects the whole can only be understood in the context of the way that the entire system affects, controls and regulates the activities of the parts.'

This is not in itself a new idea; as Walsh notes holistic and organicist views such as this have always existed on the fringes of biological thought, but have never gained the popularity, or enjoyed the empirical and explanatory success, of mechanistic reductionism—certainly not in the last hundred or so years. Where the recent organicism differs from these earlier versions is that it takes account of important recent developments in biological science, those that focus on the 'dynamics of entire complex, self-regulating systems that constantly exchange matter and energy with their surroundings' (Walsh *ibid.*, 41). Much recent work in evolutionary biology points in this holist direction, Walsh argues.[14]

Yet Walsh accepts that the rival approach—mechanistic reduction—is not generally invalid; in fact it has been extremely successful in Twentieth Century biology:

> It is important to understand how profoundly productive this [mechanistic reductionist] approach to understanding biological phenomena has been throughout the twentieth century. It has yielded an account of the most complex and recondite organismal and evolutionary processes in terms of the mechanical interactions of complex molecules. In many ways, twentieth-century biology is the triumph of reductive mechanism. (*ibid.*, 30–31)

This has found its contemporary philosophical articulation in the explanatory and methodological program of *mechanism*, one of the most significant movements in Twenty-First Century philosophy of science. This seeks to continue (and update) the reductionist tradition of explaining wholes in terms of parts, and higher levels in terms of lower, but with the specific claim that scientific explanation, including in the biological sciences, consists fundamentally in the search for underlying causal mechanisms: structures that realise functions and produce phenomena through the operation of their constituent components and their interactions, with each component behaving the way it does in virtue of *its* component parts and their organisation, and so on (Bechtel and Abrahamsen 2002, 26). Walsh is sympathetic to this contemporary mechanistic project: '[C]ontemporary mechanism is 'reductionist' in a restricted and entirely reasonable sense: the causal capacities of a complex entity, in a context, can be exhaustively accounted for by the capacities and interactions of its parts' (Walsh *ibid.*, 30). This is a fairly liberal and moderate form of reductionism, one that doesn't deny the existence and importance of higher-level patterns and regularities, or the validity and adequacy of higher-level causal explanation (Bechtel 2006). But as

[14] Nicholson (*ibid.*) gives the examples of systems theory, niche construction theory, and theories of phenotypic plasticity, among others. These ideas are at the heart of the Extended Evolutionary Synthesis also (see Laland et al. 2015).

Rosenberg notes, mechanists share 'fundamental philosophical commitment[s]' with traditional reductionists (Rosenberg *ibid.*, 67); in my terms, they share the reductionist stance.

Thus as Walsh admits, even if he is right about holism being vindicated with respect to certain domains and systems, (reasonable) reductionism remains valid with respect to other domains and systems. Holism is required *as well as*, not *instead of*, mechanistic reduction.

3.1 Pluralism and the Units of Selection

Another example of this pluralism can be seen by considering again the units of selection issue. Traditionally, the issue of reductionism vs. holism has been closely associated with the question of the level at which selection acts. Genic selectionists—the view that all selection is at the level of the gene, that the gene is the only unit of selection—have been regarded as reductionists ('crude' reductionists by their opponents) while those advocating higher-level selection, or more generally multi-level selection theory have been regarded as holists ('woolly' holists by their opponents). This seems broadly accurate. If genic selectionism was true as a matter of objective fact, holism as a universal factual or methodological claim would seemingly be refuted, while if multi-selectionism was true as an objective fact, reductionism as a universal factual or methodological claim would seemingly be refuted. There is an alternative view however, discussed in Chap. 3—'model pluralism'—on which there are no facts of the matter concerning the level at which selection acts, and the levels of selection question is entirely pragmatic and interests-relative. This view corresponds to the no-fact-of-the-matter view about holism and reductionism mentioned above: whether genic selectionism (reductionism) or multi-selectionism (holism) holds is a nonfactual matter of convention or pragmatics.

Elsewhere I have argued against model pluralism, in both its global and local versions (Boucher 2023). I hold that multi-selectionism is true, interpreted in a realist sense. This implies that (on my view) one form of reductionism—genic selectionism, or the gene's eye view—fails. But the question of holism and reductionism can be seen in a different light—one that vindicates local reductionism and local holism—when we consider the distinction between multi-level selection types 1 and 2, as mentioned in Chap. 3.[15] Recall that in MLS1, a form of reductionism holds. The nature and fitness of the collectives, and the change in the population over time, are the result of selective processes operating at the particle level. Collective fitness is reducible to—defined in terms of—average particle fitness. There are no irreducible collective-level properties or processes involved. The 'wholes' are to be understood in terms of properties, processes, and causation at the level of the 'parts'. Thus a form of reductionism can be vindicated even in the context of multi-level selection theory, i.e. even if we reject the gene's eye view.

In MLS2 on the other hand, a form of holism obtains. The nature and evolution of the collectives is the result of processes operating at the collective level, and these selective processes are not reducible to selection at the particle level.[16] The fitness of the collectives is not reducible to the fitness of the particles that compose them. Heritability is also defined at the collective level, independently of the particle level.

A stronger form of holism is involved in some formulations of MLS2. As we saw in Chap. 3, some think that it is a necessary condition for MLS2 processes that they operate on *emergent*, not just *aggregate* properties of collectives, i.e. properties that are not reducible to the properties of the particles (Vrba 1984, 1989). The existence of such emergent properties would add a stronger (or at least an additional) layer of holism to MLS2 processes. I argued however that emergent properties are not a requirement of MLS2 processes. All that is required for the latter is emergent (that is, irreducible) *fitness* at the collective level. MLS2 processes can operate on aggregate collective properties. Okasha suggests that this means the holism of MLS2 processes has nothing to do with emergent properties (Okasha *ibid.*, loc. 1827–1834), but this may be an exaggeration. Though emergent properties are not required for MLS2 processes,

[15] This point is also made by Okasha (2006, loc. 1821–1827) but I wish to make rather more of it than he does.

[16] More precisely, as I explained in Chap. 3, holism applies only when we have MLS2 processes in which the character-fitness covariance is causal. Mere MLS2-sorting, where the differential success of the collectives is a by-product of fitness and selection at the particle level, is still subject to reductionism.

MLS2 selection on emergent properties (for instance in species selection) may still take place, and may represent some of the paradigm cases of MLS2 processes. For such phenomena, a holist approach is even more clearly necessary. But Okasha is correct that the holism of MLS2 processes is secure whether emergent properties are involved or not, indeed whether or not emergent properties exist at all. Thus we should distinguish the units of selection issue from the independent debate about the existence of emergent properties in nature, whether at the level of the organism, the species, or the eco-system. In this debate holists think there are emergent group or collective properties, while reductionists think there are not (there are only aggregate group or collective properties). One could think there are emergent properties of species, for instance, without thinking that there is species-level selection on them just as, I have claimed, there can be MLS2 selection at a level without there being emergent properties at that level.

The case of MLS1 and MLS2 shows that with respect to certain processes reductionism holds, and with respect to different processes, holism holds. If we accept, as Okasha has convincingly argued (2006), that MLS1 and MLS2 are real, causal processes that take place in nature, and not just alternative ways of describing the same facts, it follows that reductionism is, in some contexts, objectively correct, and holism is, in other contexts, objectively correct. Thus both reductionism and holism understood as universal factual or methodological theses are false. But the case of MLS1 and 2 shows more than this. What we would ideally like is a principled reason *why* reductionism and holism understood as universal factual or methodological theses are false. Thus Novick, in arguing that functionalism and structuralism understood as universal factual theses are both false, notes that there is a principled reason why this is the case (Novick 2023). The reason, she suggests, is that the course of evolutionary history is contingent: 'There is no one way that evolution occurs, common to all life. As these differences may affect the relative importance of structuralist or functionalist explanatory strategies, there is no reason to expect either strategy to be universally dominant' (*ibid.*, 50).

MLS1 and 2 are not just different processes, as work on the major transitions in evolution shows (Calcott and Sterelny 2011; Clarke 2014; Bourrat 2015; Currie 2019; Ryan et al. 2016; O'Malley and Powell 2016;

Wu and Banzhaf 2011; Okasha 2022).[17] MLS1 and MLS2 processes should, in the light of the major transitions, be understood as (for the most part[18]) two temporal stages in the one historical process of the emergence of individuality and independent fitness at higher levels of biological organisation. The major transitions are characterised by the gradual transition from (1) particles acting entirely independently, to (2) particles interacting co-operatively, and forming pseudo-, or proto-collectives, in which the fitness of the 'collective' is defined as average particle fitness (MLS1); to (3) a stage in which the fitness of the emerging collectives can be defined independently of the fitnesses of the particles but is nonetheless determined by them, and finally to (4) the emergence of genuine collectives, with collective-level functions, overall integration, cohesion and coordinated activity, in which the interests and activities of the particles are subordinated to the interests and activities of the collective, and in which the fitness of the collective is not determined by the fitnesses of the particles—MLS2 (Okasha 2006). Fitness has at this final stage been transferred, or 'exported', from particle to collective. As Michod discusses (1999, 2005), the transition from 2 to 3 to 4 depends crucially on the evolution of mechanisms for the suppression of competition among the particles, reducing the possibilities of the collective breaking down due to subversion from within by selfish free-riders.

It is the fact that MLS1 processes evolve into MLS2 processes, that explains why reductionism sometimes holds, and holism sometimes holds. Integrated, cohesive collective entities whose fitness is defined at that collective level and is independent of the fitness of the particles that compose them—that is, entities, for which holism is true—evolve from looser, less integrated collectives whose fitness *is* reducible to the fitness of the particles that compose them—that is, entities for which reductionism is true. Thus holism and reductionism don't just apply to different contexts or domains—they apply to different stages of the evolutionary process. Entities for which holism is true are not just a *sui generis* feature of nature—they have been produced by evolutionary processes. Holism, we

[17] This passage draws on my (2023).

[18] I have argued elsewhere that some MLS2 processes do not evolve from MLS1 processes (Boucher 2023, 8). But collective MLS2 processes do.

may say, evolves.[19] This echoes Okasha's point that work on the major transitions in evolution shows that hierarchy is not a pre-given feature of the natural world; it has itself evolved.

This also explains why, on the new organicist view, a holistic or emergentist perspective is required with respect to organisms. MLS2 processes operate at the multi-cellular organism level (Michod 2005), among other levels. But multi-cellular organisms haven't always existed: one of the most important of the major transitions in evolution was the origin of multicellularity. The process by which this happened exemplified the four-stage process mentioned above. There was thus a stage during which collective or proto-collective entities composed of cells did not possess irreducible collective-level fitness, weren't subject to collective-level selection, and did not reproduce as collectives: they were loose aggregates of particles (cells) that retained their autonomy and independent reproductive interests. The whole was reducible to the properties and interactions of the parts. The holistic, emergent features of multi-cellular organisms, in other words—and the associated fact that, as Nicholson (2014) puts it, such organisms may be *explanans* and not just *explananda* with respect to evolutionary phenomena—are not a pre-given feature of nature, but are the result of a historical process of evolution.[20] The same goes for single-celled organisms: they too, of course, are the product of evolution. The evolution and diversification of single-celled eukaryotes only became possible because of one of those other major transitions, the origin of the eukaryotic cell and the development of its intracellular organelles, effected

[19] This argument for (limited) holism and anti-reductionism thus differs from the standard type of argument against reductionism in biology, which appeals to multiple-realisability (Sober 1999); see below.

[20] Huneman (2010) mentions the major-transitions explanation for the origin of multi-cellular organisms as one possible response that orthodox modern synthesis biologists can make (and have made) to the organicist challenge: organisms as *explananda* rather than *explanantia*. I don't see it as necessarily a means of defending orthodoxy and rejecting the new organicism. One can accept many of the claims of the latter, while still insisting that the fact that organisms are the product of evolutionary processes means that organicism is at best a limited claim, it doesn't apply to all of evolution. I find odd Huneman's suggestion that the major transitions framework is an expression of, or particularly in keeping with, modern synthesis biology, not least because one notable effect of work on the major transitions has been to entrench a multi-level perspective as against a purely genic one. It seems to me entirely possible—and necessary—to accept *both* the insights to be found in the major transitions program, *and* the insights to be found in the new organicism (suitably constrained). We need not choose between them.

through the symbiotic union of previously independently existing prokaryotes. And prokaryotes, in turn, have their own origin stories.

The new organicists may not be altogether happy with this glossing of their views in terms of MLS1–2; they offer quite different examples of and arguments for organismal (and other forms of) holism and emergence. But the general point—that biological entities and systems for which holism obtains are that way because they got that way by evolution—is one they should endorse.[21] Yet some of them, such as Walsh and Nicholson, do often talk as if the existence of organisms was a pre-given feature of the natural world. What we tend to get in their discussions is a false dichotomy between an approach to evolution in which organisms are permanently marginalised or ignored (modern synthesis biology) and one in which they and their agency and purposes are permanently foregrounded and accorded primacy in the evolutionary process *as such* (e.g. the 'situated Darwinism' Walsh (2015) advocates). An intermediate, and, I'd suggest, more defensible picture is one in which organisms, once they have evolved, are important agents in influencing certain evolutionary processes, but (a) they haven't always existed, being themselves the product of evolutionary processes to which they did not causally contribute, and (b) there are agents at other levels of the biological hierarchy that are arguably just as important in other evolutionary processes—genes, cells, groups, species, etc. (Gould 2002). The expansion of evolutionary agency to include entities above and below the level of the organism was one of the great advances in late Twentieth Century biology and philosophy of biology (Wilson 2005). In some recent attempts to re-establish the centrality of the organism there is the danger, it seems to me, that this insight could be lost in a kind of over-correction. Again, there is a middle position between the complete neglect of the organism as agent, and the neglect of agency at other levels: we can think of organisms as one important agent among others. But fully exploring this issue and fully defending the above claims would require more space than I can devote to it here, and must be left to future work.

---

[21] As D'Arcy Wentworth Thompson famously said, 'Everything is what it is because it got that way.' (Thompson 1917).

3.2 Pluralism and Instrumental Reductionism

Thus far I have spoken as if the systems for which holism is true demand holist scientific methods or tools to understand and explain them. But these things can come apart. Even if holism obtains for certain systems as a matter of fact, it may be that they are best approached *scientifically* by employing reductionist methods; in such cases, a reductionist approach would be instrumentally justified. For instance, Developmental Systems Theory (DST)—a set of ideas about evolution and development centred on the notion that it is the *whole life-cycle* that is, in every case, the unit of selection and evolution (Gray 1992, 2001; Griffiths and Gray 1994, 1997, 2001, 2004, 2005; Oyama et al. 2001) —is a strongly holist approach. One criticism of DST has been that this holism makes developmental systems empirically intractable: if every component and every causal factor in all their complex and reciprocal interactions have to be considered all at once, it would be hard to make progress in understanding the system; where would one even begin (Sterelny et al. 1996, 382–3)? Godfrey-Smith (2001b) responds to this objection by distinguishing two aims that a program like DST might have: it may aim to provide a literally true description of the target system (as such it is aiming to be a contribution to 'philosophy of nature'); or it may aim to satisfy the desideratum of empirical and heuristic fruitfulness. The unmanageable-holism objection only carries weight if the aim of DST is purely the latter. But DST *qua* philosophy of nature doesn't aim to satisfy this desideratum: all it aims at is literally true description.[22] If the description it offers is too holistic to be of any practical help to the empirical researcher, this is beside the point. Griffiths and Gray (1997, 88), in their defence of DST, similarly accept that a full theoretical articulation of DST may be strongly holist, and thus of limited use to the empirical scientist. But they suggest that DST also comes in a more empirically manageable form, in which the holism is toned down such that certain causal factors can be treated as background conditions while others are placed centre stage. So long as we remember that this is a distortion of the real situation, and that the

[22] Here and elsewhere 'literally true description' should be understood in terms of the explanatory interpretation of holism and reductionism I defined above.

distinction between foreground and background causal factors is an artificial one made purely for pragmatic or instrumental reasons, this is acceptable, they claim. They also offer a *tu quoque* argument, suggesting that the selfish replicator view favoured by some of those advancing the holism charge (Sterelny et al. 1996) also comes in two forms, a full theoretical form and a more user-friendly form, and that the full theoretical version would be scarcely less holistic than the full version of DST (Griffiths and Gray *ibid.*).

This might seem to be grist to the mill of the pluralism I am recommending. We have, again, the legitimacy of both holist and reductionist approaches, it's just that here they are being applied to the same systems. Perhaps some systems (like developmental systems) are such that they can only be fully and accurately described in holist terms; but making progress on understanding them scientifically might require, in practice, abstracting away from this holism, to consider certain parts of the system in relative isolation while keeping others in the background, in a more reductionist mode.[23] Yet these considerations can be generalised in two ways that I want to resist, as they would spell trouble for the form of pluralism I wish to defend.

Firstly, the holist may claim that *all* justified reductionism is instrumental reductionism. In this view holism applies across the board, such that reductionism is always a distortion of the true situation. It never captures the true relations between a system and its parts (or between different levels of reality) and to the extent that reductionist approaches are legitimate, it is only as convenient fictions, useful heuristics for making

[23] I am here discussing the idea that holism may be true of systems as a matter of objective fact, but reductionist methods are required to make progress in understanding them, given human limitations. But for some systems or phenomena, the reverse might be the case. Some have claimed, for instance, that Mendelian genetics, in principle, reduces to molecular genetics, such that there is, in principle, a complete and accurate description of all classical genetic phenomena in molecular terms; but that the molecular description would be so long and complicated that it could not be formulated or grasped by minds such as ours', making 'higher-level' descriptions unavoidable in practice for us. In that case reductionism would be true of these phenomena as a matter of objective fact, but holist methods, theories and explanations would be required to make progress in understanding them. This would be a kind of instrumental holism (or non-reductionism). See Rosenberg (1994).

progress in understanding the target system.[24] They would be models that (like many models in science), in the interests of tractability, present simplified pictures of the target system, abstracting away from the complexity and detail and mutual causal interaction among factors that would need to be included if the model was to be a truly accurate and complete representation of the system.[25]

Secondly, the reductionist may claim that holist approaches are *bound* to be scientifically barren, that reductionism is always needed in practice to make scientific progress, and to the extent that holist interpretations are legitimate it is only as scientifically-irrelevant 'literally true description', i.e. pure metaphysics, where the literal truth is all that counts. This view, that whatever the complete and final metaphysical truth of the matter, reductionist approaches are in practice our only means of advancing scientific explanation and understanding, is an old and influential one. These generalisations could even be combined in a kind of 'division of labour' or 'non-overlapping domains' compromise: the realm of metaphysics and literally true description belongs to the holist philosopher, while the realm of empirical progress and scientific advancement belongs to the reductionist scientist.

In fact I suspect most reductionists and holists would be dissatisfied with this compromise. Holists typically think holism is not just a matter of literally true description devoid of practical scientific benefits, that at least a lot of the time holist approaches are empirically fruitful and are means of advancing scientific knowledge. The holism of the new organicists mentioned above, for instance, is presented as a viable and

---

[24] There is also a generalised version of the view described in the previous footnote (i.e. reductionism as objectively true, holism as instrumentally justified). It is an old idea that, given the truth of physicalism, a complete and accurate description of everything in the language of completed fundamental physics must be possible in principle, and may be accessible to a mind of much greater power than our own, but given human cognitive limitations, we have no choice when trying to understand macro-level phenomena but to avail ourselves of 'higher-level' descriptions and explanations, in the language of chemistry, biology, psychology, sociology and so on. In that case reductionism would be objectively true across the board, and holism would be at best instrumentally justified across the board (Rosenberg *ibid.*). I reject this for the same reason that I reject the opposite view that holism is objectively true across the board, and reductionism as at best instrumentally justified across the board; see below.

[25] Some holists I am discussing in this chapter such as Levins and Lewontin (1985) seem to hold this view.

potentially fruitful empirical research program for biology, not just a piece of abstract metaphysics. And conversely, reductionists typically think reductionism at least often captures the actual relations between a system and its parts (or between different levels of reality), in other words, that it often succeeds as accurate and literal description, such that reductionism is often not purely instrumentally justified. For instance, with respect to the issue of the gene's eye view vs. the hierarchical (multi-selectionist) view, while a position according to which the hierarchical view captures the true situation while the gene's eye view is instrumentally justified is theoretically possible (perhaps the view that the genic perspective is useful for 'keeping the books' on evolutionary change, but fails to capture the causes of selection operating at higher levels (Sober and Lewontin 1982), could be interpreted in this way), in practice multi-selectionist think hierarchical and holist methods of analysis and tools of inquiry are required for making progress on understanding a hierarchically-structured reality, while genic selectionists think that the genic level really is the focal level for all or most selection, with selection at higher levels being, as a matter of fact, either entirely spurious, or rare enough to be unimportant and thus disregarded in evolutionary theory.[26]

I think that reductionists and holists are both right in rejecting the compromise view. Consider again MLS1–2. In cases of MLS2, it is not that holism correctly describes the metaphysics of the entities and processes in question, with reductionist methods being needed to investigate them empirically. Approaching them in terms of, say, reductionist MLS1—using MLS1 methods of analysis—would be mistaken, would give the wrong results. And to the extent that holism applies here it is not unmanageable or unwieldy; it's a tractable holism. Conversely, in MLS1, reductionism is true of the entities and processes involved: a holist description—in terms of MLS2, for instance—would be factually inaccurate. The reductionism involved in MLS1 is not just instrumental.

Thus both ways of generalising the division of labour suggested for DST fail, I claim. Holism is not true globally as literal description, and

[26] There are many other examples of reductionism intended non-instrumentally. When biologists explain regularities at one level by identifying mechanisms operating at a lower level (as emphasised by contemporary mechanists), this can be and typically is intended quite literally, not as a convenient fiction.

reductionism does not apply globally as an unavoidable methodological assumption for scientific progress to be possible. The pluralism I am defending—pluralism for both factual claims and methods—still stands. This is of course consistent with there being *some* cases where holism gives the literal truth, but forms of methodological (instrumental) reductionism are required (or at least helpful) for scientific fruitfulness, and perhaps developmental systems are one such case. And there may be more; Walsh, for instance, argues that the modern synthesis separation of inheritance, development, adaptive evolution, and the generation of evolutionary novelty was a distortion and simplification of the true holistic situation (where all these things interact in complex ways), but was nonetheless quite empirically and heuristically fertile, in allowing biologists to investigate each of these things in isolation from the others and make progress on understanding them; but 'expedience is one thing, and veracity quite another' (Walsh *ibid.*, 87).

4 Weakening the Positions?

In the previous chapter I argued that weakening metaphysical punctuationism and gradualism so they become factual claims about what happens *most of the time* is an unpromising move. The same can be said for the corresponding weakening of reductionism and holism understood as general factual or methodological claims, as the case of MLS1–2 shows. MLS1 and 2 are both important and widespread processes in nature. The former requires and vindicates a reductionist approach, the latter a holist one. To ask the question of which happens 'more often' or 'is more important' in the evolutionary process, as a means of establishing whether reductionism or holism is true as a claim about dominant relative frequency, borders on nonsensical.

The holist may object that her view is even weaker than the claim that holism enjoys a dominant relative frequency, i.e. that in most cases wholes are more than the sum of their parts (or something like this). Holism is merely the view, she may say, that at least *some of the time* wholes are not reducible to their parts and their interactions (or higher levels to lower), that global reductionism—the view that in every case wholes are reducible to or explainable in terms of their parts—is false. Holism would thus be akin to libertarianism about free will, which is the view that *there are at least some* uncaused, free actions, that determinism does not hold universally. If there is even a single uncaused, free action in the history of the universe, libertarianism would be true, on this understanding of the doctrine (in fact of course libertarians typically believe that there a lots of uncaused, free actions (whenever an agent acts freely) but their view as such does not commit them to this). I think in fact that this limited claim about holism—that it applies in some cases, to some systems, in some contexts—*is* what holists typically argue for, and they would be satisfied if they could establish this. But I take it this is too weak to count as an interesting holist view understood as a general, factual claim about nature and its workings. To return to MLS1–2 once again, it would be odd indeed to claim that the fact that MLS2 processes vindicate holism (locally) shows that holism as a *general* doctrine about the units of selection, or biology more generally, was thereby vindicated. After all, MLS1 vindicates reductionism! The limited holist view is in fact, I would argue, a version of pluralism. Holism applies in some cases, to some systems, but reductionism applies in other cases, to other systems. Neither dominates overall. It is this pluralist view that holists would be content, generally, to establish. The pluralist view is, I have argued, correct. But I see this not as a victory for holism, but rather as a defeat for both reductionism and holism understood as global factual claims about the biological world. Pluralism is a victory for pluralism.

If holists are really arguing for pluralism, not holism, in what sense are they holists? They are arguing for versions of *local* holism, to be sure. But this doesn't seem to capture what we might call their deeper philosophical (ideological?) commitment to holism. By now it will be clear how I think this commitment, this worldview, should be captured and characterised: it is a philosophical stance. Adopting the holist or reductionist stance is consistent with accepting pluralism about actual biological explanations and actual biological systems and phenomena (just as, as we've seen, adopting the punctuationist stance is consistent with adopting pluralism about the modes and tempo of change in different systems). Of course, one could accept pluralism whilst adopting no general stance at all, as some 'eclectic' or 'opportunist' scientists with neither reductionist nor holist commitments or sympathies do. There is nothing wrong with this, on my view: I argue stances can be valuable, I don't claim they are absolutely necessary.

5 Why Stances?

But why postulate stances at all? I have argued that holism and reductionism, understood as global factual or methodological claims (universal or majority-of-cases) are both false. Local forms fare better: the MLS1–2 case shows that each is true of certain contexts or phenomena. Why not stop there? Why not say there is not or should not be any *general* reductionist or holist position or stance at all, that local forms of the views are all that we require, and all that can be justified? This is the view Novick has defended with respect to functionalism and structuralism (2023). Local functionalist (structuralist) views, or explanatory strategies, pertaining to particular domains, may succeed, she suggests, but that is all there should be: general, global functionalist or structuralist positions, including stances, are unjustified and unnecessary.[27] Nobody should be a functionalist or structuralist in any more than a local sense: 'I … hope, if perhaps in vain, for a future biology free of structuralist and functionalist stances. Evolutionary biology needs structuralist and functionalist explanations, and even structure/function disputes, but … it has no need of structuralists and functionalists' (*ibid.*, 59). Perhaps we should also hope for a future biology free of reductionist and holist stances, free of reductionists and holists? This hope might be supported by the observation that many biologists today are already 'opportunists' (Rosenberg *ibid.*, 6), adopting whatever methodology (reductionist or holist) they think will work for the problem at hand, switching between methodologies as required, and are free of deep philosophical or methodological commitments to reductionism or holism.

[27] See my (2025) for a reply to her argument about functionalism and structuralism.

5.1 Explanation and Epistemic Values

To see one potentially important role for such general stances, recall I mentioned above that one way to undermine both global holism and global reductionism is to claim that there is no objective fact of the matter concerning whether reductionism or holism holds, rather this is a question of our interests, values, explanatory priorities, or tastes. The example of MLS1–2 was not, on my view, an instance of this, since I follow Okasha (2006) in regarding these as real processes in nature, not optional perspectives on the same facts. If that is correct, the no-fact-of-the-matter view does not apply universally: sometimes reductionism is true (locally) as a matter of objective fact, and sometimes holism is. But there are plausibly other cases where the view does apply. Take the question of whether Mendelian genetics is reducible to Molecular genetics, the question that has probably been the main focus of the reductionism issue in biology in the last 60 years.

Hull (1974), argued that it wasn't reducible due to the *multiple realisability* of Mendelian categories. An indefinite number of different molecular arrangements can give rise to a single Mendelian category, such as a gene. Connections between the molecular and Mendelian level are thus 'one-many' or 'many-many':

> Even if all gross phenotypic traits are translated into molecularly characterized traits, the relations between Mendelian and molecular predicate terms express prohibitively complex, many–many relations. Phenomena characterized by a single Mendelian predicate term can be produced by several different types of molecular mechanisms. Hence any reduction will be complex. Conversely, the same types of molecular mechanisms can produce phenomena that must be characterized by different Mendelian predicate terms. Hence, reduction is impossible. (Hull 1974, 39)

Kitcher extended this analysis in his famous (1984). He argued that cytological explanations are independent of the molecular level (Sterelny and Griffiths 1999, 140). If we are interested in understanding and explaining Mendelian laws, such as the law of independent assortment,[28] the 'gory details' concerning the precise molecular realisation of the relevant cytological phenomena will not be helpful. All we need to know in order to explain the law of independent assortment is that during meiosis chromosomes pair up, with one of each pair going to each gamete (and which member of a particular pair goes a gamete is independent of which members of the other pairs go to it). So we have multiple realisability again: this cytological process can be realised in many different molecular ways, such that the precise molecular details are irrelevant to the explanation, and we can (and should) abstract away from them.

> Why do genes on nonhomologous chromosomes assort independently? Cytology provides the answer. At meiosis, chromosomes line up with their homologues. It is then possible for homologous chromosomes to exchange some genetic material, producing pairs of recombinant chromosomes. In the meiotic division, one member of each recombinant pair goes to each gamete, and the assignment of one member of one pair to a gamete is probabilistically independent of the assignment of a member of another pair to that gamete. Genes which occur close on the same chromosome are likely to be transmitted together … but genes on nonhomologous chromosomes will assort independently. This account is a perfectly satisfactory explanation of why our envisaged law is true to the extent that it is … In charting the details of the molecular rearrangements the derivation would only blur the outline of a simple cytological story, adding a welter of irrelevant detail. (Kitcher 1984, 347)

In an almost equally-famous response to the anti-reductionist position, Waters (1990) argued that if we think of genes not as the direct causes of traits, but as *difference-makers*, then all instances of a Mendelian gene *will* have something molecular in common. Moreover, we saw that Kitcher claimed that we have an adequate explanation of the law of independent assortment once we understand that during meiosis chromosomes pair up, with one going to each gamete (and which member of a particular pair goes a gamete is independent of which members of the

[28] This is the law that genes for different traits segregate independently during the formation of gametes.

other pairs go to it). But we might question whether this explanation is sufficient. It doesn't, for instance, explain why this latter process happens (Sterelny and Griffiths 1999, 145), but that surely is a legitimate question we may be interested in, and presumably to answer it adequately we need to descend to the molecular level.

The anti-reductionist philosophers are, Waters suggested, out of step with the biologists: 'Practicing geneticists believe that the classical theory can be systematically explained at the molecular level, I suggest, because they have a firm grasp of the explanatory power and structure of molecular biology' (Waters *ibid.*, 135). This classic debate has its echoes in more recent debates between mechanists and their opponents concerning whether lower-level causal mechanisms are always required to explain higher-level regularities (Rosenberg *ibid.*, 59).

It has appeared to some people that there is a sense in which both sides in this debate are right, relative to different aims, goals, interests or values.[29] Sometimes, it seems, we do want to abstract away from the gory molecular or physical details in order to provide a general account of patterns and regularities instantiated at the higher level. But sometimes we want the gory details! It all depends on our goals and interests (Sterelny and Griffiths 1999, 140–141). The way Sober (1999, 550) puts this is that explanations can be broad (i.e. general, applying to a large range of cases) or deep (i.e. detailed, giving specific and detailed information about the explanans and its relation to the explanandum). There is a trade-off between these two aims; it may be hard to satisfy both at once. The Kitcher anti-reductionist view involves sacrificing depth for the benefits of generality; the Waters reductionist view involves sacrificing generality for the benefits of depth. Sober goes on to suggest that whether one prefers explanations with depth or with breadth is just a question of values and taste: neither criterion is objectively superior to the other. Neither approach fits the facts better than the other.[30]

---

[29] Or neither is: 'Our conclusion after reviewing both the 'antireductionist consensus' and the 'reductionist anticonsensus' is that nobody wins' (Sterelny and Griffiths 1999, 147).

[30] Sober's more general claim (1999) is that the multiple-realisability argument against reductionism (according to which broad explanations are objectively superior to deep (but non-broad) ones), has been greatly oversold. I think he is probably right about this, thus the arguments I offer in this chapter for (limited and local) holism do not rely on multiple-realisability (and nor, it seems, do many of the most significant recent arguments for holism-organicism in biology).

It is here, I'd suggest, where stances can come in. It is plausible that part of what defines the reductionist stance is a set of epistemic values that includes the valuing of explanatory depth, even when this conflicts with other epistemic values. The holist stance by contrast values the kind of explanatory breadth that is afforded by higher-level explanations and generalisations, even at the expense of the kind of depth valued by reductionists. (This doesn't entail of course that one cannot value *both* depth and breadth in explanations, seeking to maximise both as much as possible, and when trade-offs are necessary perhaps sometimes prioritising depth over breadth and sometimes the reverse. This is just another way of saying that one need not adopt either the reductionist or the holist stance.) The idea that part of what is involved in the Kitcher-Waters dispute is a clash of epistemic values is initially quite plausible I think; Sober's interpretation, cashed out in terms of stances, explains the sense in which this is the case.[31]

To recap, the picture I have presented is as follows. We have the reductionist stance and the holist stance. Advocates of both stances ought to recognise that in some contexts reductionism is objectively correct (e.g. MLS1) and in other contexts holism is (e.g. MLS2). But there are some cases where there is no fact of the matter about whether a reductionist or a holist explanation is superior (for instance, where the criterion of depth conflicts with the criterion of breadth). In such cases, one may opt for the explanation that manifests the epistemic values associated with one's favoured stance.

5.2 Classifying Thinkers by Their Philosophical Commitments

Thinking in terms of stances also helps us to classify thinkers by reference to their philosophical commitments. Consider the units of selection again. Gould argues (2002, 31) that the gene selectionism of Williams and Dawkins does not represent a fundamental break with orthodox Darwinism ('radical conceptual revision') but rather a 'furthering and

[31] This also helps to explain (and overcome) the somewhat stalemated character of the debate, with persuasive arguments on both sides.

intensification of Darwin's (reductionist) intent.' It is Gould's own hierarchical view which, Gould suggests, is the radical conceptual revision. We can understand this in terms of shared stance: Williams and Dawkins share Darwin's reductionist stance. If we didn't consider the question of general reductionist stance, it would be hard to claim that gene selectionism is a 'conservative adumbration' (*ibid.*) of traditional Darwinism, while the hierarchical view is the radical conceptual revision, since on the face of it, each approach departs somewhat from traditional Darwinism, and it is not obvious that the hierarchical view is the greater heresy of the two. Indeed some have argued precisely the opposite: it is gene selectionism that is genuinely revolutionary, while the hierarchical view represents in the first instance a defence of traditional Darwinism against gene selectionism, even if it goes on to posit selection at higher levels than Darwin allowed. Sterelny (2001, 4–5) tends to see matters this way. He distinguishes an 'American' and a 'British' tendency within evolutionary biology, with the latter subscribing to gene selectionism, and the former defending traditional organismic selection against gene selection, and only secondarily defending selection at higher levels. Dawkins would certainly not agree with this assessment: he argues (1982, 6) that gene selectionism and Darwin's organism-centred view are in fact equivalent, and are comrades in arms opposed to the genuine heresy, group and higher-level selection.[32] Even in *The Selfish Gene* (2006, 12), where Dawkins does argue that the genic view is factually correct, he insists that his view 'is, in substance, orthodox, even if it is expressed in an unfamiliar way'. And in the preface to the 1989 edition he says, 'The selfish gene theory is Darwin's theory, expressed in a way that Darwin did not choose but whose aptness, I should like to think, he would instantly have recognised and delighted in. It is in fact a logical outgrowth of orthodox neo-Darwinism, but expressed as a novel image' (2006, viii).

But at the level of gene selectionism vs. individual selectionism vs. multi-level selectionism, this dispute about whether Dawkins' or Gould's revisionism is the greater apostasy is difficult to adjudicate. Gould's claim

[32] Gould disagrees with this. He rejects Dawkins' model pluralism/conventionalism according to which gene selectionism and Darwin's view are equivalent. He regards them both as instances of reductionism, but denies they are 'equally correct'.

can be interpreted as the suggestion that at the different level of general philosophical orientation (stance) gene selectionists are united with Darwin as reductionists (they share Darwin's reductionist 'spirit' (Gould *ibid.*, 31)), opposed to advocates of the hierarchical view who subscribe to an anti-reductionist stance. The stance approach is useful here because it can help to justify our intuition that while the question of which modern revisionist version of Darwinism is the greater departure from orthodoxy is difficult to answer at the level of theory, there does seem a clearer sense in which, philosophically, the gene selectionist has more in common with Darwin than either has with the multi-selectionist.

5.3 The Positive Role of Stances

This only shows that thinkers *do* adopt stances that permit us to classify them as reductionists or holists; it may be wondered whether they *should*. Novick accepts that there are in fact functionalist and structuralist stances and biologists who adopt them; her claim is the normative one that there *shouldn't be*, that they play a pernicious rather than a positive role in biology. I have argued against the normative claim about functionalism and structuralism elsewhere (Boucher 2025) and I claim it is also implausible in the case of reductionism and holism.

The reductionist stance, which has long been dominant in the physical sciences and to a lesser but still significant extent in the biological sciences, has of course given rise to or been associated with spectacularly successful scientific research programs and theories in biology, from Darwin's individualist theory of natural selection through to molecular biology. A positive role that the holist stance can play (and arguably has played) has already been indicated. In a scientific culture dominated by reductionism, the holist stance can suggest, encourage or foster alternative modes of explanation and ways of understanding certain systems or phenomena. Even if the result of this is at most pluralism—establishing holist and hierarchical forms of explanation and understanding as valid and legitimate alongside more standard reductionist forms, rather than replacing reductionism with holism across the board—this is or would be a not inconsiderable achievement. While in principle such an achievement may be possible in the absence of anything like the holist *stance*, in practice it is likely to be those with holist philosophical commitments and general anti-reductionist sympathies—partisans of the holist stance with its characteristic epistemic values, goals and approaches in other words—that will be largely responsible for it. In general, as with the other metaphysical stances covered in Part II, these stances may play a constructive role in science in guiding scientific work, opening up otherwise neglected (by those in the grip of rival perspectives) avenues for inquiry, making certain issues visible, foregrounding certain overlooked or marginalised phenomena, and making progress on otherwise intractable problems. It is tempting (as Novick does for functionalism and structuralism) to try to deflate or move beyond large-scale philosophical disagreement and stance-conflict in science, and replace it with a set of tractable local empirical problems, local approaches and local solutions, to, as Novick says, take 'our debates smaller and more local, away from stances and general theories …' (Novick *ibid.*, 62). But this is, I'd suggest, unrealistic and in fact undesirable. Large-scale stances and deep philosophical and metaphysical commitments can and often do play a positive role in science, I claim.

6 Stances and Values

As with the other stances discussed in Part II, the holist and reductionist stances would be understood to be styles of thinking, perspectives or attitudes that (a) would be logically distinct from particular local, empirical reductionist or holist theses or explanations, (b) would be values-laden, (c) would influence science in the context of discovery rather than the context of justification, (d) would permit pluralism both in the sense that adopting either would be consistent with accepting the (local) validity of both reductionist and holist explanations and empirical hypotheses, and in the sense that adopting either would be consistent with accepting that the existence of a plurality of metaphysical stances (reductionist and holist) within biology is or would be a positive thing, and (e) would be pragmatically justified to the extent that the explanations and hypotheses suggested by the stance were vindicated.

It may be worth saying a little more about (b) the value-ladenness of the stances. We have seen that the holist and reductionist stances seem to embody different epistemic values (the values of explanatory depth vs. breadth for instance). But they are often associated with different non-epistemic values as well. Sober and Wilson (1998, 8) note that it is often supposed that people endorse anti-reductionist, holist views such as group and multi-level selection theory 'because they *want* the world to be a friendly and hospitable place.'[33] That is, people are inclined toward these views because they fit with their picture of how the world should be; they correspond to their social, political and ethical values. Equally, it may be said that defenders of reductionist views (such as the gene's eye view) may be inclined toward them in part because they cohere harmoniously with a set of social and ethical values in which competition, individualism, egoism and self-interest loom large. Sober and Wilson (*ibid.*, 9) note that it could also be argued that defenders of such views are attracted to them precisely because they do make the world appear harsh, unfriendly and competitive, and their proponents are pleased to think of themselves as hard-headed realists prepared to look this unpleasant reality in the face, and able to do without the comforting illusions fostered by their opponents. (Note that in this case it's not the belief itself which has pleasing effects, but the belief that one has the belief.)

Of course, as Sober and Wilson say, everyone would agree these would be bad reasons for endorsing particular factual or empirical hypotheses such as multi-level selection theory or genic selection. These ought, other things being equal, to be accepted or rejected on the basis of the evidence, not on the basis of whether they cohere with our social, political and moral values. (Even pluralists who deny that these views are rival factual claims between which the evidence can decide stop short of suggesting that this means one can adopt the view that best fits with one's social and political values.) But stances are not factual or empirical claims, and they are not adopted on the basis of the evidence. Hence such non-epistemic values can form a legitimate part of the set of values that help to define or

[33] Similarly, G.C. Williams notes that '[t]he history of science shows the inevitability of widespread emotional dissatisfaction with reductionist interpretations of the world. I suppose there are people who think that the theory of sexual selection robs the grossbeak's song of its music or that kin selection makes a mockery of brotherly love' (quoted in R. Wilson (2005, 41)).

characterise a stance in our sense, and it is permissible to adopt a stance in part because we judge that it expresses or involves the non-epistemic values we endorse (Magnus 2018).

One problem here is that certain stances seem to be associated with different non-epistemic values at different times and in different contexts. In biology and philosophy of biology, there is a very rough connection between holist/anti-reductionist positions and progressive politics, and between reductionist views and more conservative politics. Of course there are exceptions, and there is clearly no direct or simple relationship between, say, the gene's eye view and conservative, biological determinist justifications of the social or political status quo (despite some claims to the contrary), or between multi-level selection theory and leftist or progressive values. But it would be naïve to deny that the conflict between reductionism and holism in biology has often—as a matter of fact—possessed a political or ideological dimension. Reductionists have typically been more sympathetic to sociobiology and evolutionary psychology than holists and anti-reductionists, for example,[34] a topic where politics, ideology and social values have been as much to the fore as evidence and scientific reasoning.

In other contexts however, holism has been associated with reactionary politics, and reductionism (and associated views) with progressive politics. Scott and Rief (2021) have analysed the philosophical commitments and worldviews of rival schools in Germanic social science, especially in the inter-war years. Anti-holist, analytical and empiricist views and methods were adopted by Marxist social scientists (the proponents of 'empirical social science') and the so-called 'left Vienna circle', while holist, metaphysical, organicist, 'synthesising' and 'totalising' views and methods were associated predominantly with Catholic and fascist intellectual currents. In later social science, under the influence of critical theory, empiricism, reductionism and related views came to be associated with

[34] As I noted above, Sterelny (2001) identifies two broad schools in contemporary evolutionary biology, the 'British' school represented by thinkers such as Dawkins, Dennett and Cronin, and the 'American' school, represented by thinkers such as Gould, Lloyd and Lewontin. This was over 20 years ago so is rather out of date, but the general divide he is indicating arguably still holds. The Brits are sympathetic to gene selectionism, adaptationism, reductionism, extrapolationism, and sociobiology or evolutionary psychology (and the Americans are not). Very roughly, the Brits are, at least in some respects, on 'the right', and the Americans on 'the left'.

technocratic domination, and the rejection of them equated with the progressive critique of capitalist institutions and ideology.[35]

Thus, which stances are thought to embody or express which non-epistemic values can vary, it is not fixed for all time. Holism is not by definition the stance of choice for those with leftist political leanings. These connections are contingent and contested. This doesn't, it seems to me, undermine the general claim that stances may embody non-epistemic values. It just means there is no necessary connection between certain stances and certain non-epistemic values. There may be a closer connection between certain stances and certain *epistemic* values.[36] The metaphysics stance seems to necessarily involve the value of explanatory power for instance, while the empiricist stance may necessarily involve the value of minimising epistemic risk. The reductionist stance may necessarily involve the value of explanatory depth. But stances also involve beliefs, and as we have seen different beliefs are associated with different stances at different times and in different contexts. There can be reasonable disagreement about which beliefs cohere with which stances. The same may be true for the relation between non-epistemic values and stances. Indeed one may adopt a stance entirely independently of any non-epistemic considerations.

I wish now to consider a lesser-known but interesting variant of the holist stance: the *dialectical* stance.

7 Dialectics of Nature[37]

'Dialectics of nature' is the attempt to apply the Marxist doctrine of dialectical materialism to the natural world, an attempt associated in particular with Engels (1972). Dialectical materialism is a doctrine about the nature of change, and more generally the nature of reality. It says, roughly, that reality is a differentiated totality in which the totality is in some sense

---

[35] Thanks to Alan Scott for helpful discussions on these matters.

[36] In my (2018b) I argued that certain values (unlike beliefs) were necessary for certain stances, but didn't in that paper distinguish epistemic from non-epistemic values.

[37] This section draws on my (2018a) and (2019).

primary (the whole is prior to the parts), that change is a constant and pervasive feature of reality and that change takes place as the result of conflict and contradiction. The doctrine is notoriously metaphysical in the sense of being apparently immune from falsification and devoid of obvious empirical content. It has thus been viewed with suspicion or outright hostility by empiricist-minded philosophers, even those otherwise attracted to Marxism. Rogers (1999, 275) notes, for instance, that A.J. Ayer, although sympathetic to Marx's theory of history, 'argued against dialectical materialism, contending that the claim that all movement in nature is born of conflict was simply too general to be properly verified.' But I will show that, as with metaphysical punctuationism and gradualism, if interpreted as a van Fraassian stance, it may be potentially interesting and valuable, even from the standpoint of empiricism.

The Marxist biologists Richard Levins and Richard Lewontin (1985) defend a dialectical approach to biology, inspired by dialectical materialism.[38] What, specifically, do Levins and Lewontin mean by 'dialectics of nature'? They expound their position in the Conclusion of their (1985). They first summarise the outlook they are opposed to, 'Cartesian reductionism', which they see as the dominant stance in western biological and non-biological thought. This is the view that the parts of a system are 'ontologically prior to the whole; that is, the parts exist in isolation and come together to make wholes'[39] (*ibid.*, 269). The world that Cartesian reductionism describes is 'the *alienated* world, the world in which parts are separated from wholes ... causes separated from effects, subjects separated from objects' (*ibid.*, 270). They counterpose to this their dialectical outlook, which involves the ideas of totality, change and contradiction as mentioned above, but is more specific. It has several components[40]:

[38] See Godfrey-Smith (2001a), for one of the few discussions of dialectics in mainstream analytic philosophy of biology.

[39] The view has other components, but this is probably the central commitment.

[40] The following have the appearance of being factual beliefs about what the world is like. But we will see that on the stance interpretation, they should rather be understood as expressing a certain mode or style of thinking—what Levins and Lewontin call 'habits of thought' and 'forms of questioning'.

- A whole is composed of heterogeneous parts that do not exist independently of the whole (*ibid.*, 273).
- The parts have the properties they do in virtue of being parts of the whole (*ibid.*).
- Subjects and objects, and causes and effects, are 'interchangeable'. Nothing is purely passive object or purely active subject; one and the same entity (such as the organism in evolution) is frequently both subject and object, mover and moved (*ibid.*, 274).
- Change is a pervasive feature of every aspect of reality (*ibid.*, 275). The world described by dialectics 'is constantly in motion …' (*ibid.*, 279).
- Change results from contradiction (ontological, not just epistemic), which has a number of aspects: 'self-negation', which is similar to logical contradiction (*ibid.*, 282); the 'interpenetration of seemingly mutually exclusive categories' (such as 'deterministic' and 'random'), which refers to the tendency for opposing properties to involve and give rise to one another (*ibid.*, 283–4); and 'the coexistence of opposing principles', which refers to the tendency for conflicting standards (such as 'use value' and 'exchange value') to produce different effects when combined from those they would produce when taken separately (*ibid.*, 285).

The first two are fairly standard holist principles (though the claim that parts are heterogenous serves to distinguish their holism from more monistic forms of the doctrine); the other three express distinctively dialectical ideas. Levins and Lewontin note that a commitment to a strong form of holism is shared by much ecological and New-Age thought. Dialectics is distinguished from these views however by its emphasis on conflict and contradiction, rather than 'harmony, balance, or 'oneness' with nature' (*ibid.*, 274). Rees (1998, 248) suggests that the totalities posited by non-dialectical holists tend to be static, while dialectical totalities are internally contradictory, making possible change and development. As well as being a holist view, dialectics is (unlike standard forms of holism) also punctuationist in its conception of change, as noted in the previous chapter. Punctuationism is, it seems, necessary but not sufficient for dialectics: dialecticians accept the punctuationist stance, but punctuationists need not accept dialectics.

7.1 Pluralism Again

Above I argued that holists should, and typically do, recognize that reductionism is true when applied to certain systems, and reductionism as a method is often successful, undermining the plausibility of holism as a general factual thesis or general methodological recommendation in science. Levins and Lewontin accept that Cartesian reductionism has been extremely useful, as the metaphysical inspiration for major scientific theories and discoveries that count as successful on any measure, and in particular as scientific methodology: '... Cartesian reduction as a method has had enormous success in physics, in chemistry, and in biology, especially molecular biology ...' (*ibid.*).[41] Whatever the dialectical view is, it has to be consistent with this acknowledgement of the widespread and extremely impressive success of the reductionist method.

However Cartesian reductionism 'is more than simply a method of investigation; it is a commitment to how things really are. The Cartesian reductionist method is used because it is regarded as isomorphic with the actual structure of causation ... Cartesian reduction as a method has had enormous success ... and this has been taken to mean that the world is like the method' (*ibid.*, 2). Note the two-way inference: from truth of the worldview to use of the method, and conversely, from success of the method back to truth of the worldview. 'In actual practice, reduction as a methodology and reductionism as a world view feed on and recreate each other' (*ibid.*, 2). Levins and Lewontin reject the inference from method to worldview. They suggest that reductionism as a method can be and has been successful despite issuing from or being associated with a picture of the world that is false. Methodological reductionism has been successful *despite* the falsity of ontological reductionism. This is a version of instrumentalism about reductionism mentioned above.

I think what the dialectician *should* say is it's not the case that ontological reductionism is entirely incorrect (i.e. is true of no or very few systems or phenomena), just as it's not the case that the reductionist method is

[41] Although they think that success has been overstated; it has seemed more impressive than it is, they claim, because reductionist methods have been chiefly applied to precisely those problems most likely to yield to reductionist approaches (*ibid.*, 2–3).

entirely ineffective—but rather that these are *limited*, they are not *universally* valid, that there are important systems or phenomena they cannot cope with, and for which we require a different set of ideas or methods or tools. In other words, arguments for dialectics, like arguments for holism we considered above (e.g. Walsh's), are or should be arguments *against* reductionism as a universal claim or universally-valid method, not arguments *for* dialectics/holism as a general claim or universally-valid method. They are or should be arguments for pluralism, directed against reductionistic monism. As I argued above however, pluralism in this sense should not be confused with holism or dialectics understood as a general factual or methodological claim. If this kind of pluralism is correct, it means that both reductionism and dialectics as *general, global factual or methodological claims* are false. But adopting either as a stance is consistent with accepting this pluralism.

Even when we turn to the work of the 'dialectical biologists' themselves, we see the at best limited truth and applicability of the dialectical worldview and method. Maynard-Smith (1993, 33) has pointed out for instance, that much of Lewontin's own scientific work doesn't appear particularly 'dialectical'. Lewontin's key contributions to theoretical biology have probably been in the field of population genetics. This work has been thoroughly incorporated into mainstream (non-dialectical) biology.

It is true that Lewontin's more explicitly dialectical ideas and proposals have had an influence on parts of contemporary evolutionary biology. Royle (2014) and Sullivan (2015) have argued recently, for instance, that the theory of niche construction,[42] is an example of genuinely *dialectical biology*. Lewontin (Levins and Lewontin *ibid.*), and more recently, in much greater detail, Odling-Smee et al. (2003), argue that organisms *construct* their environment, rather than merely becoming adapted to independently existing environments.[43] Thus niches do not, as orthodox Darwinism would have it, exist independently of organisms, but rather are created (and indeed defined) by, the species that inhabit them. In rejecting the idea of the organism as the passive object of external evolutionary forces, Royle, Sullivan and Lewontin affirm what they regard as a

[42] Sullivan also suggests that the relationship of predator and prey exemplifies dialectics.

[43] For discussion see Sterelny (2005), Okasha (2005), and Griffiths (2005).

dialectical relationship holding between the organism and its environment. Thus on this view, the organism becomes, in Lewontin's famous phrase, the 'subject and object' of evolution (exemplifying point three in the above general characterization I offered of dialectical principles).

Royle and Sullivan argue this is a case of scientific theories that are themselves dialectical, in that they embody the principles of dialectics of nature, and indicate that, insofar as the theories are correct, dialectical materialism is true of certain parts of the world: '… the reciprocal influence of organism and environment seems to be an example of the pattern of causation that dialectical biology contends is present in biology.' (Sullivan *ibid.*[44]). But even if we grant this, those, such as Royle and Sullivan, who appeal to phenomena such as niche construction as evidence that dialectics as a factual belief is true of the world, must explain the fact that such phenomena appear to be exceptions rather than the rule. It would appear that at most a small number of phenomena in nature exemplify dialectical processes and relations: 'Whilst some of the world is constituted by dialectical relations, some of it is not. In fact, I would say that the vast majority of the world is not …' (Sullivan personal communication). In other words, phenomena such as niche construction can perhaps be used to show that a non-dialectical, reductionist perspective or method is *limited*, that there are phenomena, processes and systems for which a dialectical approach is needed (or is at least helpful). But that is all; there is no support here for the truth of dialectics of nature as general factual thesis about the whole (or very much) of nature.

If we turn to Levins' work in ecology and the philosophy of ecology, we find again an explicit acknowledgement of the limited truth and applicability of reductionism. In fact he argues that the problem with standard forms of holism is precisely that they have not understood this, they have been too one-sidedly holistic, an error just a serious as dogmatic reductionism. This leads some interpreters (e.g. Lefkaditou and Stamou 2006) to suggest that Levins' and Lewontin's dialectics is not a version of holism at all, but a kind of compromise or intermediate view

---

[44] Sullivan's suggestion (*ibid.*) that dialectics is a 'heuristic' that can assist scientists in the process of scientific discovery is quite congenial to the stance position; however he also clearly regards it is as a factual belief that correctly describes the biological world (or certain aspects of it), and is borne out by empirical investigation.

between traditional holism and reductionism, trying to do justice to the insights in both, with its emphasis on the priority of wholes, and its equal stress on the parts and their interrelations (applying dialectical categories to dialectics itself, reductionism and traditional holism would perhaps be thesis and antithesis, with dialectics being the synthesis). I would put it somewhat differently. Dialectics is, I'd suggest, unambiguously a form of holism—see for instance Levins' four strongly holist principles listed in Lefkaditou and Stamou (2006, 328)—but a non-standard form. As I noted above it rejects static wholes, encouraging us to see wholes as conflict-ridden and imbued with contradictions; not stable but in constant flux. After all, if the dialectical view was *nothing more than* the pluralistic claim that reductionism as a universal claim is false, and holism as a universal claim is false, that the truth lies somewhere in between, it would be correct but not very interesting (or dialectical). Dialecticians (like reductionists) *should* accept this pluralistic claim (as I would put it, that reductionism is true in some contexts, holism in others), this rejection of both reductionism and holism as universal factual or methodological theses. But there is more to being a dialectician (or reductionist) that this pluralism. This more is, of course, the adoption of the relevant stance. The dialectical stance is a non-standard form of the holist stance.

Levins argues, in fact, that in a sense dialectics gives us a more thoroughgoing and consistent holism that standard holism does. He notes that apparently holistic ecosystem theory shares much in common with reductionism, since it treats these systems as static and unchanging machines (Lefkaditou and Stamou 2006, 327–8). On this interpretation standard holism is the flip-side of reductionism, sharing its mechanistic assumptions; both describe an 'alienated' world, and only dialectics is able to transcend both and provide a genuinely different way of looking at the world.

Let us now consider the nature and function of the dialectical stance.

7.2 The Dialectical Stance

Lewontin and Levins are perhaps unlikely to fully agree with stance interpretation of their metaphysical commitments. They seem to hold that dialectics is both a method and a worldview (i.e. factual belief).[45] But several things they say are quite congenial to the stance construal. For instance, they claim that dialectics gives one a certain useful *perspective* on the biological world; it leads one to see it in a certain way, or approach it in a certain way, with a tendency to focus on some features rather than others, and a tendency to interpret it in a certain way. Thus they talk of dialectical 'habits of thought', 'forms of questioning' and 'ways of thinking' (1985, 267), just as Gould talks of metaphysical punctuationism as a 'style of thinking'.

Levins and Lewontin also point out (*ibid.*, 277) that while the dialectician can accept that some features of reality may be constant and unchanging, where she differs from the reductionist is in the fact that she *expects* change to characterise the systems she is studying. To *approach* the world with an *expectation* of finding change is different from holding the factual belief that change does in fact characterise all or most systems, just as we saw (in the previous chapter) that the expectation of finding punctuational patterns in phenomena, is different from and independent of the belief that punctuational change characterises all or most systems. And it seems clear that an *expectation*, being non-propositional, cannot stand in an inductive or deductive relation to particular scientific hypotheses or beliefs. This notion of dialectics as *perspective*, and *expectation*, rather than as factual belief, chimes with the stance view.

On the stance view, the dialectical perspective may be pragmatically justified on the grounds that it makes it easier to notice, and leads us to expect to find, phenomena in nature that appear to instantiate the kind of relations dialectics describes. *If* one has a dialectical outlook—adopts the dialectical perspective—one may be *more likely* to produce certain sorts of science, i.e. to access certain facts (about, say, niche construction). To that extent dialectics may be a *useful heuristic with epistemic benefits*. It makes it more likely that one will hit on certain truths about nature.

[45] They argue (*ibid.*, 2) explicitly that Cartesian reductionism is both a worldview (factual claim) and a method, and that endorsement of the method is not the same thing as endorsement of the worldview. One assumes they hold the same view about dialectics.

This approach ought to appeal to defenders of dialectics. In general, it is a major virtue of the stance approach that it enables us to see how metaphysical perspectives may be justified. The prospects for direct epistemic verification (or falsification) of general metaphysical perspectives such as dialectical materialism, or metaphysical punctuationism, appear dim. But if the stance view is right, they may be indirectly pragmatically justified in terms of their effects on empirical inquiry.

Consider again, as an analogy, selfish gene theory. Defenders of the selfish gene perspective argue that it forces on one's notice certain problems that are hard to see from other perspectives, e.g. 'the problem of the organism' (Dawkins 1982; Sterelny and Griffiths 1999, 75). Why should organisms exist at all, from the perspective of replicators? Why did replicators build organisms? This question can be asked by non-selfish-geners, but it is not as obvious or 'in your face' from perspectives that treat the existence of the organism as a given (Sterelny et al. 1996, 395).[46] This supports the conception I have outlined of metaphysical perspectives as justified in terms of their ability to draw attention to and highlight certain features of the empirical world that are accessible, but harder to see, from other viewpoints. The empirical facts are, then, independent of and separable from the metaphysics, but the metaphysics can make certain empirical facts easier to locate[47] (see Boucher 2014). Similarly, as I suggested, metaphysical punctuationism may make it easier to hit on the truth of punctutated equilibrium, and dialectics may make it easier to hit on the truth of niche construction and so on. Levins and Lewontin recognise this point when they suggest that while commitment to a metaphysical stance or 'ideology' is not necessary for the 'construction of a particular model of nature', such a commitment 'strongly predisposes us to see some things in the world and not others' (*ibid.*, 268). This is part of the effect and function of a stance/perspective. The facts are, theoreti-

[46] Neo-organicists like Walsh (2015) regard this question as pathological.

[47] Similarly Sober and Wilson argue that the multilevel selection theory they favour enables us to 'see the whole [evolutionary] stage' rather than being narrowly focused on a small part of it (1998, 332). It is, they argue, a powerful 'way of seeing' with which to view and make sense of facts that are nonetheless accessible (in perhaps less vivid and more obscure and confusing form) from other standpoints. So it is distinctive and valuable not in virtue of its uncovering of new facts, but in virtue of the orientation it provides on facts that can be represented from alternative points of view.

cally, there for all to see; but some metaphysical perspectives make it easier to pick out certain classes of facts, while others have the effect of obscuring them. Thus the influence of dialectics on empirical biology takes place in the context of discovery, not the context of justification.

Kitcher (2001, 408, 413) has argued that dialectics fails to offer anything concrete that scientists can use in their day-to-day work. Similarly, Daly suggests that Lewontin (and his ally Gould) have failed to effect the paradigm-shift in biology they have desired because they have 'no alternative research program to offer.' This is why '[s]ufficient research to fill a first issue of Dialectical Biology has yet to materialise' (Daly, quoted in Dennett 1995, 249). On the stance view I am considering, Kitcher and Daly may be right to be sceptical about the prospects for dialectical science, but wrong to thereby dismiss dialectics altogether. They do not recognise that it may play a valuable role as a metaphysical stance in the context of discovery.

We have seen that an implication of the stance view is *pluralism* with respect to rival metaphysical stances, and that Gould defends pluralism with respect to metaphysical gradualism and punctuationism, arguing that metaphysical gradualism is not *false*, and has been useful in influencing and inspiring good science. Similarly, as we've seen, Levins and Lewontin accept that Cartesian reductionism has also been extremely useful and successful. Levins and Lewontin (unlike Gould), are not however pluralists in any more than the weak sense of conceding the usefulness of different metaphysical perspectives: they do in fact regard Cartesian reductionism, however fruitful it has been as a metaphysic and as a method, to be, or involve, a factual belief about what the world is like, and is moreover a *false* factual belief. But the stance theorist, as we have seen, defends the stronger pluralist position that there is no question of truth or falsity with respect to such perspectives. On the stance view, asking whether the world is dialectical is asking the wrong question: since dialectics is not a factual claim, the question of whether it is true of the world simply doesn't arise. Adopting a dialectical perspective may help one to notice and understand certain features of reality (e.g. niche-construction), but that's all there is to it. Dialectics may be a useful heuristic, in emphasizing otherwise neglected features of reality, making certain facts and phenomena easier to see, and so on, but we shouldn't ask whether the stance is 'true', or whether the world 'really is' dialectical.

We have seen that an idea that fits naturally within this pluralist view is that of a metaphysical perspective being a useful *corrective* in a context or culture where rival perspectives have dominated and gone unchallenged. It is healthy for a scientific field to have a variety of perspectives to draw on. Levins has argued that dialectics, and associated views, could provide such a useful corrective to the dominance of reductionism: 'our emphasis on the holistic and the qualitative is intended not as a complete program for science but as a corrective for the one-sided analytical quantitative approach that is still dominant' (Levins, quoted in Lefkaditou and Stamou *ibid.*, 326). As we saw in the previous chapter, the idea of a perspective or approach being a 'corrective' to the prevailing views or approaches fits ill with the idea of the perspective being 'true' and its rivals 'false'. But it fits very well with the pluralistic stance interpretation.

The justification of the dialectics of nature perspective by way of a consideration of its epistemic fruits is something Levins and Lewontin lay stress on (see Maynard-Smith 1993). They note that approaching the biological world from a dialectical standpoint—with a *dialectical attitude*—helps one to avoid certain errors, prompts certain questions, suggests connections, and illuminates certain features of what they are studying. It thus has considerable epistemic benefits.

Hence the dialectical approach may be justified in terms of its epistemic fruits. Can it also be justified in terms of its coherence with a set of values? The approach coheres with certain holist, non-reductionist epistemic values. What about non-epistemic values? We saw above that holism in general seems only contingently related to non-epistemic values, being variously linked to both progressive, and conservative, social and political views and values. Dialectics of nature is, of course, more consistently associated with a particular political outlook, being developed initially by Marx's collaborator Engels as an application of Marxist ideas and categories (initially developed to account for human society and history) to the natural world, and later adopted (in arguably a rather crude and bastardised form) by Stalinist Marxism. Lewontin and Levins accept that their dialectical approach involves this political dimension, that it reflects and expresses a revolutionary anti-capitalist ideology (*ibid.*, 286–7). By contrast reductionism, they argue, is a product of modern western culture and industrial society, and has its rationale in the way it reflects the ideological world-picture of the bourgeoisie under capitalism. This emphasis on the non-epistemic-value-ladenness of both reductionism and dialectics fits well with the stance interpretation of these positions.

8 Coda to Part II: Higher-Level Stances?

Stephen Jay Gould used to write on the blackboard at the beginning of his lectures on Evolution at Harvard:

Gradualism; Reductionism; Adaptationism; Extrapolationism

These were, as he saw it, the philosophical pillars undergirding orthodox thinking in Darwinian evolutionary theory. Three of these—gradualism, adaptationism (a modern form of functionalism), and reductionism, represent one side in each of the three disputes we have discussed in Chaps. 8, 9, and 10 of Part II. Extrapolationism—roughly the view that natural selection operating on organisms in local populations is responsible for the full panoply of life's history, and that all macroevolutionary phenomena (evolution above the species level) is reducible to, and determined by, microevolutionary processes (evolutionary change in local populations)—is closely related to all three of the other views. Gould himself adopted the opposing views on all of these topics: punctuationism, holism/hiearchical thinking, structuralism, and anti-extrapolationism.

This leads to an obvious question: might we be able to identify higher-level stances that incorporate the metaphysical stances discussed in Part I? That is, is there a higher-level stance that includes the more specific stances of functionalism, gradualism and reductionism, and an opposing higher-level stance that includes structuralism, punctuationism, and holism? Are there certain styles of thinking, sets of attitudes, general approaches, perspectives on the world, (connected to certain values), that functionalism, gradualism and reductionism on the one hand, and structuralism, punctuationism, and holism on the other, are particular expressions or manifestations of? Gould seems to have thought so. In his (2002) he defends forms of the latter set of views, and argues explicitly that they are not unconnected, but are united as expressions of a more general worldview or attitude or orientation, as are the set of opposing views. It is not that punctuationists *must be* structuralists, or functionalists *must be* reductionists, that there are logical or even inductive relations between them. Just as we have seen that, for Gould, at the lower level, specific punctuationist hypotheses are logically and inductively independent of one another, but are united as expressions of the punctuationist stance (Chap. 9), so he suggests that at the higher level, the stances are logically and inductively independent of one another, but are united as expressions of a higher-level stance.

While he may be right about this, I am sceptical not so much of the existence of the higher-level stances, but of the explanatory role they may play.[48] I accept Gould's point that for the most part, the stances do tend to go together—if someone adopts one they are likely to adopt others in the set—they *reliably cluster together*, to use an expression from the natural kinds debate—and that this isn't just a contingent historical accident.[49] But it seems to me that as we ascend to the level of the higher-level stances, things begin to appear a little murky, we lose our bearings a bit, and it is difficult to know if there is anything very definite, or theoretically or explanatorily fruitful, we can say.

Firstly, how are we to characterize the higher-level stances? It is fairly straightforward to characterise the punctuationist, or reductionist, or functionalist stance. But what do functionalism, gradualism and reductionism (and structuralism, punctuationism, and holism) have in common? Gould thought he could answer this. The views and stances he advocated (structuralism, punctuationism, holism and other views) shared a 'common underlying vision': a commitment to 'complexity, interaction, multiple levels of causation, multidirectional flows of influence, and pluralistic approaches to explanation in general' (2002, 31). But while this might be true of the holist stance, it's less clear that it captures what holism has in common with punctuationism or structuralism:

[48] In my (2018b) I was more sympathetic to positing these higher-level stances, though I tended to see the lower-level views included within them (such as gene-selectionism, empirical adaptationism or sociobiology) as theories or beliefs rather than *themselves* being stances.

[49] I have noted, for instance, the close connection between the dialectical and the punctuationist stances.

why must the functionalist or the gradualist deny this thesis about complexity, interaction and pluralism? For instance, Gould argues that whereas functionalists deny interaction and complexity by positing a 'unidirectional flow from an external environment', structuralists accept that natural selection and internal constraints interact in complex ways (*ibid.*, 32). But the opposing view to functionalism here is not structuralism, but a *pluralistic* view that sees function and structure as *interacting*, and both important (Boucher 2025). A 'pure' structuralist perspective would, for Gould, be just as simplistic, just as guilty of ignoring interaction and complexity, as a pure functionalist view. We have seen that a natural pluralistic position accepts the validity of both holist *and* reductionist factual claims and explanations, both gradualist *and* punctuationist factual claims and explanations: why should 'pluralistic approaches to explanation' be part of the stance that includes holism and punctuationism but excludes reductionism and gradualism?

Perhaps there *are* certain attitudes, values, ways of thinking and approaches to explanation that are common to all and only the stances in each set, but until we can identify what precisely these are, and say something meaningful about them, gesturing vaguely towards the higher-level stances will not be very helpful, and worse, will only play into the hands of those stance-sceptics who doubt that identifying stances ever does any valuable theoretical or explanatory work, that it ever goes beyond vague hand-waving.[50]

Secondly and relatedly, while I have argued in Part II that the metaphysical stances I have discussed may play, and have played, a constructive role in guiding and informing scientific theory and practice, it is not at all clear how this could be true of the higher-level stances. They are too inchoate, too far in the background, too distant from and unrelated to actual conceptual, theoretical or empirical scientific work, for them to play the kind of positive role, even in the context of discovery, that I have argued the more specific stances can and do play. Gould admits that at this level, we are talking more about vague and inchoate social and psychological influences and impetuses—a *zeitgeist*—than anything directly

[50] Gould admits that the paragraphs in which he discusses these higher-level stances are 'the most vague and impressionistic … of the entire book' (*ibid.*, 30).

relevant to theoretical or empirical scientific advance (*ibid.*, 29–32). Just as Williams (1966) argued the concept of adaptation in evolutionary biology needed to be *disciplined*, and deployed sparingly, lest it be applied everywhere and to everything in a haphazard manner and lose its theoretical and explanatory value (explaining everything and nothing), the notion of a stance also needs to be disciplined, and not necessarily be applied to just any general, large-scale idea or worldview or zeitgeist, lest it be denuded of the sort of theoretical and explanatory value and import I have claimed it can offer when applied judiciously.

References

Bechtel, W. 2006. *Discovering cell mechanisms: The creation of modern cell biology.* Cambridge: Cambridge University Press.

Bechtel, W. & Abrahamsen, A. 2002. *Connectionism and the mind: Parallel processing, dynamics, and evolution in networks.* Second Edition. Oxford: Basil Blackwell.

Boucher, S.C. 2014. What is a philosophical stance? Paradigms, policies and perspectives. *Synthese*, 191(10): 2315–2332.

——— 2018a. Stances and epistemology: values, pragmatics and rationality. *Metaphilosophy* 49 (4): 521–547

——— 2018b. What is the relation between a philosophical stance and its associated beliefs? *dialectica* 72 (4): 509–524.

——— 2019. An Empiricist Conception of the Relation between Metaphysics and Science. *Philosophia*, 47 (5): 1355–1378.

——— 2020. Methodological naturalism in the sciences. *International Journal for Philosophy of Religion*, 88 (1): 57–80.

——— 2023. An Argument for Global Realism about the Units of Selection. *Biology and Philosophy* 38 (5): 1–22.

——— 2025. Functionalism and Structuralism in Biology: Stances or Explanatory Strategies? *Synthese*, 206 (1): 1–23.

Bourrat, P. 2015. Levels, Time and Fitness in Evolutionary Transitions in Individuality. *Philosophy and Theory in Biology* 7:e601, 1–17.

Calcott, B. & Sterelny, K. (eds.). 2011. *The Major Transitions in Evolution Revisited.* Cambridge, MA: MIT Press.

Clarke, E. 2014. Origins of Evolutionary Transitions. *Journal of Biosciences* 39 (2): 303–317.
Currie, A. 2019. Mass extinctions as major transitions. *Biology and Philosophy* 34 (2): 1–24.
Dawkins, R. 1982. *The Extended Phenotype, 1st ed.* Oxford: Oxford University Press.
——— 2006. *The Selfish Gene*. 30th anniversary ed. Oxford: Oxford University Press.
Dennett, D.C. 1995. *Darwin's Dangerous Idea*. New York: Norton.
Dupré, J. 1993. *The Disorder of Things: Metaphysical Foundations for the Disunity of Science*. Cambridge MA: Harvard University Press.
———2021. *The Metaphysics of Biology*. Cambridge: Cambridge University Press.
Engels, F. 1972. *Dialectics of Nature. 5th edition*. Moscow: Progress Publishers.
Ernst, W.G. (ed.). 2000. *Earth systems: Processes and issues*. Cambridge University Press.
Fang W. 2022. *The Explanatory Autonomy of the Biological Sciences*. Routledge.
Godfrey-Smith, P. 2001a. Organism, Environment, and Dialectics. R.S. Singh, C.B. Krimbas, D.B. Paul, and J. Beatty (eds.). *Thinking about Evolution: Historical, Philosophical, and Political Perspectives, Volume 2*. Cambridge: Cambridge University Press. 253–266.
——— 2001b. On the Status and Explanatory Structure of Developmental Systems Theory. S. Oyama, P. Griffiths, and R. Gray (eds.). *Cycles of Contingency: Developmental Systems and Evolution*. Cambridge, MA: MIT Press. 283–297.
Gould, S.J. 2002. *The Structure of Evolutionary Theory*. Cambridge, MA: Harvard University Press
Gray, R.D. 1992. Death of the Gene: Developmental Systems Strike Back. P. Griffiths (ed.). *Trees of Life*. Dordrecht: Kluwer Academic Publishers. 165–209
——— 2001. Selfish Genes or Developmental Systems? R.S. Singh, C.B. Krimbas, D.B. Paul, and J. Beatty (eds.). *Thinking about Evolution: Historical, Philosophical, and Political Perspectives, Volume 2*. Cambridge: Cambridge University Press. 184–207.
Griffiths, P. 2005. Review of 'Niche Construction'. *Biology and Philosophy* 20:11–20.
Griffiths, P.E. and R.D. Gray. 1994. Developmental Systems and Evolutionary Explanation. *Journal of Philosophy*, 91 (6). 277–304.
——— 1997. Replicator II—Judgement Day. *Biology and Philosophy*, 12 (4). 471–492.

——— 2001. Darwinism and Developmental Systems. S. Oyama, P. Griffiths, and R. Gray (eds.). *Cycles of Contingency: Developmental Systems and Evolution.* Cambridge, MA: MIT Press. 195–218.

——— 2004. The Developmental Systems Perspective: Organism-Environment Systems as Units of Evolution. K. Preston and M. Pigliucci (eds.). *Phenotypic Integration: Studying the Ecology and Evolution of Complex Phenotypes.* Oxford: Oxford University Press. 409–431.

——— 2005. Discussion: Three Ways to Misunderstand Developmental Systems Theory. *Biology and Philosophy*, 20. 417–425.

Hein, H. 1969. Molecular biology vs. organicism: The enduring dispute between mechanism and vitalism. *Synthese* 20, 238–253.

Hull, D. 1974. *Philosophy of Biological Science.* Englewood Cliffs, NJ: Prentice-Hall Prentice Hall

Huneman, P. 2010. Assessing the Prospects for a Return of Organisms in Evolutionary Biology. *History and Philosophy of the Life Sciences*, Vol. 32, No. 2/3: 341- 371

Kim, J. 1999. Making Sense of Emergence. *Philosophical Studies* 95: 3–36

Kitcher, P. 1984. 1953 and All That: A Tale of Two Sciences. *Philosophical Review* 93: 335–73

——— 2001. Battling the Undead: How (and How Not) to Resist Genetic Determinism. R.S. Singh, C.B. Krimbas, D.B. Paul, and J. Beatty (eds.). *Thinking about Evolution: Historical, Philosophical, and Political Perspectives, Volume 2.* Cambridge: Cambridge University Press. 396–414.

Lakatos, I. 1970. Falsification and the Methodology of Scientific Research Programmes. I. Lakatos and A. Musgrave (eds.). *Criticism and the Growth of Knowledge.* Cambridge: Cambridge University Press. 91–196.

Laland, K.N., Uller, T., Feldman, M.W., Sterelny, K., Müller, G.M., Moczek, A., Jablonka, E., and Odling-Smee., J. 2015. The extended evolutionary synthesis: its structure, assumptions and predictions. *Proc. R. Soc. B* Vol. 282 Issue 1813.

Lefkaditou, A. and Stamou. G. P. 2006. Holism and Reductionism in Ecology: A Trivial Dichotomy and Levins' Non-trivial Account. *History and Philosophy of the Life Sciences*, Vol. 28, No. 3: 313–336.

Levins, R. and R. Lewontin. 1985. *The Dialectical Biologist.* Cambridge, MA: Harvard University Press

Longino, Helen E. 1990, *Science as Social Knowledge: Values and Objectivity in Scientific Inquiry.* Princeton: Princeton University Press.

Magnus, P.D. 2018. Cautious realism and middle range ontology. *Metascience* 27: 365–70
Maynard-Smith, J. 1993. Molecules are Not Enough. J. Maynard-Smith. *Did Darwin Get it Right? Essays on Games, Sex and Evolution*. Penguin. 30–38.
Michod, R. E. 1999. *Darwinian Dynamics: Evolutionary Transitions in Fitness and Individuality.* Princeton NJ: Princeton University Press
——— 2005. On the transfer of fitness from the cell to the multicellular organism. *Biology and Philosophy*, 20(5), 967–987.
Mitchell, S. D. 1992. On Pluralism and Competition in Evolutionary Explanations. *American Zoologist* Vol. 32, No. 1: 135–144.
——— 2003. *Biological Complexity and Integrative Pluralism*, Cambridge: Cambridge University Press
Mossio M, ed. 2024. *Organisation in Biology*. Springer
Nicholson, D.J. 2014. The Return of the Organism as a Fundamental Explanatory Concept in Biology. *Philosophy Compass* 9/5: 347–359.
Nicholson, D.J. and Dupré, J. (eds.). 2018. *Everything Flows: Towards a Processual Philosophy of Biology.* Oxford University Press.
Novick, R. 2023. *Structure and Function*. Cambridge, Cambridge University Press
Odling-Smee, F, Laland, K.N. & Feldman, M.W. 2003. *Niche Construction: The Neglected Process in Evolution*. Princeton University Press.
Okasha, S. 2005 On Niche Construction and Extended Evolutionary Theory. *Biology and Philosophy* 20: 1–10.
——— 2006. *Evolution and the Levels of Selection*. Oxford: Clarendon Press. Kindle edition.
——— 2022. The Major Transitions in Evolution—A Philosophy-of-Science Perspective. *Frontiers in Ecology and Evolution* Vol. 10: 1–10.
O'Malley, M., and Powell, R. 2016. Major problems in evolutionary transitions: how a metabolic perspective can enrich our understanding of macroevolution. *Biology and Philosophy* 31, 159–189.
Oyama, S., P. Griffiths, and R. Gray. 2001. Introduction: What is Developmental Systems Theory? S. Oyama, P. Griffiths, and R. Gray (eds.). *Cycles of Contingency: Developmental Systems and Evolution*. Cambridge, MA: MIT Press. 1–11.
Rees, J. 1998. *The Algebra of Revolution: The Dialectic and the Classical Marxist Tradition*. London: Routledge.
Reiss M.J., & Ruse M. 2023. *The New Biology*. Harvard University Press.
Rogers, B. 1999. *A.J. Ayer: A Life*. New York, NY: Grove Press Books

Rosenberg, A. 1994. *Instrumental Biology, or The Disunity of Science.* University of Chicago Press

——— 2020. *Reduction and Mechanism.* Cambridge University Press.

Royle, C. 2014. Dialectics, Nature, and the Dialectics of Nature. *International Socialism*, Issue 141, Winter: https://www.marxists.org/history/etol/newspape/isj2/2014/isj2-141/royle.html.

Ryan, P. A., Powers, S. T., and Watson, R. A. 2016. Social niche construction and evolutionary transitions in individuality. *Biology and Philosophy* 31: 59–79.

Scott, A. and Rief, S. 2021. Reactionaries of the lectern: Universalism, anti-empiricism and corporatism in Austrian (and German) social theory. *European Journal of Social Theory.* DOI: https://doi.org/10.1177/1368431021992205.

Sober, E. 1999. The Multiple Realizability Argument Against Reductionism. *Philosophy of Science* 66: 542–564.

Sober, E. and Lewontin, R.C. 1982. Artifact, Cause and Genic Selection. *Philosophy of Science*, Vol. 49, No. 2: 157–180.

Sober, E. and D. S. Wilson. 1998. *Unto Others: The Evolution and Psychology of Unselfish Behavior.* Cambridge, MA: Harvard University Press.

Steffen, W., Richardson, K., Rockström, J. *et al.* 2020. The emergence and evolution of Earth System Science. *Nat Rev Earth Environ* 1: 54–63.

Sterelny, K. 2001. *The Evolution of Agency and Other Essays.* New York: Cambridge University Press.

——— 2005. Made By Each Other: Organisms and Their Environment. *Biology and Philosophy* 20: 21–36.

Sterelny, K. and P. Griffiths. 1999. *Sex and Death: An Introduction to Philosophy of Biology.* Chicago: The University of Chicago Press.

Strerelny, K., K.C. Smith, and M. Dickison. 1996. The Extended Replicator. *Biology and Philosophy*, 11 (3). 377–403.

Sullivan, T. 2015. Dialectical biology: A response to Camilla Royle. *International Socialism*, Issue 145, Winter: https://www.marxists.org/history/etol/newspape/isj2/2015/isj2-145/sullivan.html.

Thompson, D.W. 1917. *On Growth and Form.* Cambridge University Press.

Vermeij, G.J. 2010. *The Evolutionary World: How Adaptation Explains Everything from Seashells to Civilisation.* New York: Thomas Dunne Books.

Vrba, E. 1984. What is Species Selection? *Systematic Zoology* 33, 318–28.

——— 1989. Levels of Selection and Sorting with Special Reference to the Species Level. *Oxford Surveys in Evolutionary Biology* 6, 111–68.

Walsh, D.M. 2015. *Organisms, Agency and Evolution*. Cambridge: Cambridge University Press.

Waters, C.K. 1990. Why the Anti-Reductionist Consensus Won't Survive The Case of Classical Mendelian Genetics. *Proceedings of the Biennial Meeting of the Philosophy of Science Association*: 125–139.

Williams, G.C. 1966. *Adaptation and Natural Selection: A Critique of Some Current Evolutionary Thought*. Princeton, NJ: Princeton University Press.

Wilson, R.A. 2005. *Genes and the Agents of Life: The Individual in the Fragile Sciences, Biology*. Cambridge: Cambridge University Press.

Wu, S. and Banzhaf, W. 2011. Evolutionary transition through a new multilevel selection model. *Proceedings of the 11th European Conference on Artificial Life*. Paris, France. ASME. https://doi.org/10.7551/978-0-262-29714-1-ch132.

11

Conclusion

1 The Relation Between Epistemic Stances and Metaphysical Stances

I have in this book explored and defended two quite different paradigms for the concept of a philosophical stance, and applied them to a range of issues in the metaphysics of biology.[1] On the epistemic stance conception, the epistemic attitude one takes to ontology, in particular, one's judgments as to the epistemic legitimacy of ontological beliefs of varying degrees of removal from empirical evidence and empirical vulnerability, depends on the prior adoption of a stance. I focused mainly on epistemic stances that at least aspire to be naturalistic, i.e. those that see ontology as a matter of the drawing and endorsing of 'metaphysical inferences' from science and empirical data, with a priori reasoning assigned a relatively marginal role. On the metaphysical stance conception, certain metaphysical positions, in particular those that play a role in scientific inquiry, are reinterpreted as stances, something which, while able to be accepted on

[1] In previous work (e.g. Boucher 2014) I had not recognised that these represented two very different ways the stance idea could be developed, and thus had tried to shoehorn them into a single, unified account (arguing, for instance, that what I am calling metaphysical stances could be thought of as epistemic policies or strategies for forming beliefs), as others have tried to do. I now regard epistemic stances and metaphysical stances as fundamentally distinct, with neither being reducible to the other, although both are legitimate for different purposes.

S. C. Boucher, *Empiricism, Naturalism and the Metaphysics of Biology*,
https://doi.org/10.1007/978-3-032-00555-7_11

its own merits and whatever one's prior philosophical commitments, ought to be particularly attractive to the empiricist who wishes to account for the role of metaphysical assumptions and frameworks in science, while staying true to empiricist scruples about traditional metaphysics. While the focus of Part I was primarily on the metaphysical inferences from science that philosophers draw, the focus of Part II was metaphysical presuppositions and their role *within* science. My claim is certainly not that epistemic stances are only of interest to philosophers and never influence science or scientists, or that metaphysical stances are only of interest to scientists. Clearly scientists have adopted epistemic stances such as realism or empiricism and this has influenced their work (Forbes 2017; Boucher and Forbes 2024); and equally clearly philosophers can adopt and have adopted metaphysical stances such as holism, reductionism and functionalism. But I think it is true to say that issues related to epistemic stances in the sense of attitudes towards scientific ontology and the legitimacy of inferences from science to metaphysics are, today, primarily matters that philosophers, rather than scientists, concern themselves with; and while philosophers (or anybody else for that matter) may adopt and make use of metaphysical stances, it is the role of the latter within science that I have wanted especially to explore in this work.

Unlike Part I, the issues discussed in Part II are not, for the most part, matters of ontology or ontological commitment. The punctuationist and the gradualist, or the functionalist and the structuralist, do not really disagree about *what exists*, at least not in the way that realists and anti-realists about species, or units of selection, or teleology, do. The topics I discussed in Part II are not realism issues, at least not directly. Of course we *could* construe some of them in these terms if we wanted to: perhaps the holist affirms, while the reductionist denies, that emergent properties or irreducible higher-level causal processes exist. But the central disagreement between the holist and the reductionist concerns what can be explained in terms of what, not what entities we should have in our ontology.

One clear difference between the approaches concerns where the stances come in. On the metaphysical stance approach, the relevant metaphysical views *are* stances, while on the epistemic stance approach, metaphysical views (i.e. views on scientific ontology) are *beliefs*, but are accepted or rejected against the background of a stance that specifies the types of beliefs and inferences that are considered admissible. So the stances come in in different places. See Fig. 11.1.

Epistemic stances as we have understood them—at least, those that aspire to be naturalistic—are fundamentally epistemic policies with respect to scientific ontology. They are, or include, sets of V-prohibitions and V-permissions with respect to beliefs, and they each involve a characteristic trade-off between explanatory power and epistemic risk, or if not a trade-off, at least a choice as to which of these to value. Metaphysical stances are very different. They are not policies for generating beliefs, at least not in the same way; they do not involve V-prohibitions and V-permissions with respect to ontological beliefs or commitments. They are, as I put it, ways of looking at or approaching the world—metaphysical *perspectives* on or *styles of thinking* about the world. A metaphysical stance is often characterised by a tendency to focus on, emphasise, and find meaning and significance in certain features or aspects of the world, and de-emphasise others. We have seen that in some forms, they are defined by a set of values and preferences that identify certain phenomena as the most interesting, striking or problematic features of the natural world—the phenomena which above all others cry out for explanation.

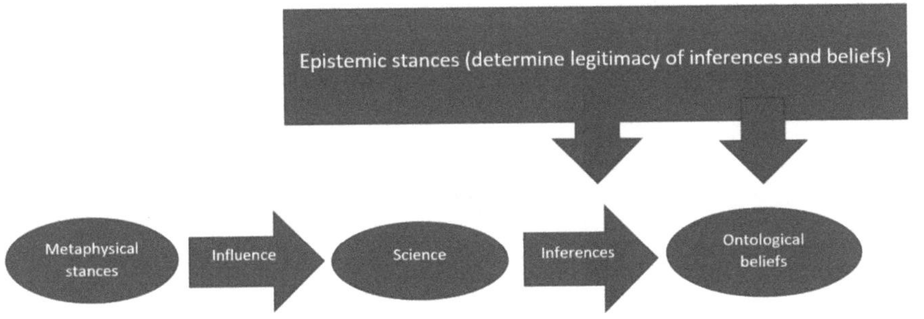

Fig. 11.1 The role of metaphysical stances and epistemic stances with respect to science and ontology

And they tend to influence science in the context of discovery, as a potentially fruitful source of empirical hypotheses.[2]

Yet epistemic and metaphysical stances clearly have things in common. They are both clusters of beliefs, values, attitudes, goals and commitments. They are different ways of developing van Fraassen's original concept of a stance, which was itself neutral between metaphysical and epistemic stances (though his central examples of empiricism and materialism pointed in the different directions the idea could be taken). An analogy that may be helpful is the relation between the two dominant views with respect to biological functions, the etiological view and the Cummins-causal-role view. These are different 'paradigms' (as Millikan 2002 puts it) for developing and making rigorous the general idea of function and purpose in biology. As I noted in Chap. 6, it is now widely held that neither account is 'the' correct account of biological functions, but rather each is valid in certain contexts, and each is applicable to certain areas of biology. Neither is reducible to or more fundamental than the other. But neither are they entirely unrelated. They are different ways of developing the one *general* idea of function in biology, an idea that doesn't entirely dissolve when we disambiguate between the etiological and the causal-role meanings. Kitcher (1993) argues the two approaches are united by the concept of *design*: the function of T is what T was designed to do; something applicable to functions in both senses.[3] While this is controversial (see e.g. Neander 2016), ideas of goal-directedness, of component parts contributing to the ends of a larger system, and of parts or traits being defined by what they do, by their role, rather than their material constitution, are common to functions in both senses. Similarly the epistemic stance and the metaphysical stance conceptions are different elaborations or applications of the core notion of a stance as originally proposed.

---

[2] Thus, the arguments we have considered for treating these views as stances clearly do not apply to the ontological beliefs that were the subject of Part I. Neither the historical argument—the continuity of the metaphysical positions across changes in theory—nor the epistemic argument—the manifest falsity of the views considered as factual claims combined with the evident fruitfulness of the positions for science—apply, at least in any straightforward way, to realism or anti-realism about species, natural kinds, teleology, and so on.

[3] See Mossio et al. (2009) for a different attempt to combine the two approaches within a general, unified account of functions.

In this book I have defended the general metaphysical stance and epistemic stance approaches, that is, I have argued that the positions in question—empiricism and the various grades of realism for epistemic stances, and functionalism, punctuationism etc. for metaphysical stances, *are* stances, and considered what follows from that. I have not taken a position on *which* stances (from either category) one ought to adopt. I have my views on this, but this book was not the place to defend them (to the extent that it is possible to rationally defend stances). It was the meta-level stance interpretation that I was concerned to defend here.[4]

What views could or should those with empiricist or realist philosophical commitments hold with respect to these questions? Both can and should accept both the epistemic stance and metaphysical stance general conceptions, to the extent that my arguments are persuasive, since those arguments, and those conceptions, do not presuppose either realism or empiricism. As I have said, the metaphysical stance conception is particularly congenial to empiricism; we may go further and claim that the conception is close to mandatory for the empiricist, if she is to do justice to the role of metaphysics in science without compromising her empiricism. But if it is necessary or almost necessary for empiricism, it is not sufficient for it; the metaphysical stance conception is perfectly consistent with realism. Likewise, the general meta-level epistemic stance conception is compatible with both realism and empiricism, as is shown by the fact that the conception has been accepted by those from across the realism-empiricism spectrum (van Fraassen 2002; Chakravartty 2017; Psillos 2011).[5] That is, both the realist and the empiricist can and should accept the meta-level claim that realism and empiricism are epistemic stances.

The situation for the realist and the empiricist is this. The empiricist ought, *qua* empiricist, to accept the metaphysical stance conception. She should also accept the epistemic stance conception, though this conception is not any more congenial to empiricism than to realism. And she ought, of course, at the lower level, to accept empiricism as an epistemic stance. The empiricist stance is a standpoint from which one

[4] See Bhakthavatsalam (2023) for a pragmatist account of stance-justification quite congenial to the views I defend.

[5] One does need, I have argued, to accept stance voluntarism to accept the stance idea, so to that extent there is theoretical commitment involved.

recommends, in the realm of scientific ontology, endorsing only those metaphysical inferences that do not take us beyond belief in observable objects, events and processes. It is *also* a standpoint from which one advocates construing certain metaphysical views as metaphysical stances, rather than beliefs.[6] This is consistent, since when it comes to scientific ontology, the 'metaphysical' beliefs the empiricist stance licenses are hardly metaphysical in any strong sense, being limited to the observable. The metaphysical views the empiricist wants to interpret as metaphysical stances, on the other hand, are metaphysical in a much stronger sense (or would be if construed as metaphysical *beliefs*). Note that the empiricist does not need to accept the strong claim that *all* metaphysical positions are stances. She can accept that ontological views we have discussed—realism about species, realism about teleology, and so on—are beliefs (albeit ones she does not accept). It is the large-scale, high-level metaphysical positions, worldviews and orientations we have focussed in Part II, such as reductionism, punctuationism and functionalism, that she is concerned to reinterpret as metaphysical stances, especially with respect to their role in science.

The realist ought to accept the metaphysical stance conception if my arguments are persuasive, but unlike the empiricist, is not compelled to adopt this view *qua realist*. The realist does not typically share the empiricist's concerns about metaphysical influences on science, and so may, at least as far as her realism is concerned (depending perhaps on the strength of her realism, i.e. on how strong a realist stance she endorses) be happy to accept the traditional conception of functionalism, punctuationism and the rest as metaphysical beliefs or theses rather than stances. The realist should also accept the epistemic stance conception, though this conception is not any more congenial to realism than to empiricism. And she ought, of course, at the lower level, to accept realism (of one form or another) as an epistemic stance.

[6] Notice the stances coming in at two levels: from the standpoint of the empiricist stance, certain metaphysical positions should be construed as stances. We can avoid confusion if we keep in mind the difference between epistemic and metaphysical stances.

Table 11.1 Realist and empiricist options at the first-order level and the meta-level

	Realist	Empiricist
Meta-level metaphysical stance conception	Accept if arguments persuasive, but not *qua* realist	Accept if arguments persuasive, and accept *qua* empiricist
First-order metaphysical stances	Can accept any, *qua* realist	Can accept any, *qua* empiricist
Meta-level epistemic stance conception	Accept if arguments persuasive, but not *qua* realist	Accept if arguments persuasive, but not *qua* empiricist
First-order epistemic stances	Accept one or other realist stance	Accept empiricist stance

The realist's and empiricist's options at the first-order level and the meta-level are summarised in Table 11.1 (I am assuming that both can accept any of the metaphysical stances we have considered[7]):

Note that here realism and empiricism are both epistemic stances at the first-order level, *and* standpoints from which one accepts or rejects the meta-level stance conception. E.g. one may accept the empiricist epistemic stance at the first-order level, then, from that standpoint, go on to accept, at the meta-level, that the empiricist epistemic stance *is* an epistemic stance. This is a little disorienting, but it is coherent. Van Fraassen, in his (2002), defended the idea that empiricism was a stance, and he clearly did that, in large part, from the standpoint of empiricism itself. A set may contain itself as a member. As I argued in my (2012), there is a sense in which the idea that empiricism, and other philosophical positions, are stances was *itself*, initially, an empiricist idea, or at least, was highly congenial to empiricism. But I have argued in this book that its appeal can and should be much broader. Although I have suggested that the *metaphysical* stance idea is particularly helpful to the empiricist, I have made the case that one need not be an empiricist to accept either the epistemic stance conception or the metaphysical stance conception. This book has in fact been an extended argument for the claim that, whatever one's philosophical commitments or inclinations, each of these

[7] In Chap. 10 I suggested that there may be affinities between reductionism (in certain forms) and empiricism (in certain forms), but to the extent that there are such affinities they are likely to be loose and defeasible.

conceptions are both coherent and plausible in themselves, and fruitful in helping us make sense of, and provide a consistent and illuminating metaphilosophical underpinning for, various metaphysical debates in and about science.

2 The Pro-stance Stance

When it comes to epistemic and metaphysical stances, we can distinguish the descriptive or factual question—do these stances in fact exist, and play a central role in science and philosophy?—from the normative question—is it *desirable* that they exist and play this role? Note that the normative question here is not whether the stance *idea* is a useful tool of analysis for understanding, say, the metaphysics of biology, but whether the stances *themselves* and the role they play, are on the whole a positive and productive thing for science and philosophy. One may, after all, agree with my descriptive or factual claim about the impact and influence of stances while regarding this situation as regrettable or even disastrous. In this book my focus has primarily been on the descriptive question, though I have addressed the normative question here and there, for instance in Chap. 10, Sect. 5.3.

With respect to epistemic stances, the relevant normative question is whether the stance-dependence of views on and arguments relating to scientific ontology undermines the rationality of the latter. If the view one takes with respect to particular questions of scientific ontology are downstream from stance choice, and the latter is understood in voluntarist terms, do we court the kind of incommensurability, irrationality, and relativism that has troubled the peace of philosophers of science since Kuhn? I argued in Part I that this worry is misplaced. Recognising the inevitable limits on the effectiveness of rational argument between positions undergirded by rival, voluntarily-chosen epistemic stances need not commit one to the strong and unattractive claim that it is permissible to believe anything one likes so long as it chimes with one's values, preferences or interests. Evidence, argument and reason can play a critical role in matters of scientific ontology even if they are constrained by value-laden stances.

With respect to metaphysical stances, we saw in Chap. 10 that Novick (2023) accepts my descriptive claim about the existence and influence of the functionalist and structuralist stances, but bemoans this as a decidedly non-optimal situation, entrenching ideological or philosophical disagreements, encouraging heated and unproductive polemics, and stymying scientific advance. She advocates taking 'our debates smaller and more local, away from stances and general theories…' (Novick *ibid.*, 62). The opposing view is articulated by Hein (1969), who, we saw in Chap. 10, argues for something similar to the stance-interpretation of holism and reductionism on the basis of what I called the historical argument. She disagrees with Novick on the normative question, arguing that the clash of large-scale worldviews or 'meta-theoretical commitments' such as holism and reductionism is on the whole likely to be conducive to scientific progress. This is in large part because each side is forced to clarify its claims, sharpen its arguments, and improve its concepts, in response to the challenge continually presented by the opposing side (apologies for the sexist language, this was 1969):

> …I should like to point out the intellectual benefit which results from the clash of these two positions. I do not know if there is such a thing as scientific progress, but if there is, I think it would not occur if men of conviction were not challenged by opposing convictions to constantly re-examine, and, where necessary, revise their own beliefs. (Hein *ibid.*, 250)

Even if we favour, say, the reductionist view, and its contemporary manifestation in molecular biology, we should recognise that that research program was unlikely to have achieved what it has achieved, or to have attained its current state of theoretical and methodological maturity, without the constant prods provided by the opposing holist or vitalist camp: 'That very adventurousness which lends molecular biology its current glamour and popularity may well be the consequence of vitalistic stimulation which is perpetuated by the organicist biologists' (*ibid.*). Notice that for the benefits of meta-theoretical convictions to be realised, for Hein, it is essential that they be embodied in rival, competing traditions or standpoints. They would not, it seems, be realised if a single stance or set of meta-theoretical convictions dominated a scientific field to such an extent that its proponents did not feel under any pressure to respond to challenges presented by an opposing stance or viewpoint.[8] But in all the cases we have discussed in this Part II of this book we do find such rival stances in operation, thus allowing for the benefits that Hein indicates.

Hein concludes by noting that the fact that the meta-theoretical convictions and commitments

> are not necessarily rational in origin or due to the examination of evidence does not seem to me to be a mark against them. Nor is it a failing that the same prejudices reassert themselves in constantly new guises. Rationality demands that we submit our beliefs to critical scrutiny… It does not require that we be super-humanly free of prejudice to begin with. (*ibid.*)

I, of course, side with Hein here rather than Novick. But Novick makes a further, telling point that it seems appropriate to end the book on (*ibid.*, 62). She suggests[9] that the views I am attributing to her and to Hein are themselves value-laden, higher-level stances. One stance (call it the 'pro-stance stance') extols the virtues and value of entrenched philosophical positions and large-scale theoretical conflict in science. It sees science as riddled with clashes of worldviews, and views this is at least often a positive thing. The opposing stance that she adopts (call it the 'anti-stance stance'[10]) rejects the value (though not, as we've seen, the existence) of large-scale theoretical conflict and stance-disputes in science, and sees clashes of worldview as generally inimical to scientific progress.[11] While it would be implausible to suppose that such stances (or factual claims associated with them) must be entirely unresponsive to evidence—it should

[8] And indeed it is likely that at least some of the *disadvantages* that Novick sees in the role played by metaphysical stances in science would also disappear in the absence of significant stance *conflict*.

[9] This is my gloss on what she says here, she doesn't put it in quite this way.

[10] We may of course wonder whether the anti-stance stance is self-undermining, as it seems to be opposed to itself.

[11] In the Coda to Part II I expressed scepticism about the explanatory value of positing *some* higher-level stances, for instance a general stance that includes the punctuationist, functionalist and holist stances. I think Novick's proposal about higher-level stances—that there is, roughly, a pro-stance philosophical stance and an anti-stance philosophical stance, and that there is some explanatory value in recognising them—is more fruitful and illuminating, but I am not able to fully defend that claim here.

be possible, after all, to marshal evidence from the history of science in support of the claim either that stance-conflict promotes, or that it hinders, scientific progress—it would be equally mistaken to suppose that this dispute can be entirely resolved by appeal to empirical evidence. The reason is that the positions involve not just factual claims, but also goals, attitudes, values and commitments. They are philosophical stances.[12] And when it comes to stance choice, at the end of the day, to quote Lenin's quoting of Napolean, 'On s'engage et puis… on voit' (roughly, 'You commit yourself, then you see') (Lenin 1965).

References

Bhakthavatsalam, S. 2023. Revisiting stance voluntarism: in defense of an active stance pluralism. *Synthese* 202, 184

Boucher, S.C. 2012. *Empiricism, Metaphysical Stances and the Philosophy of Biology*. Ph.D. dissertation, University of Melbourne

———. 2014. What is a philosophical stance? Paradigms, policies and perspectives. *Synthese*, 191(10). 2315–2332

Boucher, S.C. and Forbes, C. 2024. The Pragmatic Turn in the Scientific Realism Debate. *Synthese*, 203 (4). 1–23.

Chakravartty, A. 2017. *Scientific Ontology: Integrating Naturalised Metaphysics and Voluntarist Epistemology.* Oxford University Press

Forbes, C. 2017. A pragmatic, existentialist approach to the scientific realism debate. *Synthese* 194: 3327–3346

Hein, H. 1969. Molecular biology vs. organicism: The enduring dispute between mechanism and vitalism. *Synthese* 20, 238–253

Kitcher, P. (1993). Function and Design. *Midwest Studies in Philosophy* 18: 379–397

Lenin, V.I. 1965. Our Revolution. *Collected Works, 2nd English Edition*. Progress Publishers, Moscow, Volume 33: 476–80

Millikan, R.G. 2002. Biofunctions: Two Paradigms. André Ariew, Robert Cummins, and Mark Perlman (eds.), *Functions: New Essays in the Philosophy of Psychology and Biology*. Oxford: Oxford University Press

[12] This is similar to the point made in the previous section that the idea that empiricism is a stance is itself an expression of the empiricist stance.

Mossio, M., Saborido, C., & Moreno, A. 2009. An organizational account of biological functions. *Br J Philos Sci* 60(4): 813–841

Neander, K. 2016. Kitcher's Two Design Stances. Mark Couch, and Jessica Pfeifer (eds.), *The Philosophy of Philip Kitcher*. New York: Oxford University Press

Novick, R. 2023. *Structure and Function*. Cambridge, Cambridge University Press

Psillos, S. 2011. Choosing the realist framework. *Synthese* 180 (2): 301–316

Van Fraassen, B.C. 2002. *The Empirical Stance*. New Haven, CT: Yale University Press

Author Index[1]

[1] Note: Page numbers followed by 'n' refer to notes.

S. C. Boucher, *Empiricism, Naturalism and the Metaphysics of Biology*,
https://doi.org/10.1007/978-3-032-00555-7

Subject Index[1]

[1] Note: Page numbers followed by 'n' refer to notes.

S. C. Boucher, *Empiricism, Naturalism and the Metaphysics of Biology*,
https://doi.org/10.1007/978-3-032-00555-7

E

F

G

FSC
www.fsc.org
MIX
Papier | Fördert gute Waldnutzung
FSC® C083411

Kolibri 360 GmbH
Ferdinand-Jühlke-Straße 7
99095 Erfurt, Deutschland
produktsicherheit@kolibri360.de